Postharvest Management of Dryland Fruit Crops

The Editors

Dr J. K.Dhemre is presently working as Associate Professor of Horticulture and Officer Incharge, Onion & Grape Research Station, Pimpalgaon Baswant, Dist. Nashik, Mahatma Phule Krishi Vidyapeeth, Rahuri. He has about twenty years of teaching, research and extension experience. He has to his credit 01 variety, 09 research recommendations, 12 books, 21 book chapters, 06 folders, 03 booklets, 02 pocket diary, 48 research papers, 210 popular articles and 67 technical papers published in journals of national and international repute, magazines and newspapers. He has awarded Mahatma Gandhi Excellence award-2017 at New Delhi in recognition of contribution for research in Postharvest Technology and for Extension in increasing the economy of farmers.

Dr V. P. Kad is presently working as Associate Professor in the Department of Agricultural Process Engineering, Dr. Annasabeb Shinde College of Agricultural Engineering and Technology, Mahatma Phule Krishi Vidyapeeth, Rahuri. He has about eighteen years of teaching, research and extension experience. He has to his credit 10 research recommendations, 09 books, 23 book chapters, 01 edited volume, 56 research papers, 251 popular articles and 110 technical papers published in journals of national and international repute. He has received "Engineering Achievement Award" by the Institution of Engineers; "Agro-care Best Teacher Award" by Balidharma Charitable Society, Nashik and "Fellow Award" by Society of Krishi Vigyan, Ludhiana.

Dr S. A. Ranpise is presently working as Head, Department of Horticulture and Director, Centre of Advanced Faculty Training in Horticulture(Fruits) in Mahatma Phule Krishi Vidyapeeth, Rahuri. He has about thirty-three years of teaching, research and extension experience. He has to his credit 16 varieties, 39 research recommendations, 17 Books, 20 Book chapters, 110 research papers, 31 technical papers and 85 popular articles published in journals of national and international repute, magazines and newspapers. He has been awarded with VC Rawal highest Marathi Pustak Vikri Purskar; Best paper presentation in Poster Session by Ornamental Society of India in National Symposium and Late Randive first award for best paper on mango technology.

Postharvest Management of Dryland Fruit Crops

– Editors –

Dr J. K. Dhemre

Dr V. P. Kad

Dr S. A. Ranpise

2019

Daya Publishing House®

A Division of

Astral International Pvt. Ltd.

New Delhi – 110 002

ISBN: 9789389569124 (Int. Edition)

Published by : **Daya Publishing House®**
A Division of
Astral International Pvt. Ltd.
– ISO 9001:2015 Certified Company –
4736/23, Ansari Road, Darya Ganj
New Delhi-110 002
Ph. 011-43549197, 23278134
E-mail: info@astralint.com
Website: www.astralint.com

Digitally Printed at : Replika Press Pvt. Ltd.

Foreword I

India is the fruit and vegetable basket of the world. India, being a home of wide variety of fruits and vegetables, hold a unique position in production among other countries. In India, horticulture crops occupy 9.0 per cent of cropped area, however contributes to approximately 30.0 per cent in agriculture GDP. Diverse agro-climatic zones in the country make it possible to grow almost all varieties of fruits and vegetables. Area under horticulture crops grew by about 3.0 per cent per annum with a rise in production to 7.6 per cent per annum. This higher growth rate was due to improvement in productivity in horticulture crops from 8.8 MT/ha in 2001-02 to 12.11 MT/ha in 2014-15. Fruits that dominate India's export market, include mangoes, bananas, citrus fruits, pomegranate, apples, guava, papaya, pine apple and grapes. Production of these fruits in India makes a sizeable contribution on global scale.

Although, India ranks second in the world with respect to production of fruits and vegetables, with more than 283.5 Mt of horticultural produce annually which is higher than the food grain production since 2012-13, it is disheartening to note that postharvest losses to the extent of 20-30 per cent of total produce occur every year. This is mainly because of improper and inadequate pre and postharvest management practices. It is indeed imperative to reduce these losses that otherwise arc estimated to save over Rs. 40,000 crores annually. These losses can be significantly reduced, if the integrated approaches arc adopted in pre and postharvest management of horticultural crops. The concept of integrated approaches encompasses the greater planning of entire demand driven production activity, appropriately linked with pre and postharvest practices including handling, transportation, storage, packaging, processing, value addition, distribution and marketing network for effectively spacing product in the hands of consumer. In view of these considerations, a Model Training Course on "Advances in postharvest management of dryland fruit crops" was organized by the Course Director, Model Training Course and Head,

Department of Horticulture, Mahatma Phule Krishi Vidyapeeth, Rahuti (MS) from 14th to 21st October, 2016 which was sponsored by the Directorate of Extension (DOE), Ministry of Agriculture, Government of India, New Delhi.

The lectures delivered during this programme by various experts from this University are brought together in the form of compendium, which was edited by Dr. J. K. Dhemre, Dr. V. P. Kad and Dr. S.A. Ranpise. The scientific information incorporated in Ibis compendium will be of immense use to UG and PG students, research workers, teachers and extension personnel to upgrade and advance their knowledge in pre and postharvest management for enhancing fruit production.

This University is indebted to the Director (Extension and Training), GOI, Ministry of Agriculture,Department of Agriculture and Cooperation, Krishi Vistar Bhavan, IASRI Campus, Pusa, New Delhi for sanctioning the Model Training Course, through which this training was possible.

Dr. K. P. Viswanatha

Vice-Chancellor

Mahatma Phule Krishi Vidyapeeth

Rahuri, Dist. Ahmednagar, M.S.

Foreword II

Indian agriculture is characterized by overwhelmingly small holdings due to higher population density and nearly two-third of its population residing in the rural areas coupled with unabated land fragmentation due to the inheritance laws of the country. India is one of the most drought prone countries in the world. Over 75 per cent cropped area in the semi-arid tropics in the country (131 million of 174 million ha). Ninety nine districts in 14 states are declared as drought prone district by the Central Water Commission (2002). Most of them are concentrated in Andhra Pradesh, Maharashtra, Tamil Nadu, Karnataka, Rajasthan affecting 265 million people in the rural areas. Low and erratic solar radiation makes these the most vulnerable region in India. Thus the arid and semi-arid region faces serious challenges due to lack of sufficient food security. Bio-diversity and food security are directly related. Crop diversification and intercropping systems are the means to reduce the risk of crop failure due to adverse weather events crop pest and insect attack. Crop improvement programmes can play a significant role in this direction. The significant achievements made in last couple of decades in breeding and crop production in different fruit crops by MPKV, University needs to be disseminated to the researcher, teachers and extension workers.

In view of these considerations, a Model Training Course on "Advances in post harvest management of dryland fruit crops" was organized by the Course Director, Model Training Course and Head, Department of Horticulture, Mahatma Phule Krishi Vidyapeeth, Rahuti (MS) from 14th to 21st October, 2016 to disseminate the latest developments regarding production technologies and post harvest management of different fruit crops which were standardized by our University.

The book entitled "Postharvest Management of Dryland Fruit Crops" "of lectures delivered during the said training programme by different experts in the field of Horticulture. This book will be useful to the researchers, teachers and extension workers. I congratulate Dr. J. K. Dhemre, Dr. V. P. Kad and Dr. S. A. Ranpise for bringing out this well documented publication.

This University is indebted to the Director (Extension and Training), GOI, Ministry of Agriculture, Department of Agriculture & Cooperation, Krishi Vistar Bhavan, IASRI Campus, Pusa, New Delhi for sanctioning the Model Training Course, through which this training was possible.

Dr. K. D. Kokate
Director of Extension Education
Mahatma Phule Krishi Vidyapeeth
Rahuri, Dist. Ahmednagar, M.S.

Preface

India is the fruit and vegetable basket of the world. India, being a home of wide variety of fruits and vegetables, hold a unique position in production among other countries. In India, horticulture crops occupy 9.0 per cent of cropped area, however contributes to approximately 30.0 per cent in agriculture GDP. Diverse agro-climatic zones in the country make it possible to grow almost all varieties of fruits and vegetables. Area under horticulture crops grew by about 3.0 per cent per annum with a rise in production to 7.6 per cent per annum. This higher growth rate was due to improvement in productivity in horticulture crops from 8.8 MT/ha in 2001-02 to 12.11 MT/ha in 2014-15. Fruits that dominate India's export market, include mangoes, bananas, citrus fruits, pomegranate, apples, guava, papaya, pine apple and grapes. Production of these fruits in India makes a sizeable contribution on global scale.

Although, India ranks second in the world with respect to production of fruits and vegetables, with more than 283.5 Mt of horticultural produce annually which is higher than the food grain production since 2012-13, it is disheartening to note that post harvest losses to the extent of 20-30 per cent of total produce occur every year. This is mainly because of improper and inadequate pre and post harvest management practices. It is indeed imperative to reduce these losses that otherwise are estimated to save over Rs. one lakh crores annually. These losses can be significantly reduced, if the integrated approaches are adopted in pre and post harvest management of horticultural crops. The concept of integrated approaches encompasses the greater planning of entire demand driven production activity, appropriately linked with pre and post harvest practices including handling, transportation, storage, packaging, processing, value addition, distribution and marketing network for effectively spacing product in the hands of consumer. In view of these considerations, we take an opportunity to present this standard treatise entitled as "Postharvest Management of Dryland Fruit Crops". The object of this book is to present the subject matter in most concise, compact, to the-point and lucid manner.

Although every care has been taken to check mistakes and misprints, yet it is difficult to claim perfection. Any errors, omissions and suggestions, for the improvement of this treatise, brought to our notice will be thankful acknowledged and incorporated in the next edition.

Dr. J. K. Dhemre

Dr. V. P. Kad

Dr. S. A. Ranpise

Contents

Chapter 1

Strategic Approaches to Upscale the Postharvest Management of Dryland Fruit Crops

K.D. Kokate

Director of Extension Education,
Mahatma Phule Krishi Vidyapeeth, Rahuri – 413 722, Dist. Ahmednagar, M.S.

Indian agriculture is characterized by small holdings due to higher population density and nearly two-third of its population residing in the rural areas coupled with unabated land fragmentation due to the inheritance laws of the country. Nearly 62 per cent of the estimated 142 m ha area is rainfed. The present population of India is about 1.32 billion, which is 17.74 per cent of total world population of about 7.2 billion. Though India has only 2.3 per cent of total land area on the planet earth, it supports the survival of a substantial portion of world population. India's well-orchestrated Green Revolution began in the mid 60's. It was ushered in through the adoption of higher and balanced doses of the biological, chemical and mechanical inputs together with the timely intervention of the Government. The later ensured the availability of the required inputs of high yielding seed varieties, fertilizers, pesticides water and improved power sources and equipment. The Government provided the minimum support price, easy access to procurement markets, rural roads and other infrastructure which helped to trigger the green revolution in selected areas of the country. Resultantly gross food grain production increased from 50.8 million tonnes in 1950-51 to 283.4 million tonnes in 2017-18 and land productivity rose from 0.58 tons/ha/year to more than 2.14 tons/ha/year. Whereas the quantum jump in production and productivity was brought about by a combination of factors, farm mechanization was often at the centre of controversy due to its impact on employment of human labour in a labour abundant economy.

Every human being on planet earth wants prosperity, happiness and security. To achieve this goal, diversification and modernization of agriculture for higher productivity and equitable distribution of food commodities and other necessities of life are needed. In developing countries, it could be achieved to a great extent through selective mechanization of agriculture and appropriate postharvest management and value addition in the production catchments, leading to employment generation in the rural sector and minimization of losses of the harvest biomass. Thereby, enhancing per capita availability of food, fiber and other essential commodities of life from the same area without any lapse of appreciable time, which otherwise, would have been difficult to get through the production process taking 4-6 months of time, energy and inputs.

It may be noted that India is a predominantly an agricultural economy and 52-55 per cent of its population live in village/rural sector and earn their livelihood through agriculture. Rural population of India was 91 per cent in 1901 and now in 2011, it is around 60-65 per cent and may reach to 50 per cent by 2020. Rural people migrate to urban areas for employment and letter amenities of life. Such opportunities are presently not available in rural areas but could be created through selective mechanization of agriculture and appropriate postharvest management and value addition to the harvested biomass in the productions catchments.

Agro processing could be defined as set of techno economic activities carried out for conservation and handling of agricultural produce and to make it usable as food, feed, fibre, fuel or industrial raw material. Hence, the scope of the agro-processing industry encompasses all operations from the stage of harvest till the material reaches the end users in the desired form, packaging, quantity, quality and price. Ancient Indian scriptures contain vivid account of the postharvest and processing practices for preservation and processing of agricultural produce for food and medicinal uses. Inadequate attention to the agro-processing sector in the past put both the producer and the consumer at a disadvantage and it also hurt the economy of the country. Agro-processing is now regarded as the sunrise sector of the Indian economy in view of its large potential for growth and likely socio-economic impact specifically on employment and income generation. Some estimates suggest that in developed countries, more than 18 per cent of the total work force is engaged in agro-processing sector directly or indirectly. However, in India, only about 4-5 per cent of the work force finds employment in this sector revealing its underdeveloped state and vast untapped potential for employment. Properly developed agro-processing sector can make India a major player at the global level for marketing and supply of processed food, feed and a wide range of other plant products.

India is one of the largest producers of food grains besides being second in the world in the production of fruits and vegetables, and first in milk production and livestock population, which makes this sector further attractive. India produces about more than 500 million tons of raw food material of plant and animal origin which are refined, stored and transformed into various usable products using congenital and modern postharvest and food processing technology. India is said to have 52 per cent cultivable land as against the world average of 11 per cent. It

.najor climates of the world and has 46 out of 60 types of soil, making it the .nber one contender for becoming a global sourcing hub.

Currently of the total production, only 4.4 per cent of fruits and vegetables, 26 per cent of marine, 6 per cent of poultry, 20 per cent of buffalo meat and 35 per cent of milk is processed. This is quite low in comparison to the food processed in developed countries.

While, some part of the produce is consumed raw and unprocessed, a large part rots in the warehouses in absence of proper transporting system, terminal markets, distribution channels, *etc.* As much as 30-40 per cent of the total fruits and vegetables production is wasted every year. The way out is food processing. Increased food processing can lead to minimal wastage.

The present levels of post-production losses are:

- 5-15 per cent in durables
- 20-30 per cent semi-perishables
- 30-50 per cent in perishables

Present level of production of different types of food commodities and their estimated postharvest losses along with monetary values are given. It shows that various food items, to the tune of around 65-70 million tons, worth Rs.1,00, 000 crores per annum, are lost during postharvest phase. A substantial amount of these losses could be prevented if appropriate agro-processing centers having backward linkage with farmers to ensure constant supply of quality raw food materials are established and operated. Profit generated through value addition must also be shared with farmers who are only in the world to create wealth every year in the form of food, fiber and other commodities necessary for human survival. It can bring a sea of change in rural areas, where 52-55 per cent population lives, in respect of economy, health and happiness.

The sunshine industry, as food processing industry is popularly known as, is witnessing newer heights. It holds immense potential for small-scale entrepreneurs much more than it does for large corporate. With the shift in consumer needs, a rapid evolution in the global food landscape can be witnessed. Hectic schedules have made ready-to-cook (RTC) and ready-to-eat (RTE) food a new favourite among the younger generation. There is an increasing demand for processed and convenience food. However, the supply side has failed to keep up with the demand. The imparity between the current demand and supply offers immense scope for the manufacturers, retailers and suppliers of processed food.

The share of the agricultural sector in the Gross Domestic Product (GDP) in the Indian economy declined from nearly 55 per cent in 1951/52 to 16-18 per cent in 2010-11. However, this structural change in output has not been accompanied by corresponding change in the occupational structure of the country's labour force. Nearly 60 per cent of the country's population lives in rural areas.

Postharvest Technology involves operations like cleaning, grading, drying, storage, milling packaging, transport and marketing and utilization, At the end of each operation, value is added to the product. The lowest and highest monetary values of a food commodity are, respectively, when it is in raw and fresh from when it is in processed and ready to consumed/eat form. Postharvest and food processing and technology are commodity and location specific. It enhances and augments per capita food availability from a unit arable land and other recourses by preventing avoidable postharvest losses and adding value to the fresh agro –products. It also creates opportunities for employment and income generation. Integration of production agriculture with on farm better quality and products for domestic and export markets. It, therefore demands establishments of agro processing is needed to have higher and sustainable production. Productivity and establishments of agro processing centers in the production catchments itself to facilitate backward linkage with farmers, have fresh and best quality raw food materials for processing and value addition, minimize material movements, check migration of rural people to urban areas for jobs and thereby reducing pressure on public utilities in urban areas. Such center would be a very strong tool for rural reconstruction and its upliftment. It would help in reducing rural- urban disparity and ensuring household food and nutritional security for all at an affordable cost.

The Proposed Action Program is as Under

- ✫ Continued monitoring of status of processing and utilization of different crops/commodities, processes, quality assurance, energy audits and economic competiveness.
- ✫ Development and promotion of farm storage and rural warehouses for perishables, semi-perishable as well as durables that minimize storage and associated losses enable growers/herdsmen/fishermen having capacity to negative with forces of marketing, provide off-season goods to local people at cheaper rates and raw materials to entrepreneurs/industry with dividends to both.
- ✫ Study of postharvest physiology, senesces, ripening, respiration, *etc.* of different crop varieties and commodities as influenced by time, temperature, humidity, mechanical injurious/interaction, *etc.*
- ✫ Development of HACCP and food safety measures for different commodities and products and developing mechanisms to apply and certify them, setting up of referral laboratories and human recourse development capacities in good/biosafety.
- ✫ Harnessing biotechnology and genetic engineering that enhance shelf life, quality and nutritive value; provide protection against postharvest insects,pests and mycotoxins; improve products market appeal and product recovery,and ultimately make the products cheap and globally competitive.
- ✫ Modernization and cost reduction of cleaning, grading, sorting, milling, processing and packaging equipment for food grains, oilseeds, horticulture crops, animal products and fish, noise and pollution control/effluent treatment.

- ✰ Modernized packages of processing and utilization of agricultural commodities/crops and their residues and that of underutilized plant in to products in demand as well as developing processes and pilot plants for high value futuristic products or their intermediate stages needed by industry.
- ✰ Develop and commercialize diet convenience and specialty foods and agro process.

Looking into the present agricultural production and post-production scenario in India, the most appropriate action for development and employment generation in rural sector is primary processing and value addition in production catchments. Technology for establishing such agro-processing complex in rural areas are available in India and requires to be adopted and should be demonstrated through pilot plant. This will not only help in reducing post production losses, generating employment opportunities in rural sector and higher income to farmers but also providers better quality products to the consumers at reduced rate. This will also leads to the economic utilization of by products enriching animal health through feed, and soil via composting. Thus, agro-processing in production catchments will benefit people, livestock and the mother earth.

Integrated intensive farming system (IIFS) including rural based postharvest activities are recommended for the overall development of the rural sector. IIFS would create employment and income generation opportunities and thereby enhance the living standard of the people.

Agro-food processing industries have the potential to generate directly significant employment in production activities and also indirect employment through its forward and backward linkages. This employment will be in rural areas where these industries have to be located near the source of raws materials, especially perishable agricultural products. These industries would help in reducing postharvest losses and wastes as well as in using by-products more efficiently. This can increase farmers' income by getting them better prices, to avoid fluctuations of prices when glut in market and also consumer welfare by increasing the availability of agricultural consumer goods. The available vast potential in our country could be sufficiently exploited through:

a) Selection of appropriate scale and technology of production;
b) Up-gradation of technology of existing units
c) Establishment of suitable linkages between products and consumers at home and abroad; and
d) Establishment of suitable institutional arrangements.

The term "agro-food processing industries" covers a wide range of activities utilizing farm, animal and forestry based products as raw materials. There are certain traditional agro-based industries such as rice and flourmills, sugar, khandsari and pulp manufacture, edible oils and the processing of plantation crops like tea, coffee and cashew nuts. There are also some relatively modern food processing industries such as dairy products, confectionery, marine products, horticultural and vegetable products as well as meat and poultry products.

Although most of the technology is readily available in the country, it is not being extensively adopted because economic incentives are often missing, or institutional arrangements for collection, processing and marketing of the byproducts may be lacking. Since the processing of byproducts in the cereals sector constitutes introduction of new manufacturing activities, or the expansion of existing ones, such activities would generate additional employment.

The next category of food processing concerns processing and packaging in order to provide easy transportability and marketability of some food products. The processing of milk and milk products falls in this category. This would increase the income of farmers, especially small farmers and landless agricultural labourers in the rural areas. It would also promote consumer welfare. The third category relates to processing activities which would help in extending the storage life of seasonal food products. Fruits and vegetables belong to this category. Processing of fruits and vegetables would help in reducing postharvest losses and would also provide stable income to the growers by eliminating the seasonal fluctuations in income.

The food processing sector is highly fragmented industry, it widely comprises of the following sub-segments: fruits and vegetables, milk and milk products, beer and alcoholic beverages, meat and poultry, marine products, grain processing, packaged or convenience food and packaged drinks. A huge number of entrepreneurs in this industry are small in terms of their production and operations, and are largely concentrated in the unorganized segment. This segment accounts for more than 70 per cent of the output in terms of volume and 50 per cent in terms of value. Though the organized sector seems comparatively small, it is growing at a much faster pace.

The organized sector is relatively more modern in technology requiring larger investments as compared to the unorganized sector. However, processing in the organized sector helps in achieving higher efficiency in the use of raw materials and by products. Processing in the organized sector generates additional employment in trade and transport activities which may be quite substantial as compared to direct employment created in processing activities. At the same time, consumer welfare is enhanced by modern processing which ensures better quality and clean packaging in an uncontaminated environment, also improves export potential.

India's food processing mainly involves primary processing which accounts for 80 per cent of the value. As much as 42 per cent of the food industry is in the unorganized sector, 25 per cent of the food industry is in the organized sector and 33 per cent in the small scale, tiny and cottage sectors (Figure 1.1). The value addition to agricultural commodities is less than 10 per cent. Food habits in India are traditional in nature and varied across the country. Busier schedules and a growing number of working women have collectively led to an increase in the demand for ready-to-eat traditional and or newer foods. As a segment of the food industry, traditional foods are the largest, both in terms of quantity and value. Most of the operations are manual, even in relatively large-scale units, causing variation in quality.

Structure of the Indian Food Processing Industry

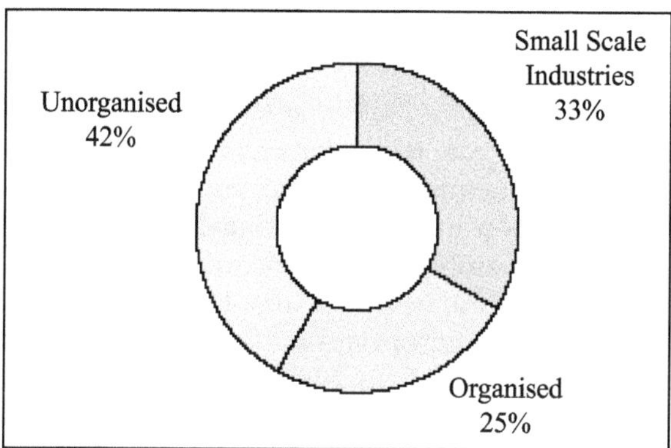

Figure 1.1: Fruits and Vegetable Processing Units in Organized Sector (numbers): 5293.

(*Source*: Ministry of Food Processing Industries, Annual Report 2003-04)

Present Scenario

The food processing industry is a catalyst connecting both growers and consumers. If the commodity is processed properly then farmer will get price for graded and sorted material and use proper varieties which produce such quality raw material. Even the use of harmful chemical, fertilizers and pesticides will be reduced if proper care is taken of soil health and GAP and GMP are followed. Similarly as our consumers is within 12 miles of the food production they will get the international quality food at much lower price as compared to what we pay for imported food items. We should make effort in establishing viable integrated postharvest industry which will take care of whole value chain and each and every component of the biomass produced by the mother earth is salvaged and made into value added product.

Agriculture in India is expected to meet the food, feed, fibre, fuel and other requirements of the society. Green revolution in the seventies led to the quantum jump in agricultural production. However, poor infrastructure and technological backstopping resulted into unacceptably high postharvest losses (6-18 per cent).

Another major question relates to the level of processing of various commodities in the country. While it is true that the processing levels are generally low in comparison to those in developed countries, it is necessary to determine the thresholds and the reasons for the low levels.

Indian agriculture is characterized by a large number of tiny, marginal, and small farm holders who could not derive adequate monetary benefits from production activities alone. Most of these farmers are hardly employed for 150-200 days in a year. Therefore, while their incomes are abysmally low; they are inadequately employed. Unless the profitability of agriculture and the income of farmers improve substantially, food, nutritional and environmental sustainability of the country is at stake. One of the NSSO surveys has clearly indicated that given the opportunity, 40 per cent of farmers would leave farming. Agricultural produce management and appropriate level of value addition in production catchments could meet these twin objectives of providing additional off-farm employment and income.

Supply chain from farm gate to markets and industries is highly fragmented. There is inadequate infrastructure for storage, packaging, handling and transport from the place of production to the markets and processors. As a result, huge postharvest losses take place, cutting into the income of the farmers and deterioration of quality of the raw materials for the processors. There is an essential requirement of postharvest management and appropriate value addition in production catchments to minimize the quantitative and qualitative losses. While people consume a substantial fraction of the agricultural produce directly with minimum processing, the surplus needs to be stored and directed towards processing industries for greater value addition and trade.

Processing and Value Addition

The two major goals of postharvest technology are loss prevention and value addition to the raw food commodities through preservation and processing. Raw food materials are cleaned, graded and then conditioned either for storage or processing. Processing is done to make raw commodities edible through primary and secondary processing and ready to eat through tertiary processing. At every stage of processing, value is added to the raw food materials through primary and secondary/tertiary processing in India are 75 per cent and 25 per cent respectively. It, therefore, shows that primary processing has a greater role to play in improving the economic benefits to the farmers.

A wide range of tools, machineries and equipment are used in post-production agriculture the globalization of trade may compel many agro processing industries to rely on an imported technology at a high cost and the advantages of such technology may be available by a few who have an exclusive business interest. Majority of agriculture-based enterprises will continue to depend upon indigenous technology and therefore, R&D through public support has to be strengthened to become globally competitive and serve small-scale food processing sector of the country.

Agricultural Processing

All the raw food materials are processed to improve their palatability, nutritional value and shelf life. Processing is carried out on household, micro, small and large scales either for home consumption and/or for trade. Conventional postharvest operations, which can be improved, take better advantages of raw materials and modern technology.

In crop production and utilization system, the post –production loss occurs at all stage of operations. Right in the field with standing crops; in transport, threshing, cleaning, drying, storage, milling, packaging, preparation and utilization. But these losses can be avoided by adopting improved postharvest practices and equipment evolved through R&D post – harvest processing.

Fruits and Vegetables

Presently, the processing of fruits and vegetables is estimated to be around 4.4 per cent of the total production in the country. The major processed items in this segment are fruit pulps and juices, fruit based ready-to-serve beverages, canned fruits and vegetables, jams, squashes, pickles, chutneys and dehydrated vegetables. The new arrivals in this segment are vegetable curries in retortable pouches, canned mushroom and mushroom products, dried fruits and vegetables and fruit juice concentrates.

The fruits and vegetables processing industry is rather fragmented. A large number of units are in household and small-scale sector, having low capacities of up to 250 tonnes per annum. From the year 2000 onwards the industry has seen a significant growth in ready-to-serve beverages, pulps and fruit juices, dehydrated and frozen fruits and vegetables products, pickles, processed mushrooms and curried vegetables and units engaged in these segments are export oriented.

The immense diversity in agro-climatic conditions across the different regions enables India to produce a large variety of fruits and vegetables that are generally grown under sub-tropical and temperate climatic conditions.

The value of fruits and vegetables production in 1992/93 was of the order of Rs. 20.94 crores accounting for 12.3 per cent of the value of output of all crops. India produced nearly 32 million tonne of fruits in 1993 which accounted for 8.6 per cent in world production of fruits. The production of vegetables in 1993 is estimated at 71 million tonnes which formed 15 per cent of world production. The total area under vegetables cultivation is 6.2 million hectares which is about 3 per cent of the total area under cultivation.

Consumer Foods

Consumer food industry includes pasta, breads, cakes, pastries, rusks, buns, rolls, noodles, corn flakes, rice flakes, ready to eat and ready to cook products, cocoa products, biscuits, soft drinks, beer, alcoholic beverages (non-molasses based), mineral and packaged water. Bread and biscuits constitute the largest segment of consumer foods. Their production is about 4.00 million tons per year. Out of the total production of bread, 40 per cent is produced in the organized sector and the remaining 60 per cent in the unorganised sector. Similarly, production of biscuits in the organized sector is about 80 per cent and quantity of biscuits produced in the unorganised sector is about 20 per cent.

Bevergaes – Alcoholic and Non-Alcoholic

India is the third largest market for alcoholic beverages in the world. Liquor manufactured in India is categorised as beer, country liquor and Indian Made Foreign Liquor (IMFL). IMFL production includes wines, whisky, rum, vodka, gin and brandy. Canned beer is an even more recently introduced new beverage. Country liquor is consumed in rural areas and by low-income groups in urban areas. IMFL is consumed by the middle and high income groups, primarily in urban areas.

The soft drinks constitute the 3rd largest packaged foods regularly consumed after packed tea and packed biscuits. The aerated soft drinks industry in India comprises over 100 plants across all States. Installed capacity of sweetened/aerated water as on 01.01.2006 is reported to be 29.60 lakh tons per annum. There is vast scope in natural beverages prepared from different fruits and vegetables. Now a day consumers are health conscious preferring such type of natural beverages. In coming era, only natural beverages were preferred by the consumers.

Rural Based Agro-processing

There is a convergence of through at all levels that development and application of appropriate postharvest technologies leading to the establishment of agro processing centers (APCs) in the production catchments and owned and operated by targeted beneficiaries, individually or collectively, are must. It has greater capacity for employment and income generation than production agriculture. It can help for stabilizing market prices through removing vulnerabilities of perishables by transforming them into semi-perishable agricultural commodities for table use. De-urbanization has come to be a development goal in order to improve lot of the rural people.

The present scenario of declining public funding and emphasis on privatization, weak public-private linkages and lack of confidence between them, adjustments in public funding and audit procedures impede the development process towards agro processing more so on farm agro-processing that can directly increase income and employment to the rural people for their socio-economic development. There are production and productivity related constraints too that impede acceleration in agro-processing. Many of our crop varieties tailored for table grade use when used for processing the product recoveries tend to be low, much lower than the strains and varieties available with the world leaders. The major goals of postharvest technology at rural threshold should be:

- ☆ Minimization of harvest and postharvest losses, improving net availability and net returns to the growers.
- ☆ Capacity to the handle and hold the produce without excessive losses at affordable cost to negotiate with the forces of marketing avoiding distress sales.
- ☆ Transform the perishables in to semi-perishables or durables for better marketing and also value addition.
- ☆ Creation of rural agro processing enterprises that meet needs of the rural people at the least cost and market surpluses alter value additional income and employment.
- ☆ Improve livelihood base of rural people through entrepreneurship development and up gradation of skills.
- ☆ Provide consumers fresh and processed products at reasonable rates.
- ☆ Make better use of crop residues, processing of by-products andwastes in eco-friendly and economically rewarding fashion.

- ☆ Meet hygiene and quality standard specified for domestic sand export markets for fresh and processed products.
- ☆ Appropriate packaging and marketing of minimally processed and value added products through cooperative super market and other retail outlets.

Major Issues

It is evident that there is considerable scope for expansion of processing of fruits and vegetables. Development of processing industry on modern, scientific lines would produce a variety of benefits. Firstly, it will significantly reduce the magnitude of postharvest wastage and losses. Secondly, it would generate employment opportunities on a significant scale both directly in processing activities and indirectly in trade and transport activities.

Current Status of Technology

The technologies currently used for different postharvest operations are:

1. Harvesting: Mostly by hands or neither the help of hand tools such as clippers, scissors, *etc.*
2. Sorting/Grading: Very limited. Mostly by visual inspection. Some large marketing agencies use weight based grading systems.
3. Precooling: Limited facilities (forced draught type) for grapes, strawberries, *etc.* and mainly for export purposes.
4. Packaging: mostly bamboo baskets and wooden boxes for domestic markets and corrugated fibre board (CFB) for exports. Controlled atmosphere (CA)/Modified Atmosphere (MA) packaging technology is available, but not used commercially.
5. Transportation: Mostly open vehicles. Limited use of refrigerated vehicles and trains which are mainly earmarked for exports.
6. Storage: Mostly ventilated storage used at the farm level. Use of cold storage at mandis is very limited.

Steps to Promote Agri-industries and Agro-processing Industries

During last two five year Plans, Government of Maharashtra has given emphasis on the development of horticulture along with Agriculture. All out efforts in this regards have reflected through increase in area and production. Maharashtra plays a major role in horticulture development under scheme EGS, which linked horticulture plantation and its production of fruits and vegetables. The export from Maharashtra for fresh vegetables and fruits to the Indian basket is 30 per cent and that of processed food products is nearly 50 per cent.With the availability of varied range of agricultural produce, the state provides great potential for establishment of agri-business. The Government needs to take fiscal steps, which would be conducive to sustain the investment. With this broad view in mind, state government proposed to develop infrastructure to speed up agri-business.

The infrastructure comprises postharvest facilities like mass collection centers, sorting grading, packing houses, store houses, refrigerated vans, pre-cooling, cold storage, establishment of Gramin Information Centers for market of higher value, residue testing laboratories for fruits and vegetables. In the case of agro-processing for food grains and oil seeds, dal mill, soyabean processing and oil processing units *etc.* at the village level.

Agriculture Export Zone (AEZ)

Already the state has started implementation of Agri Export Zones (AEZ). The State has announced eight (8) Agri Export Zone for important crops being implemented through various Agencies in the State. The concept of Agri Export Zones attempts to take comprehensive look at particulars produce/products located in a continuous area for the purpose of developing and sourcing the raw material, their processing/packing, leading to final export.

Promotion of Exports

As a leading agricultural State in the country, today the state of Maharashtra has a major contribution in country's agriculture sector and continues to play a role in export of agriculture produce and products from the country. Large portion of all horticulture produce exported from the country are from Maharashtra.

Similarly the state has major contribution to exports of cashew nut and mango from the country today and it has a potential to increase exports of agricultural produce and value added processed foods from the State.

Growth Drivers of the Sector

- Increasing urbanization
- Increasing nuclear families and working women
- Increased disposable incomes
- Organized retail and private label penetration
- Demand for functional food
- Increased spend on health food

Opportunities

The opportunities are ample in sectors like bakery, convenience food, snack, RTS, RTC, RTE *etc.* of the two segments of the Indian food processing industry - commodity-based processing and value-added processing, the former contributes to two thirds of the processed food market. And although the latter comprises a smaller segment of the market, it is experiencing a radical growth. In fact, India is moving towards convenience-foods that evolve through value added processing.

Value-Added Processing: The New Future

Currently, the demand and consumption of commodity-based processed products is higher. However, this trend is expected to change significantly on account of changing life-styles and increasing income levels. The change will soon become more prominent in urban India with an increased demand for value-added products, particularly in dairy, poultry, fruits and vegetables, *etc.*

Future Scenario

Postharvest losses on farm and in the supply chain are unacceptably high. However, there has been increased awareness in recent years about adoption of appropriate postharvest technology. The postharvest loss scenario may be changing faster than ever. There is a need to minimize these losses through appropriate technologies. This step alone could significantly improve the availability of agricultural produce for industry and consumers.

Another major challenge is to improve the income generation for farm and rural families. Therefore, on-farm and off-farm livelihood opportunities need to be created. Processing and value addition to primary agricultural produce could, therefore, provide the right opportunity in the production catchments for additional livelihood and income. A major incentive would come by way of reduced volumes of processed or semi-processed goods for transportation since the surplus residues and by-products would remain in the production catchments.

About 70 per cent of Indian population resides in rural sector deriving sustenance essentially from agriculture and allied activities. They also form the majority of consumers of food and other agriculture-based goods. Therefore, instead of transporting the entire agricultural produce from production catchments to processors in organized sector and then transporting the processed goods back, it would be appropriate if the agricultural produce is processed in production catchments for use by the rural consumers and the surplus is made available to the organized sector industries. It cuts down on the cost of goods available to rural consumers and the transportation costs leading to better environmental sustainability.

Clearly, improved scientific handling of agricultural produce through proper supply/cold chain, storage, higher level of processing into value added products in production catchments and utilization of valuable by-products available from processing industry as well as crop residues left in the field to extract bioactive ingredients or other value added products are at the core of ensuring food security and enhanced income and employment in rural sector without detriment to environment.

Way Forward

Postharvest Technology is being recognized as one of the top priority areas for the development of food production scenario in India. Production of food grains and horticultural crops have increased manifold during the recent years and destined to double in another five years but there has not been matching progress in the development and adoption of postharvest technology resulting in huge postharvest losses. Quality, nutrition and safety of the produce have also gained utmost importance. Hence, strategies are needed to ensure these aspects apart from minimizing postharvest losses and promotion of processing and value addition. The rapid globalization, market liberalization, urbanization and increased consumer awareness have created newer challenges to be managed by postharvest technologists. In order to achieve these targets, an Integrated Postharvest Technology has to be developed with a "System Approach".

Chapter 2

Present Status and Future Prospects of Dryland Fruit Crops

R.S. Patil

Director of Research,
Mahatma Phule Krishi Vidyapeeth, Rahuri – 413 722, Dist. Ahmednagar, M.S.

In response to changing dietary patterns, the composition of agricultural production has diversified over the years. As a result, the horticulture and livestock sectors have emerged as major drivers of growth in the agricultural and allied sector. On the production and productivity front, the horticulture sector outperformed conventional food crops. Between 2004-05 and 2014-15, horticultural output achieved an annual growth of about 7 per cent as compared to around 3 per cent growth in food grain production. This increase in production has come from an increase in acreage and even larger increase in productivity. While the area under horticulture crops grew by about 2.7 per cent per annum, productivity increased by 37 per cent between 2004-05 and 2014-15. As a result, India has maintained its second rank in the global production of fruits and vegetables, next to China. A comparative picture of horticulture and food grain production over the last few years is given in Figure 2.1.

Horticulture

The horticulture sector has been a driving force in stimulating growth in Indian agriculture. India is currently producing 277.7 million tonnes of horticulture produce from an area of 23.2 million hectares, which has surpassed the estimated food grain production of 257 million tonnes. Though the production of food grains and horticultural produce are not meaningfully comparable due to fundamental differences in the nature of their farming, characteristics of produce, nature of land requirements, and most importantly, their nutritional purpose and value, it has come to light that horticultural farming is much productive and gainful. The productivity

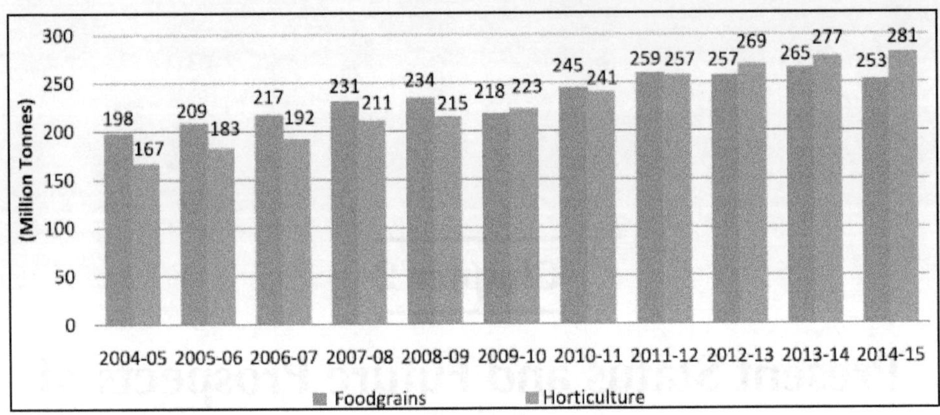

Figure 2.1: Trend in Food Grain and Horticulture Production.
*** Based on Fourth Advance Est. for Food grains and Third Advance Estimates for Horticulture.**
(*Source*: Ministry of Agriculture and Farmers Welfare).

of horticultural crops has increased by about 34 per cent between 2004-05 and 2014-15. The special attention given to the sector, especially after the introduction of the National Horticulture Mission (NHM) in the 11th Plan, has borne bumper fruit. Given the increasing pressure on land, growth strategies have been focusing on raising productivity through high density plantations, protected cultivation, micro irrigation, quality planting material, rejuvenation of senile orchards and an emphasis on postharvest management and marketing of produce for better price realization. The trend in horticultural production growth over the last 10 years is depicted in Figure 2.2 and the share of different commodities in horticulture production is given in Figure 2.3.

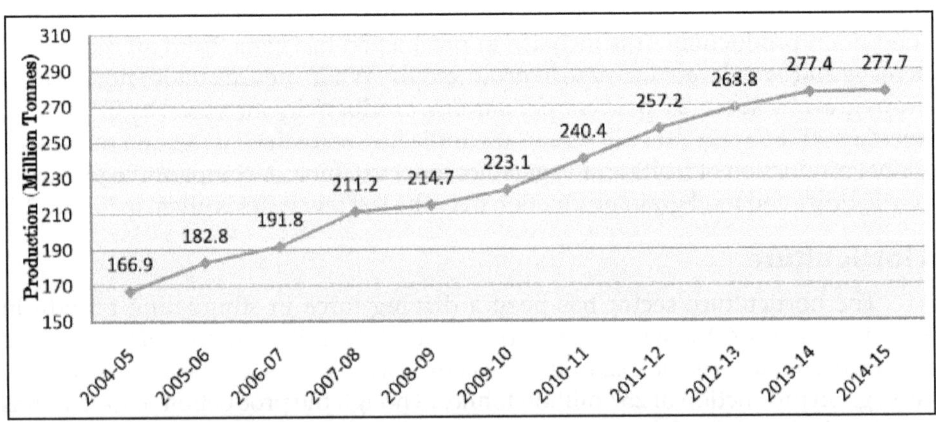

Figure 2.2: Growth Trend in Horticulture Production.
(*Source*: Department of Agriculture, Cooperation and Farmers Welfare).
: Figures for 2014-15 are based on the Third Advance Estimates.

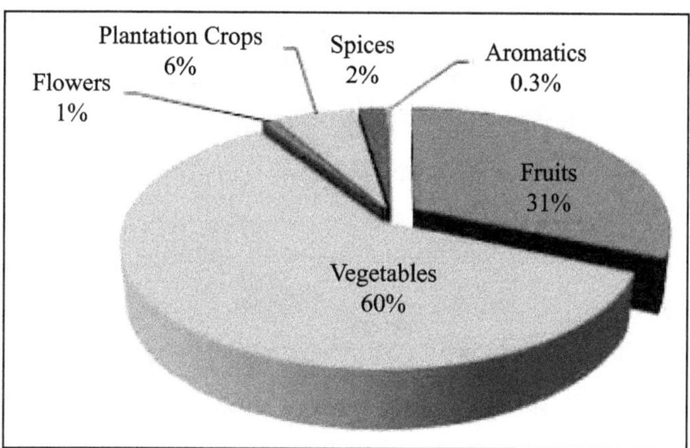

Figure 2.3: Share of Different Commodity Groups in Horticulture Production (per cent of 277.7 million MT)
(*Source*: Department of Agriculture, Cooperation and Farmers Welfare).

Agricultural Production

Fruits

With a production of 88.8 million tonnes, fruits account for about 31 per cent of total production of horticulture crops. The area under fruit crops cultivation during 2013-14 was 6.3 million hectares, which is about 27 per cent of total area under horticulture cultivation in India. The area under fruit crops cultivation has increased from 5 million hectares in 2004-05 to 6.24 million hectares in 2014-15, with a corresponding increase in production from 50.9 to 86.2 million tonnes. A large variety of fruits, such as banana, mango, citrus, papaya, guava, grape, sapota, pomegranate, pineapple, aonla, litchi, pear, plum and walnut are grown in India. India accounts for about 13 per cent of the total world production of fruits and leads in the production of mango, banana, papaya, sapota, pomegranate, acid lime and aonla.

During 2014-15, Maharashtra stood first in terms of fruit production with a 12.22 per cent share in total production followed by Andhra Pradesh with 10.57 per cent, Uttar Pradesh with 10.03 per cent, and Gujarat with 9.27 per cent and Tamil Nadu with 6.26 per cent shares. These states together contributed about 50 per cent of the total fruit production in the country. Banana is the most cultivated fruit accounting for 33 per cent of total production, followed by mango at 21 per cent, citrus at 14 per cent, papaya at 6 per cent, guava at 4 per cent, grapes at 3 per cent, apple at 2 per cent and others with a 16 per cent share in the country. In the case of the Himachal Pradesh and Jammu and Kashmir, the value of output from apples, plums, pears and stone fruits exceeds the value of output from cereal crops. The area under cultivation, production and productivity of fruit crops have registered significant increases during the last decade, as depicted in Figure 2.4.

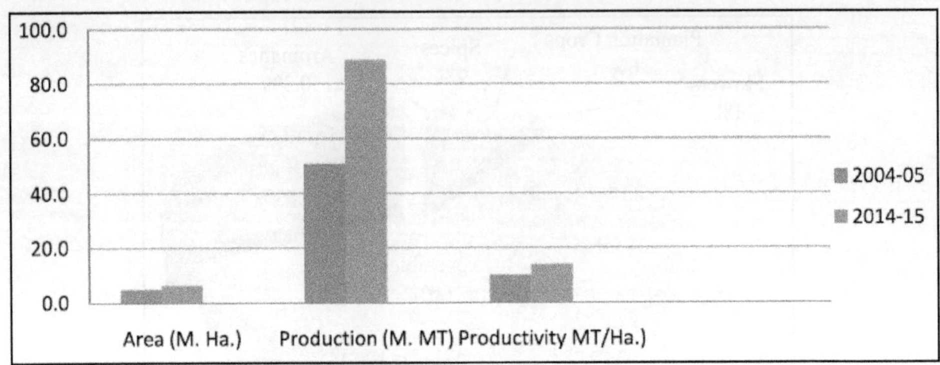

Figure 2.4: Area, Production and Productivity of Fruits.
(*Source*: Department of Agriculture, Cooperation and Farmers Welfare).

Fruit Sector

- India is the second largest producer of fruits in the world and is the leader in producing fruits like mango, banana, pomegranate, sapota, acid lime and aonla.
- India's fruit productivity is better than that of China, though China is the largest fruit producing country.
- Special efforts are being made to improve the productivity of fruit crops by enhancing the supply of quality planting material from accredited nurseries and improved package of practices.
- The per capita availability of fruit to the Indian population is 189 g/person/day and has been helping in supplementing nourishment.

Vegetables

Vegetables occupied an area of 9.5 million hectares during 2014-15 with a total production of 167 million tonnes having average productivity of 17.6 tonnes per hectare. Vegetable production registered a quantum jump of 66 per cent between 2001-02 and 2014-15. The major vegetable crops grown in the country are potato, tomato, onion, brinjal, cabbage, cauliflower, peas, okra, chillies, beans, melons, *etc.*

Vegetable Sector

- Productivity of vegetables in India continues to be low compared to world average productivity.
- Special emphasis is being given for production of vegetables under protected cultivation under Mission for Integrated Development of Horticulture (MIDH).
- Per capita availability of vegetables in India is 357 g/person/day, which is helping in fighting malnutrition.

Details of the area, production and productivity of vegetables during the last decade are depicted in Figure 2.5.

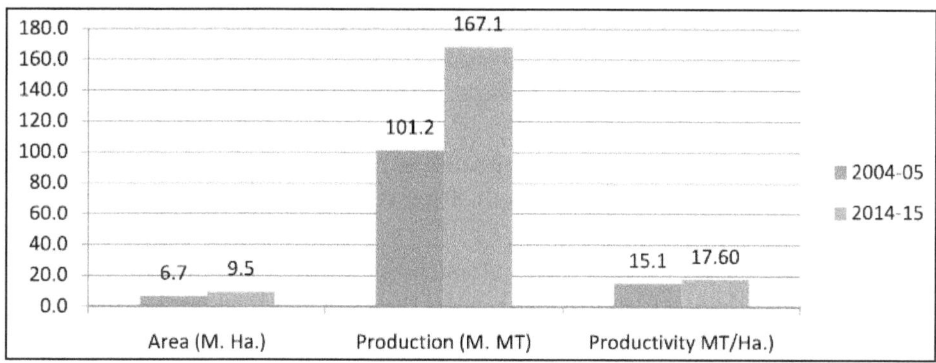

Figure 2.5: Growth in Area, Production and Productivity of Vegetables. (*Source*: Department of Agriculture, Cooperation and Farmers Welfare).

India is the second largest producer of vegetables after China and is a leader in the production of vegetables like peas and okra. Besides, India occupies the second position in terms of production of brinjal, cabbage, cauliflower and onion and the third position in the production of potato and tomato in the world.

Growth in Horticulture Exports

Commensurate with increased horticulture production, there has been a significant improvement in terms of export earnings in the horticulture sector as shown in Table 2.1.

Table 2.1: Value of Exports of Horticulture Commodities

Sl.No.	*Commodity*	*Value (Rs in crore)*		
		2004-05	*2014-15*	*2015-16*
1.	Fresh fruits and vegetables	1,725.25	7,759.72	4,727.99
2.	Floriculture products	222.92	460.76	306.95
3.	Spices	1,883.18	14,842.36	10,531.46
4.	Cashew	2,477.18	5,565.77	3,361.06
	Total	**6,308.53**	**28,628.61**	**18,927.46**

Source: Directorate General of Commercial Intelligence and Statistics (DGCIS), M/o Commerce and Industry.

The Department of Agriculture, Cooperation and Farmers Welfare is working closely with the Agricultural and Processed Food Export Development (APEDA) Department under the Ministry of Commerce and Industry and the state governments to ensure that infrastructure and institutional support is available for exports so that farmers can leverage export markets for higher incomes. Out of the 60 agri-export zones in the country, 52 focus on horticultural crops.

Achievements in Horticulture Sector during 2014-15 and 2015-16

- Attention has been paid to increasing the cold chain capacity in the country. About 1.5 million tones of capacity has been increased during last two years.
- India is the second largest producer of fruits and vegetables globally. Horticulture contributes about 30 per cent to agriculture GDP, using only 17 per cent of the land area.
- Horticulture exports were valued at Rs. 28,628 crores in 2014-15.
- Storage capacity has been increased by 66,000 tonnes by establishing 2,600 low cost onion storage units at farmers' field.
- Exports of fresh fruits like grapes, banana, mango and vegetables like peas and potato are on the rise.
- Nine Centers of Excellence (CoEs) have been established for demonstrating the latest technologies in fruits, vegetables and fruits cultivation through Indo-Israel and Indo-Dutch co-operation during the last two years.
- Integrated cold chain availability platform – an online portal providing information about all Government assisted cold chain projects – was launched.

Horticulture Development

The major planned activities taken up under the Mission for Integrated Development of Horticulture (MIDH) scheme included programmes for production of planting material, area expansion including high density planting, rejuvenation of old and senile orchards, protected cultivation, creation of water resources and promotion of INM and IPM, which are basically aimed at productivity improvement. Organic farming and good agricultural practices (GAP) are promoted to enable chemical residue free horticulture produce, besides addressing environmental concerns of soil and land degradation. Horticulture mechanization is promoted to bring in efficiency in horticulture production and harvesting operations. Production and productivity improvement programmes are complemented by the creation of infrastructure facilities for postharvest management, processing and marketing.

Towards Sustainable and Inclusive Growth in Horticulture

The major challenge for horticulture is to sustain this growth in a manner which ensures a higher income for the primary producer through better intuitional support mechanisms such as infrastructure and technology support for the entire value chain – from pre–planting to postharvest management. While the State agricultural universities, ICAR system with its research institutions, national research centres have addressed issues relating to soil health, planting material and new and adaptive varieties, the major challenge for the DAC and FW is to assure higher returns for the farmer by ensuring that what is produced is not lost in transit on account of poor handling or perishability and that the farmer is part of the value chain.

Food Processing and Value Addition

A strong and dynamic food processing sector plays a vital role in attaining higher value-addition, reduction in wastage of agricultural produce, enhancing shelf life of food products, promoting employment generation and increasing income of the farmers. The term "food processing" covers all the processes that food items go through right from the farm till the time they reach the consumer's plate. Value addition processes include i) basic cleaning, grading and also alteration of the raw material to a stage just before the final preparation; and ii) making ready-to-eat and ready-to-drink foods, such as bakery products, instant foods, flavoured and health drinks, *etc.* The Ministry of Food Processing Industries (MoFPI) acts as a catalyst for mobilizing investment into this sector, guiding investors and creating a supportive environment for the healthy growth of the food processing industry. It realizes that the food processing sector offers rural communities an opportunity for creating sustainable livelihoods and economic development. Ever-changing lifestyles, food habits and customer tastes have altered industry dynamics, and food production and consumption patterns are evolving with changes in customer needs. Food processing benefits all sections of society.

Performance of the Food Processing Sector

Growth in the food processing sector is also expected to open up a lot of opportunities for players having strong linkages in the agri-value chain. Significant investment opportunities await in supply chain management, cold storage, financing, retailing and exports. Historically, agriculture and food processing industries remained stagnant due to low public investment, poor infrastructure, inadequate credit availability and high levels of fragmentation. These issues are being resolved expeditiously by the Government, so that significant improvement possible in terms of the creation of supporting infrastructure and investment and, consequently, the overall growth of the sector. With its huge production base, India can easily become one of the leading food suppliers to the world while at the same time serving its vast growing domestic market of over a billion people. India's large market size, with growing incomes and changing lifestyles, create incredible market opportunities for food producers, food processors, machinery makers, food technologists and service providers in this sector.

The food processing sector, identified as a thrust area for development, needs huge investments in logistics for supporting the value chain from farm to plate. The Government is aware that rules and the policy regime determine the sector's performance, and has therefore exempted most food processing enterprises from the provisions of the Industries (Development and Regulation) Act, 1951 (except beer and alcoholic drinks, and items reserved for the small-scale sector). For foreign investment, automatic approval is given for up to 100 per cent equity for a majority of processed foods. Effective postharvest management not only minimizes losses but also increases the value of agricultural products marketed by transforming agricultural raw materials. Central Government policy initiatives include assistance for setting up of mega food parks, cold chains, modernization of abattoirs, development of agri-export zones, skill development and R&D activities. Various state governments are also implementing their own food processing promotion policies and schemes.

The food processing industry is one of the major employment-intensive segments of industry, and constitutes 13.04 per cent of all employment generated in the registered factory sector in 2012-13. According to the Annual Survey of Industries (ASI), the total number of persons engaged in registered food processing industries during 2012-13 was 16.89 lakhs. During the last five years ending 2012-13, employment in the registered food processing sector has been increasing at an average annual growth rate of 2.41 per cent. The unregistered food processing sector supports the employment of 47.9 lakh workers, in the year 2010-11.The overall scenario of employment in food processing sector is given in Tables 2.2 and 2.3 and Figure 2.6.

Table 2.2: Employment in Food Processing Industry: Registered and Unregistered Units

Sector	*Food Processing* Sector*	*Overall Industry*	*Share of Food Processing Sector (per cent)*
Registered# (2012-13)	16.89 lakh	129.50 lakh	13.04
Unregistered (2010-11)	47.90 lakh	348.88 lakh	13.72

* Includes food products and beverages segments.

Factory registered under Sections 2m (i) and 2m (ii) Factory Act 1948

Table 2.3: Employment in Registered Food Processing Units (2008-09 to 2012-13)

Year	*2008 -09*	*2009-10*	*2010-11*	*2011-12*	*2012-13*	*AAGR**
Persons (in lakh)	15.64	16.06	16.62	17.77	16.89	-
Growth (per cent)	3.87	2.71	3.46	6.92	-4.94	2.41

Source: Annual Survey of Industries, MOSPI; AAGR.

*: Average Annual Growth Rate for 5 years.

In terms of investment in fixed capital, the registered food processing sector has been growing at an annual average of 18.47 per cent during the five years ending 2012-13.

Major Initiatives to Promote the Food Processing Industry

- ☆ The Ministry of Food Processing Industries (MoFPI) has developed an "Investors' Portal" to disseminate information on state-specific resource potential, policy support and fiscal incentives for investors in the food processing sector. Both domestic and foreign investors can also seek guidance on specific issues by accessing the Investors' Portal (http://foodprocessingindia.co.in) or the Ministry's website (http://mofpi.nic.in).

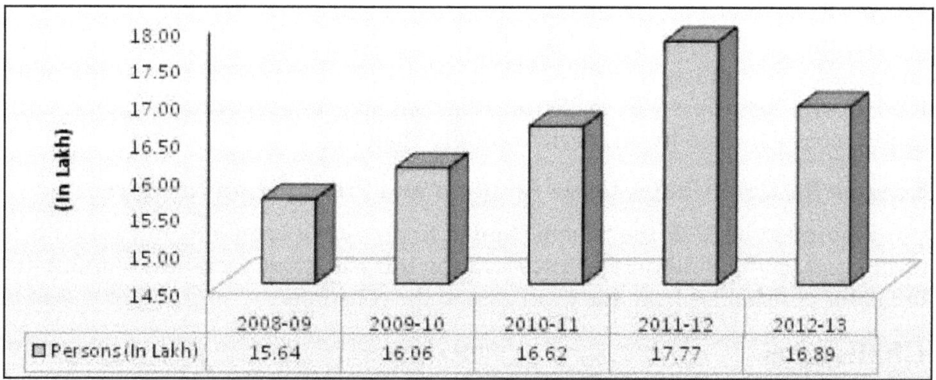

Figure 2.6: Persons Employed in Registered Food Processing Units. (*Source*: Annual Survey of Industries, CSO).

- Excise duty on food processing and packaging machinery has been reduced from 10 per cent to 6 per cent. The Union Budget 2015-16 exempted from Service Tax the services of preconditioning, precooling, ripening, waxing, retail packing and labelling fruits and vegetables.
- Government had announced setting up of special fund of Rs 2,000 crores in the financial year 2014-15 in NABARD for extending affordable credit to designated food parks and the individual processing units in the designated food parks at concessional rates. The fund is being continued in 2015-16.
- Recently, the Reserve Bank of India classified loans to food and agro-based processing units and cold chains under agriculture activities for priority sector lending (PSL), subject to an aggregate sanctioned limit of Rs 100 crores per borrower. It will ensure greater flow of credit to entrepreneurs for setting up food processing units and attract investment to the sector.
- With a view to help investors and prospective entrepreneurs identify the raw material base in different parts of the country, the Hon'ble Minister, Ministry of Food Processing Industries released "Food Maps of India" at a function organized by Yes Bank on 18 December 2014. These food maps have been uploaded to the Ministry's website.

Major Plan Schemes

Scheme for Infrastructure Development

Integrated and holistic infrastructure is extremely important for the food processing sector. Towards this end, the Ministry of Food Processing Industries (MoFPI) has been supporting the creation of modern enabling infrastructure and efficient processing facilities, such as (i) mega food parks and (ii) cold chain, value addition and preservation infrastructure.

i. Mega Food Parks Scheme: The Mega Food Parks Scheme is a flagship programme of the Ministry of Food Processing Industries under implementation since the Eleventh Five Year Plan. The scheme aims to accelerate the growth of the food processing industry by facilitating the establishment of strong food processing infrastructure backed by an efficient supply chain.

Scheme for Cold Chain, Value Addition and Preservation Infrastructure

Approved in 2008, the Scheme aimed to provide a complete, integrated cold chain, and value addition and preservation infrastructure facilities, without any break for perishables from the farm gate to the consumer.

Challenges

One of the major challenges faced by the sector is the significant postharvest wastage of agricultural produce, particularly in fruits and vegetables. Such wastage may be minimized by creating an efficient system of transportation, storage and product delivery, and by increasing the level of processing of agricultural produce. Apart from infrastructure constraints, the food processors face problems in procurement of raw materials for processing due to restrictive provisions in marketing of agricultural produce.

Another major constraint faced by food processors is the lack of availability of processable varieties of fruits and vegetables, due primarily to the inadequate linkage between production and processing. This is often addressed through a system of contract farming between food processors and farmers. Some states permit contract farming, but a number of states are yet to notify the rules.

Some corporate entities operating in the food processing sector have successfully organized farmers, and supply farm inputs and provide training on crop practices. They have also entered into arrangements for procuring agricultural produce from farmers and agreed to a price. The farmer is assured of a reasonable return, and the processer is assured of adequate, timely supply of raw materials of the required quality, with no further intermediation. This kind of collaboration is critical for the success of the food processing industry in India because of the peculiar nature of landownership. Effective tie-ups across the value chain will translate into assured supply of adequate inputs, efficient agri practices, monitoring of quality and minimization of cost and reduction of wastage. In the era of trade liberalization, the constraints of technology and skill can be overcome; what is perhaps needed is a long term partnership with farmers on mutually beneficial terms.

Way Forward

The huge need for value addition and for minimizing the reduction of wastage, and given the high employment potential of the sector, the allocation for the sector should be stepped up significantly.

Postharvest Management and Value Addition

Novel products and technologies were developed for the reduction of postharvest losses, utilization of by-products and the creation/generation of employment opportunities, besides enhancing farmers' income.

References

1. Anonymous, 2015. A Report on Horticulture Statistics at a Glance 2015 published by Horticulture Statistics Division, Department of Agriculture, Cooperation and Farmers Welfare, Ministry of Agriculture and Farmers Welfare, Government of India.
2. Anonymous, 2016. A Report on State of Indian Agriculture 2015-16 published by Government of India, Ministry of Agriculture and Farmers Welfare, Department of AgricultureCooperation and Farmers welfare, New Delhi, India.

Chapter 3

Fruit Production in India: Issues and Challenges

S.A. Ranpise

Head,
Department of Horticulture,
Mahatma Phule Krishi Vidyapeeth, Rahuri – 413 722, Dist. Ahmednagar, M.S.

India is the second largest producer of fruits and vegetables in the world after China with diverse soil and climate comprising several agro-ecological regions provides ample opportunity to grow a variety of horticultural crops which form a significant part of total agricultural produce in the country comprising of fruits, vegetables, root and tuber crops, flowers and other ornamentals, medicinal and aromatic plants, spices, condiments, plantation crops and mushrooms. It is estimated that all the horticulture crops put together cover nearly 11.6 million hectares area with an annual production of 91 million tonnes. Though, these crops occupy hardly 8 per cent of the cropped area in India with approximately 30 per cent contribution in agricultural GDP. Export of medicinal plants, fruits and vegetables have also exhibited rising trend. Horticultural crops play a unique role in India's economy by improving the income of the rural people. Cultivation of these crops is labour intensive and as such they generate lot of employment opportunities for the rural population. Fruits and vegetables are also rich source of vitamins, minerals, proteins, carbohydrates *etc.* which are essential in human nutrition. Hence, these are referred to as protective foods and assumed great importance in nutritional security of the people. Thus, cultivation of horticultural crops plays a vital role in the prosperity of a nation and is directly linked with the health and happiness of the people. India with more than 28.2 million tonnes of fruits and 66 million tonnes of vegetables is the second largest producer of fruits and vegetables in the world next only to Brazil and China. However, per capita consumption of fruits and vegetables in India is only around 46kg and 130g against a minimum of about 92g and 300g respectively recommended by Indian Council of Medical Research and National Institute of Nutrition, Hyderabad (Datta, 2013).

Horticultural Development in India

Horticultural development had not been a priority in India until recent years. In the period 1948-80, the main focus of the country was on cereals. Much planned efforts had not been made for horticultural development, except for some technical support and development efforts for specific commodities like spices, coconut and potato. During 1980-92 there was consolidation of institutional support and a planned process for the development of horticulture. It was in the post-1993 period that a focused attention was given to horticulture development through an enhancement of plan allocation and knowledge-based technology. Despite of this decade being called a "golden revolution" in horticultural production.

The horticulture sector has been a driving force in stimulating a healthy growth trend in Indian agriculture. India having current area under horticulture crops has increased from 16.6 million ha in 2001-02 to 23.7 million ha in 2012-13 with corresponding increase in production from 145.8 million tonnes to 268.8 million tonnes Over the last decade, Similarly, the Area under fruit crops has increased from 4010 thousand ha in 2001-02 to 7216 thousand ha in 2013-14 with corresponding increase in production from 43001 thousand MT to 88977 thousand MT and also increase in productivity from 10.7 thousand MT/ha in 2001-02 to 12.3 MT/ha in 2013-14. The area under horticulture grew by about 4.0 per cent per annum but production rose by 7.6 per cent per annum.

Table 3.1: Area, Production and Productivity of Horticulture Crops in India

INDIAN HORTICULTURE PRODUCTION - AT A GLANCE

Year	Fruits		Vegetables		Flowers		Nuts (Almond & Walnut)	
	A	P	A	P	A	P (Loose)	A	P
1991-92	2874	28632	5593	58532	NA	NA	NA	NA
2001-02	4010	43001	6156	88622	106	535	117	114
2002-03	3788	45203	6092	84815	70	735	117	114
2003-04	4661	45942	6082	88334	101	580	106	121
2004-05	5049	50867	6744	101246	118	659	106	121
2005-06	5324	55356	7213	111399	129	654	130	149
2006-07	5554	59563	7581	114993	144	880	132	150
2007-08	5857	65587	7848	128449	166	868	132	177
2008-09	6101	68466	7981	129077	167	987	136	173
2009-10	6329	71516	7985	133738	183	1021	142	193
2010-11	6383	74878	8495	146554	191	1031	including in fruit	
2011-12	6705	76424	8989	156325	254	1652	including in fruit	
2012-13	6982	81285	9205	162187	233	1729	including in fruit	
2013-14	7216	88977	9396	162897	255	2297	including in fruit	

Source: NHM Data Base 2015

Table 3.2: Area and Production Estimates for fruit Crops

(Area in '000 ha, Production in '000 MT and Productivity = MT/ha)

Crops	2011-12			2012-13			2013-14		
	Area	Production	Pdy.	Area	Production	Pdy.	Area	Production	Pdy.
Banana	797	28455	35.7	776	26509	34.2	803	29725	37.0
Mango	2378	16196	6.8	2500	18002	7.2	2516	18431	7.3
Citrus	915	7922	8.7	1042	10090	9.7	1078	11147	10.3
Papaya	117	4457	38.1	132	5382	40.7	133	5639	42.3
Guava	220	2510	11.4	236	3198	13.6	268	3668	13.7
Apple	322	2203	6.8	312	1915	6.1	119	2585	21.8
Pineapple	102	1500	14.7	105	1571	14.9	313	2498	8.0
Sapota	163	1426	8.7	164	1495	9.1	177	1744	9.9
Grapes	116	2221	19.1	118	2483	21.1	110	1737	15.8
Pomegranate	112	772	6.9	113	745	6.6	131	1346	10.3
Litchi	80	538	6.7	83	580	7.0	84	585	7.0
Others	1383	8224	5.9	1402	9315	6.6	1484	9872	6.7
Total	6705	76424	11.4	6982	81285	11.6	7216	88977	12.3

Source: Indian Horticulture Database 2015.

Table 3.3: Production Share of Major Fruit Crops in India (2013-14)

Name of Fruit crop	**per cent share**
Banana	33.4
Mango	20.7
Citrus Fruits	12.5
Papaya	6.3
Guava	4.1
Grape	2.9
Apple	2.8
Sapota	2.0
Pine apple	2.0
Pomegranate	1.5
Litchi	0.7
Others	11.1

Prospects in Horticulture

Growth achieved in horticulture sector so far needs to be sustained:

- ☆ Increase in demand of horticulture products due to growing population

- ✰ Increase in household income and growing sensitivity towards nutritional security.
- ✰ Enhance returns per unit of area.
- ✰ Harness existing potential for horticulture sector.

To achieve this, there is need to address critical issues and challenges.

Important Issues

1. Inadequate Availability of Quality Planting Material

The increasing demand for horticultural products for meeting the internal and external demands has resulted in the increased demand of planting material and seeds for various development programmes. Most of the planting material of the perennial horticulture crops is vegetatively propagated, produced in the nurseries and sold to the farmers. The planting material - seedlings/grafts of fruits and plantation crops show characteristics after a long gestation period. It is, therefore, essential to ensure that the planting material produced and supplied by nurseries is true- to- type, disease free and of recommended varieties.

The acute shortage of quality planting/seed materials including seeds of improved varieties is one of the major constraints limiting faster development in many horticultural crops. This is despite the fact that a large number of private and government owned nurseries and seed companies are engaged in production and supply of planting materials. Also the process involves the production of nucleus planting material by the ICAR/SAUs and its multiplication in the nurseries. While the nurseries are useful for meeting the usual demand, the large scale propagation involves micro propagation techniques like tissue culture. It is however be necessary to ensure that public sector nurseries have mother blocks for supply of true to type planting material of good pedigree to small nurseries as well as growers. The mother stock planting material should be made available from the Research Institutes, State Agriculture Universities and other sources, where productivity and quality have been tested and certified by the Horticulture Department of the State Government. The functioning of the existing units will have to be reviewed and good ones accredited for ensuring the supply of good quality planting material. This would be achieved by giving emphasis on nursery development in the private sector (Anon, 2001).The impact of using quality of planting material will become manifest in increased productivity of several horticultural crops *e.g.*, mango, citrus, grape, pineapple, banana among fruits.

2. High Prevalence of Senile Orchards

India has witnessed tremendous increase in the production of horticultural crops especially fruits since independence. Due to specialized scientific interventions and application of latest research tools, the country is now among the top fruit producing countries of the world ranking second next only to China. However, the productivity has still remained low as against area under the fruit cultivation. Several neglecting issues related to the production technology have remained unattended so far. The old and senile orchards are now reverting towards a declining trend of production because of plant age factor, non-compatible varieties and poor canopy management (Baba, 2011).The low productivity of the perennial fruits like mango, citrus, apple and the plantation crops like coconut, arecanut, cashewnut, *etc.* is mainly due to the small size of holdings, the preponderance of old and senile trees and the poor management available to them.

The average size of a fruit orchard holding is reported to be 0.21 to 0.79 ha. Apart from the small holding size, the poor condition of trees in most plantations and orchards is another contributing factor to the low yields. The thickly shaded mango orchards in the Malihabad areas of U.P. and the seedling orchards of guava throughout the country and the disease affected orchards plantations are commonly seen in large numbers in different parts of the country. These have brought down the average productivity in these crops drastically.In fruit crop about 20-30 per cent of the older plantations are of seedling origin embodying non-descript material and poor genetic potentiality which become senile.In several areas, plantations of improved varieties having good genetic potentiality have either gone unproductive or showing marked decline in productivity. This is the outcome of overcrowded and intermingling of large branches and meager foliage, allowing poor light interception and utilization by photosynthetic surface to growing shoots within the canopy. This renders them uneconomical. Such exhausted trees can be rejuvenated by heading back of branches for the production of new shoots, which can bear good crops in the years to come.

3. Pests/Disorders of National Importance

Most of the major horticultural crops, are beset with several physiological disorders, which have defied all attempts for resolution, and have contributed heavily to low productivity. Besides, there are pests and virus diseases, which have no solution. Important problems/diseases affecting productivity adversely are the following: **Mangoes**: Malformation; alternate bearing/irregular bearing; spongy tissues. **Banana**: Sigatoka Disease; Bunchy top; **Guava**: Wilt; **Citrus**: Decline/ Dieback; **Custard apple:** Fruit Set in Custard apple; **Pomegranate: Oily** Spot in Pomegranate; **Apple:** Apple Scab; **Papaya:** Papaya Ring Spot Virus.

In addition, the nutritional deficiency disorders like zinc deficiency in citrus and grapes, potassium deficiency in mango affect the growth and yield of crops. Non adoption of adequate and timely control measures against pests and diseases also cause major crop losses.(Anon, 2001).

4. Management of Canopy Architecture in Fruit Crops

In many fruit crops, increase in production with enhanced fruit quality is obtained by managing canopies of short statured trees. Small trees capture and convert sunlight into fruit production in a better way than larger ones. For commercial fruit growing, the natural form and shape of fruit trees have to be modified through the practice of pruning to achieve the target of higher yield and better quality fruits. The fruit growers desire to regulate the growth and development of their trees in a scientific manner. This is because it is not always wise to allow a plant to develop naturally, since unwanted portions may develop at the expense of those which are essential from the cultivator's point of view. The removal of unwanted parts of plant is thus known as pruning. (Singh, 2010)Canopy management depends on the nature and growth pattern of plant, number of plants/ha and pruning techniques. Pruning should be done when trees are young to establish their basic shapes and sizes, and continued as trees mature. When pruned to allow light to reach the top and lower sides of trees, the increased bearing surface results in more fruit setting. Any sprouts arising from the rootstock *i.e.* from below the graft union) should be removed. Pruning of the scion of young trees only delays growth and extends the juvenility. Pruning is done either to encourage the growth (thinning) or to reduce the tree size (heading back). Thinning of bearing trees encourages vegetative growth and removes interior branches, encouraging the outward growth into the allotted area planned for mature trees to occupy. Heading back reduces the outward canopy growth through topping and hedging of branches. To fulfill the nutritional demand of burgeoning population in the country and to meet the global challenge in the international market, it is need of the hour to accommodate almost 10 times more plants per unit area. This is only possible with the management of canopy to intercept maximum light in the innermost parts of trees. The canopy management has direct correlation with dry matter production, flower bud initiation and fruit quality.

5. Impact of Climate Change in Indian Horticulture

Global warming and climate change is the greatest concern of mankind in 21st century. The established commercial varieties of fruits will perform poorly in an unpredictable manner due to aberration of climate. A large variety of fruits are grown in India, of which mango, banana, citrus, guava, grape, pineapple and apple are the major ones. Due to rise in temperature, crops will develop more rapidly and mature earlier. For example, Citrus, grapes, melons *etc.* will mature earlier by about 15 days. Strawberries will produce more runners at the expense of fruits. Specific chilling requirements of pome and stone fruits will be affected hence dormancy breaking will be earlier. Delay in monsoon, dry spells of rains, and untimely rains during water stress period, supra-optimal temperatures during flowering and fruit growth, hailstorms are some of the most commonly encountered climatic conditions experienced by the citrus growers over the past decade or so. The climate change increases the atmospheric temperature and change of rainfall pattern, as a result, banana cultivation may suffer from high temperature, soil moisture stress or flooding/water logging (Datta, 2013).

High temperature and moisture stress also increase sunburn and cracking in apples, apricot and cherries and increase in temperature at maturity will lead to fruit cracking and burning in litchi (Kumar and Kumar 2007). Air pollution also significantly reduced the yield of several horticultural crops and increase the intensity of certain physiological disorders like black tip of mango which is induced by coal fume gases, sulphur dioxide, ethylene, carbon monooxide and fluoride. Leaf production increases by one leaf per month for every 3.3 to 3.7°C rise in minimum or mean temperature from 10-20 °C or 13.5 to 25 °C respectively. Higher temperature (31-32°C), in general, increases the rate of plant maturity in annual species, thus shortening the growth stages, during which developing fruits and suckers absorb photosynthetic products. The temperature below 10oC leads to impedance of inflorescence and malformations of bunches. Chilling symptoms on leaves are not seen immediately but it may take 2 to 4 days to appear. The production of apple has gradually increased but the productivity has fallen from 10.8 to 5.8 t/ ha (Awasthi *et al.*, 2001).

6. Development of Horticulture

Dryland

The dryland/rainfed areas of the country account for more than 70 per cent of the cultivable area in the country. These areas would require focussed attention to tap the potential of dryland horticultural crops, area under fruits like ber, aonla, date palm, pomegranate, citrus, sapota *etc.*

Wastelands

The unproductive lands belonging to the farmers, particularly those affected by salinity/alkalinity and other erosion and land degradation problem can be utilized profitably by introducing horticulture in these areas. This will not only help in increasing the production of the much needed horticultural produce but also help in rehabilitating the degraded area thereby providing remuneration to the farmer.

Island Ecosystems

The island ecosystems provide a good potential for development of horticulture because of its isolation from the mainland with minimum impact of pests and diseases. The Andaman and Nicobar Islands, located in the Bay of Bengal is a natural habitat for a variety of horticultural crops like coconut, cashew nut, jamun *etc.* In fact one island *viz.* Car Nicobar island is fully under coconut plantation. Similarly, there is scope to improve the production of horticultural crops in the Lakshadweep Islands. Development of horticulture with focussed attention would result in the supply of horticultural commodities, both for the mainlanders and the local tribal population.

7. Postharvest Losses

Out of the total horticulture crops produced in the country, approximately 60 per cent is consumed by the local population or marketed in the nearby market yards and only about 40 per cent of the produce is channelized in the regulated markets for the consumption of the urban population in the large cities. Export markets account for approximately 10 per cent of the total production. What to talk of rural marketing, the bare minimum infrastructural facilities are lacking even in the regulated markets. Handling of horticulture crops still becomes worse because of tropical climate available in most parts of India, under which physiological spoilage and microbial damages take place quite fast.(Mittal, 2006).

The horticulture produce suffers heavy postharvest losses in the absence of adequate postharvest and marketing infrastructure *viz.*, precoolingunits, packing and grading sheds, short and long term coldstorage facilities, refrigerated containers, storage and phytosanitary facilities at airports. Further, in spite of having varied agro-climatic conditions, abundance of natural resources like sunlight and water sufficient labour availability, multiple produce, our country is trailingbehind in productivity, export and processing of horticultural products ascompared to other horticulturally advanced countries. There is considerable loss in the amount of fruits and vegetables produced in India due to improper postharvest operations, these results in a wide gapbetween the gross production and net availability. Assuming an average loss of 25 per cent (though the range varies between 8-37 per cent in various crops during different stages after harvest) in all the horticultural crops together, the losses are phenomenal. To minimize these losses, it is essential to analyse the contributing factors which result into these losses.

- **Harvesting and pre-harvesting**: due to spoilage, spoilage and trimming.
- **Transport**: due to bruising, breakage and infection as a result of dust, heat, rain and humidity.
- **Storage**: due to over ripening or under ripening.
- **Processing and packing**: due to inefficiency and contamination.
- **Marketing**: due to loss of weight and quality with multi-level handling.

The problem is further complicated due to the fact that there are no storage facilities at the farm level and the farmers are forced to dispose of the entire produce immediately on harvesting. This createsa glut situation in the market. Thus, the margins of the whole salers and retailers are much higher than in the advanced countries.

8. Mechanization of Horticulture Operations

Nursery

The prices of seeds especially F_1 hybrids are very high. These needs proper environment for better germination. Greenhouses for different agro-climatic conditions need to be developed for raising the nursery. Healthy and disease free seedlings can be raised in shorter duration under greenhouse conditions.Necessary root media can be pulverized, mixed, pasteurized and filled inthe pot or tray. For this necessary equipment need development. The root media can be mixed in batches by modifying the concrete mixer. The pasteurization can be done by applying steam to the rooting media bymaintaining the media at 60°C to 82°C for the required duration. Different diseases and weeds can also be controlled in this way. Portable steamgenerator can be used to generate the steam and the aerated steam can be passed through perforated pipes buried below the media or beds.

Transplanting

The transplanting of nursery in the field is very labour intensive and seedling transplanters for different crops are required. Crops like onion, which need close spacing need bare root transplanter development. Necessary systems for transporting the seedlings, filling in the transplanter and transplanting systems needs development.

Pesticides Spraying

Mostly the spraying of fruit, vegetable and floricultural crops are being done by manual sprayers. These operations are labour intensive and can bemechanized by using power-operated sprayers. The efficient orchard sprayers need development. Tall tree sprayers are also required for old plantations of mango and for other plantation crops like coconut.

Harvesting and Transporting

Horticultural crops like fruits and vegetables give very high yield and need very careful harvesting and transporting. The high capacity harvesters for different fruit crops like mango, guava, sapota, orange, pineapple, *etc.* need tobe developed. The harvesters for reduction of labour for vegetable crops likeonion, cabbage, cauliflower, tomato *etc.* are required.

Grading

The better quality and grade of fruits or vegetables get a premium pric ein the market. For export of fruits there are different grades based on weight of individual fruits. The electronic fruit grader is required to be tested and adoptedby the export houses. The size grader for different fruits like sapota, orange,mango, pineapple *etc.* is required for local market.

Packaging

Around 30 per cent of the produce get spoiled during the marketing chain from farm to the retailer under Indian conditions. The proper packaging for fruits,flowers and vegetables for internal trade and export needs development.

9. Inadequate Processing Facilities

The weak processing infrastructure, as it exists today, has been one of the contributing factors for ineffective utilization of the raw materials resulting in huge postharvest losses. Further, serious price fluctuations commonly observed in the Indian horticulture scenario lead to glut situation in individual commodities. Insufficient technologies for commercial utilization of by-products/value added products also act as a damper for handling horticulture produce. Lack of adequate standards for quality produce also hinders the export prospects of these crops. Lack of infrastructure drying yards due to high capital investment hinders effective utilization of raw materials. As small and marginal growers predominate the production sector, individual processing facilities are not available due to lack of skill as well as investment capability

10. Inadequate Research and Extension Support

Having regard to the number of crops covered in the horticulture group, the peculiarities of the problems being faced by each one, their perishable nature and the elaborate organizational support required, the type and volume of the research are not commensurate with the situation. The R&D support available is highly inadequate despite major strengthening of the research system through establishment of National Research Centers and upgradation of centers to Institutes. Besides most of the NRCs have been started only during the VIII and IX Plan periods and they have yet to make a mark in their achievements in the areas being dealt with by them. (Anon, 2001).

Improved production technologies developed at Research Stations have not been transferred to the field to the required extent due to lack of sufficient extension personnel and infrastructure. Motivation of farmers to adopt new technology through field demonstration has not been carried out satisfactorily.

Important Challenges

Notwithstanding the critical role that horticulture plays in modern communities, horticulture globally is undergoing some very significant challenges that are impacting on both the production of horticultural crops and also on the priorities for research and development. Many of these challenges are quite recent and are of considerable magnitude. They include, but are not limited to, the following:

1. Global Population

By 2050, the global population is expected to increases to 3 billion. The challenge of feeding a growing population is daunting while all the economic sectors depend to some degree on ecosystem services like agriculture and allied activity specially like that of horticulture. Horticulture help in maintaining healthy ecosystem, water regulation, pollination, erosion control, and climate and wind regulation so also world food security issues could have been solved by paying greater importance to the role of horticulture (Kamei, 2013)

2. Urbanisation

Globally, an increasing proportion of the world's population is living in urban environments where their interest in and knowledge of farming and therefore of food production, is becoming increasingly deficient. In 1950 approximately 71 per cent of the world's population lived in rural locations while in 2010 this had declined to 50 per cent (or as low as 10, 18, and 22 per cent for the United Kingdom, the USA and France, respectively) and is projected to be as low as 30 per cent globally by 2050 (United Nations, 2007). There is, in all developed countries, a growing expectation and dependence on the supply of a low-cost, year-round supply of premium quality fruit and vegetables. Urbanisation is also impacting significantly on the availability of labour for farming activities in general (Sharma and Alam, 2013)

3. Increased "Healthy Diet"

Consumers have become more interested in understanding the sources of the foods that they eat. Current trends indicate that consumers are looking for increased variety, freshness, and healthy options in their eating choices. They are also seeking greater ease and a higher proportion of fresh produce in their diets. Those in metropolitan environments are more and more aware of and dependent on green spaces for their livelihoods and wellbeing. The future for horticulture and its foundation sciences within such an environment is, therefore, exhilarating, tricky, motivating and surely worthwhile. (Sharma and Alam, 2013)

4. High Investments and Long Gestation Period

The long gestation period that the perennial horticultural crops like mango, sapota, citrus, apple and plantation crops like coconut, arecanut, cashewnut, oil palm coffee, tea and rubber coming to the economic bearing age. This calls for liberalized credit facilities in easy installments for repayment in the form of soft loans to small and marginal farmers to be introduced if the benefits of the horticulture industry are to be fully exploited. High cost of inputs and lack of enough incentives for production of quality varieties/species, product diversification, value addition, etc. also hinder crops development.

5. Germplasm Conservation and Product Enhancement

There is a growing awareness world-wide about the need to conserve plant germplasm for the use of future generations. Furthermore, the fact that most fruit crops of horticultural significance are vegetatively propagated and not seed propagated restricts the options for their conservation other than through relatively expensive live plant collections.In the related area of germplasm enhancement and new cultivar development, there is now a major shortage of conventional plant breeders world-wide as programmes have become disproportionately focussed on molecular techniques. Formal training in plant breeding as well as the retention of conventional breeding in both tertiary education and government research institutes has also seen a major decline in the past two to three decades.

6. Sustainable Production Practices

The concepts of sustainability of horticultural crops relate both to the use of resources, such as water and nutrients, in a manner which considers future needs, and the responsible use of compounds such as pesticides which will not compromise the quality of the environment (or the safety of the produce that is produced). The most topical issue that is facing horticultural production in almost all areas of the world is the availability of water. Pressure on water resources for urban, industrial, recreational, conservation and other uses all appear to have higher priority within societies than the availability of water for horticultural crop (food) production (Sharma and Alam, 2013) Current demands on water from both river and subterranean sources are outstripping availability resulting in attention being given to water harvesting options with on-farm dams, the use 7 of waste water and to the use of more efficient application methods. Research is urgently needed to resolve issues such as the development of drought tolerant crops, the management of crops under managed water deficits, dealing with increased salinity, and the use of low quality water. Integrated crop production systems have been shown to be effective on a number of crops for reducing the number of pesticide applications and for shifting to new control compounds that have less persistence in the environment and which are of lower overall toxicity. Producers have behaved responsibly and with some urgency to ensure the adoption of new practices.

7. Year-Round Supply

The year-round provision of safe, nutritious fruit crops to consumers has, over the past 50 years, primarily been through a combination of locally-produced and imported crops that have been handled through various transport and storage regimes some crops such as apples have been stored for up to 10–12 months and shipped to both local consumers and to opposite hemispheres using such methods. The high costs of energy for transport and for storage, changes in consumer consumption preferences, and changes in consumer concerns To meet the steadily rising requirements of low-cost, year-round supply of premium quality fruit and vegetables especially in developing countries, production of fresh vegetables for export has grown rapidly in a number of countries around the world over the last decade. (Sharma and Alam, 2013)

8. Genetically Modified Crops

GMOs are now widely used in a number of countries for the production of large-scale agronomic crops, in caseof horticultural crops where concerns and debates about the safety and the wider justification of genetic modification prevail in many countries. Horticultural crops, especially perennial crops like fruits, suffer from major disadvantages in terms of application of biotechnology, including longer periods of crop durations, very high heterozygosity and inter-plant variability, less amenability to most genetic analyses and lack of genomic resources. These problems are more severe in the case of mango, coconut, banana, pomegranate, etc. which are the major fruits of India.Ringspot (papaya), bunchy top and streak (banana), yellow vein mosaic (okra) and other fungal and bacterial pathogens such asFusarium (banana) and wilts (most horticultural crops including coconut), the scenario is depressing, as there are either no resistant genes available globally or the breeding progress is slow. Innovative biotechnological interventions are the need of the hour to address these maladies.the direction of current research on horticultural biotechnology in India and a sense of urgency and requirement of streamlining of research initiatives. Also, the importance of innovation-driven horticultural biotechnology research, as deliberated, to address pre-sent and future challenges indicated the much needed emphasis in lifting up the quality of research and corresponding positivity in the enhancement of contribution of Indian horticulture to the national nutritional security while keeping in view the Millennium Development Goals.(Hanur and Aswath2013).

9. Poor Market Intelligence

For any planning process aimed at development of particular sector, an authentic and upto date statistical base is an essential pre-requisite. Marketing of horticultural produce is a major constraint in the production and disposal system and has a major role to play in making the industry viable or otherwise. Fruits and vegetables are mostly marketed through commission agents. A very small portion is handled by cooperative marketing societies. In some cases fruits, vegetables and flowers are also auction by the owners to the pre-harvest contractors. The pre-harvest contract could be for one or even three years in perennial cropsThe returns from such arrangements are very low. Such sales also result in poor upkeep of the orchards and the contractors hesitate to make further investment in the upkeep of such orchards. There is an urgent need to develop knowledge of market information (Anon, 2011).

Way Forward

- ☆ Special emphasis on establishment of mother blocks/rootstock blocks with hi-tech nurseries and tissue culture units.
- ☆ Accreditation of nurseries-All nurseries need to be accredited to ensure production and supply of quality planting material.
- ☆ Import of planting material of improved varieties and rootstocks-States need to identify variety/rootstocks and its requirement for establishment of progeny blocks.
- ☆ More attention is required for procurement of planting material of standard qualities instead of tendering system.
- ☆ Protocols and standards of vegetatively propagated fruit plants - More awareness is required on quality parameters of vegetatively propagated fruit crops.
- ☆ Concerted efforts are essential in ensuring timely availability of seeds.
- ☆ Establishment of crop based centre of excellence.
- ☆ Promotion of high density/ultra high density planting systems in potential crops.
- ☆ Tree canopy management right from establishment stage.
- ☆ Rejuvenation of old, senile and unproductive orchards.
- ☆ The management of crops under managed water deficits
- ☆ Capacity building through training and demonstration of improved technologies.

Future Thrust and Strategies

- ☆ Burgeoning population, shrinkage of arable land for agriculture due to urbanization and industrial growth and global competition have limited the possibilities for horizontal expansion. Therefore, lies in vertical growth in term of appropriate location-specific high value crops with maximum productivity and cost effectiveness.
- ☆ After attaining food security, our efforts have been focussed on achieving nutritional security by laying emphasis on production, protection and postharvest management of horticulture crops.
- ☆ We now have the mandate to increase per capita availability of these crops and meet the domestic and export needs, besides the needs of the processing industries.
- ☆ Horticultural crops have already demonstrated their role in terms of their potential in increasing income per unit area, generating additional employment opportunities, providing sustainable income to small, marginal and tribal farmers and earning sizeable foreign exchange through exports and savings through import substitution.

- Still there are several crops and areas which need strengthening, so that horticulture sector could be more competitive to face the present situation of global trade.
- The development strategy for horticulture would addresses all the issues starting from the production activities to marketing of the final produce is essential and vital to augment production and provide on farm and farm employment. Therefore, thrust should be on the integrated development of horticulture in Mission Mode to ensure adequate appropriate, timely and concurrent attention to all the links in production, postharvest and consumption chain which should maximise ecological and social benefit from investment and promise ecologically sustainable intensification, ecologically desirable diversification and employment. The following thrusts and strategies are proposed to achieve the above goals.
- Improving production.
- Improving productivity.
- Reducing cost of production.
- Improving quality of products for exports.
- Value addition.
- Marketing and Export.
- Price stabilization.
- Strengthening of organizational support.
- Human Resource Development
- Addressing relevant policy issues.

References

Anonymous (2001).Report of Planning Commission Working Group on Horticulture Development for Xth Plan, Ministry of Agriculture, GOI.

Anonymous (2011).Report of Planning Commission Working Group on Horticulture and Plantation Crops for XIIth Plan, Ministry of Agriculture, GOI.

Anonymous (2015).Area, production and productivity of horticultural crops in India. National Horticulture Database.

Awasthi RP, Verma HS, Sharma RD, Bhardwaj SP and Bhardwaj SV (2001) Causes of low productivity in apple orchards and suggested remedial measures: Productivity of temperate fruits. Jindal, K.K. and Gautam, D.R. (Eds), 1-8.

Baba, J.A., Akbar P. I. and Kumar V. (2011) rejuvenation of old and senile orchards: a review. Annals of Horticulture 4(1): 37-44.

Datta, S. (2013) Impact of Climate Change in Indian Horticulture -A Review International Journal of Science, Environment and Technology, 2 (4), pp: 661–671

Hanur, V.S. and Aswath, C. (2013) A report on the National Seminar on Horticultural Biotechnology, Current Science, 105 (4).

Kamei, P. (2013) Horticulture crops potentiality, problem and its role to alleviate the poor economy of the hill district in Manipur with focus on tamenglong district International Journal of Humanities and Social Science Invention, 2 (1): PP.01-06

Kumar R and Kumar KK (2007) Managing physiological disorders in litchi. Indian Horticulture 52 (1): 22-24.

Mittal, Surabhi (2006).Can Horticulture Be A Success Story For India? ICRIER Working Paper No 197, August 2007.

Sharma V, Alam A (2013) Current Trends and Emerging Challenges in Horticulture. J Horticulture 1:1.

Chapter 4

Recent Techniques in Dryland Fruit Crops for Quality Production

V.S. Supe

Professor of Horticulture,
Department of Horticulture,
Mahatma Phule Krishi Vidyapeeth, Rahuri – 413 722, Dist. Ahmednagar, M.S.

India is in unenviable position in terms of expertise in handling more than 120 crops ranging from Temperate to Arid zone crops, from fruits to medicinal and aromatic plants. Very few countries in the world have the competence, infrastructure and above all the natural resources including genetic resources. India is the second largest producer of fruits (63MT) and contributes 10 per cent share in global fruit production. The area under fruit crops in India is 5.81 mha with a production of 62.86 MT. Among the horticultural crops, fruits recorded a two folds increase in area and production compared to 1991-92. India occupies first place in production of mango, banana, papaya, pomegranate and aonla. The productivity of grape is highest in the world. India has higher national average productivity in banana and sapota compared to world average productivity, while in citrus, mango, apple, guava, pineapple and papaya, India has substantially low productivity in comparison to world average. Besides, pomegranate and grapes, India leads in production of other arid fruits such as ber, bael, aonla, custard apple, fig and tamarind. In addition, the important minor fruit crops of the region are timroo, pilloo, jamun, mahuwa, mulberry, kainth, karonda, lasora, khejri, ker and many others. These crops not only rich in minerals and vitamins but has medicinal and therapeutic values.

Horticulture has emerged as commercial enterprise moving from rural confine, owing to technological advancement and policy environment provided by the Government. The fruit production has increased sustainability due to strong institutional system, which has resulted in development of many high yielding varieties/hybrids and production system. Despite achievements the challenges confronting are still many. Although India is ranked as second in the world production of fruits, but the availability still continues to be much below the dietary requirements. With increase in per capita income and accelerated growth of health conscious population, demand for fruits will increase. But there is a growing competition in open economy, which demand competitive price of standard quality produce. This opens up opportunity and challenges. Hence, the potentialities which exist need to be harnessed and gains have to be sustained. India is required to enhance producing high quality produce to face the stiff competition under the liberalized economy. Attention is therefore required to tackle the issues of low productivity, poor quality and high costs of horticultural produce. Development of improved cultivars with high quality characteristics, productivity, resistance to pest and disease and tolerant to abiotic stresses, high nutritive and nutraceautical value and technologies for improving the efficiency of water and nutrients by reducing variability in yield, quality and reducing pre and postharvest crop losses are needed. The crop monitoring mechanism would need strengthening to enable the forecast of crop prospects. Efforts will also be needed to ensure timely availability of quality planting material to farmers. The development has to be seen as integrated approach, addressing important gaps, in harnessing the potential horticultural development. However, it will require sustained research efforts to improve competitiveness, enhancing efficiency of production system, assessing quality and safety and enhancing the capability of community to manage the change. Consequently, technology driven fruit production will contribute significantly for economic development especially in the area which is largely the marginal and degraded.

Research Infrastructure for Arid Horticulture

Initially the work on arid horticulture was done at Central Arid Zone Research Institute, Jodhpur with collection of germplasm and production management of ber, pomegranate, aonla, and few other fruit crops. However, recognising the importance of fruits in arid region a separate ad-hoc Scheme entitled, "Research on some selected fruits on arid and semi- arid areas in India" was started in 1976 with 14 centers (Abohar, Jodhpur, Chandan Well, Hissar, Khedoi, Rahuri, Lahau, Yercaud, Bangalore and Udaipur). The scheme was merged during the Sixth Five Year Plan to form the cell III of All India Co- ordinated Fruit Improvement Project (AICFIP) IN 1978. During the Seventh Five Year Plan, the cell III of the AICFIP was restructured to form an independent Project entitled, "All India Co-ordinated Research Project on Arid Zone fruits" at Hissar, which was later upgraded during the VIII Five year plan with its headquarter at National Research Centre for Arid Horticulture (NRCAH) in April, 1993 at Bikaner in Rajasthan which continue to monitor research programmes at different centres in the mode of Directorate. In the year 2000, NRCAH was upgraded at Central Institute of Arid Horticulture with with one regional centre at CHES Godhra. At present, the AICRP project has twelve centres located in different parts of the country. Out of twelve centres, three are located in Rajasthan, three in Maharashtra, two in Gujarat and one each in Tamil Nadu, Karnataka, Andhra Pradesh, Punjab, Haryana and Uttar Pradesh.

Looking into increasing demand of arid fruit crops, for internal consumption and export, threat created by diseases and pests, Indian Council of Agricultural Research realised that some more infrastructural facilities for research may be created for such potential crops. This resulted in establishment of National Research Centre on Pomegranate in 2005 at Solapur in Maharashtra to solve the problems of Pomegranate production technology in India.

Technological Advances in Arid Fruit Crops

Genetic Enhancement

Biodiversity

The big challenge to attain the food and nutritional security for the increasing population can be met by improvement in the productivity through genetic enhancement. Accumulation of variability in the form of germplasm is the pre-requisite for this purpose. Our country is endowed with germplasm of vide range of arid and semi-arid fruit crops. As a commitment to provide strong backing to the breeding programmes, germplasm collection efforts were strongly supported by the national research programme. Concerted efforts are therefore made for documentation, characterization, conservation and utilization of plant genetic resources in horticultural crops. India is endowed with large germplasm pool (72,000 germplasm accessions of cultivated, wild and related taxa) is being scientifically conserved in various institutes, with about 6000 fruits which includes about 1360 accessions of arid zone fruits, 25,400 in vegetables, 15,700 in species and condiment crops and 10,100 in plantation and tuber crops. National Bureau of Plant Genetic Resources (NBPGR), New Delhi holds the germplasm repository of such fruit crops at National Active Germplasm Sites.

CIAH Bikaner is the lead active germplasm centre which have 1360 accessions of the germplasm of different arid zone fruit crops. Till today, the available germplasm reservoir consists of 619 of ber, 122 of aonla, 134 of datepalm, 228 of pomegranate, 46 of bael, 23 of fig, 119 of custard apple and 69 of tamarind is being maintained at 12 AICRP centres. Besides, many collections of few arid fruits are being maintained at IIHR, Banglore; CIHS, Lukhnow and CHES, Godhra. The collections have been evaluated for many economic characters like high fruit yield, resistant to diseases, insects-pests, maturity, adaptability, processing attributes and nutritional qualities. Evaluation studies conducted over years has lead to identification of promising varieties. Till now the major emphasis on germplasm collection and evaluation has been for targeting the high yield but now keeping in view the dynamic needs of diverse stakeholders, there is need to make more explorations for collection of variability and further sourcing for resistance to diseases, insects and more particularly for processing attributes for developing cultivars for processing industry and export. Keeping in view the nutritional and health care need of the people, it has become necessary to identify accessions possessing high nutritional value and bioactive compounds which play a great role in health protection. Such collections shall be helpful in breeding varieties of arid fruits with special attributes.

Varietal Development

As a consequence of the research efforts made at various research institutes, many high yielding varieties and hybrids of arid fruits have been developed. Some of the region specific varieties developed are Gola, Seb, Umran, Banarasi, Karaka, Kaithali, Mundia, Goma Kirti, Thar Bhubraj, Thar Sevika, Narendra Ber Sel-1 and Sel-2, ZG3, Sanaur of Ber; ganesh, Dholka, Jalore Seedless, Mridula, phule Arakta, Bhagawa,Ruby, Amlidana, G-137, Jyoti, Basin seedless of pomegranate; Banarasi, Chakaiya, Francis, NA-6, NA-7, NA-10, kanchan, Krishna, Anand-1, Anand-2, Laxmi-52, BSR-1,Chakaiya, BSR-1 of aonla; Kagzi, Mirzapur seedling, Etawah, Ayodhya, NB-9, Pant Aparna, Pant Urvashi, Pant Shivani, Pant Sujata, CISH Bael-1 and CISH Bael- 2 of bael; Balanagar, Mammoth, Island Gem,APK (Ca) 1, Arka Sahan of Custard apple and Poona Fig, Dinkar, Deanna, Conardia, Excel of fig are popular amongst the farmers. Adoption of improved varieties by farmers has exhibited significant contribution in overall production of horticultural produce particularly in the arid ecosystem. There is need to develop more varieties suitable for different agro- climatic conditions and the emphasis for improvement should not only be confined to target yield but it should be need based and have ability to withstand against biotic and abiotic stresses.

No specific breeding attempt has been made for the development of varieties suitable for specific purpose. However, some of the existing varieties identified with processable characters are Medzool, Shamran, Khadrawy, Chip Chap of Datepalm for dehydration (Chhuhara); NA-6, NA-7, Chakaiya, Banarasi of Aonala for preserve; Kanchan, Krishna, NA-10 of aonla for pickle and shreds; BSR-1 of aonla for medicinal purpose; Amlidana of pomegranate for dehydration (Anardana); Umran ber for dehydration; Gola, Seb, Mundia, Banarasi, Karaka ber for preserve; Seb, Umran, Mundia ber for better shelf-life; Poona fig, Deanna, Excel fig for dehydration. In order boost up export, there is need to develop varieties with processing attributes for preparation of value added items from arid fruits. In fruit crops, most of the commercially adopted varieties are developed through selection only. There are few varieties which have been developed through selection only. There are few varieties which have been developed through planned hybridization. Some of these varieties are Thar Sevika (Seb x Katha) of ber, Arakta (Ganesh x Russian Selection), Mridula (Gul Shah red x Ganesh) of pomegranate and Arka Sahan (Iceland Gem x Mammoth) of custard apple. In case of datepalm, there is need of early variety for the arid zone of Rajasthan to escape from early rain which spoils the fruit quality. Some of the serious biotic and abiotic problems in arid region are fruit fly in ber, oil leaf spot and blackening of aril in pomegranate, fruit cracking in pomegranate and bael, frost effect on aonla and tamarind in North West India. There is need to give greater emphasis on breeding for development of resistant varieties against diseases/pests which are impediment in raising profitable crop and also improvement for nutraceutical values (vitamins, minerals and other bioactive compounds) Tolerance for abiotic stresses and resistant to insects and pests have yet to receive attention under the co-ordinated programs. Breeding through hybridization and hybrids could prove useful in combining such important genes and further use of molecular markers can shorter the period of selection of desirable types by the breeders.

Rootstocks for Enhanced Profitability

Several rootstocks with immense potential have become integral part of the fruit industry. The strong horticulture industry today based on grapes, citrus, apple and few other fruit crops would have never flourished in absence of better rootstocks. There is need to use efficient rootstocks for boosting fruit production and productivity in the country. The success of rootstock is determined by its tolerance to prevailing conditions of soil, climate and disease, while still producing high yields of good quality fruit. In ber, some of the popular rootstocks are *Z. mummularia* (drought hardy and drought stature) *Z. mauritiana* var. Tikdi and *Z. mauritiana* var. Shukhawani (Vigorous growth) *Z. rotundifolia* (Vigourous growth and drought hardy); and in sapota Khirni (*Manilkara hexandra*) has proved drought hardy behaviour. Several components of future commercial fruit growing will depend on successful use of rootstocks for better scion compatibility, canopy architecture, fruit quality and nutrient absorption, water use efficiency, biotic and abiotic stress tolerance, adaptation under the influence of climate change. Rootstocks in various arid and semi- arid fruit crops have also not been exploited properly. There is need to breed newer rootstocks for fruit growers of ber, pomegranate, aonla, custard apple *etc.* for quality yield and resistance to abiotic and biotic stresses particularly drought and salinity.

Production Technology for Planting Material Production

Use of true-to-type of quality planting material is essential for development of arid and semi-arid horticulture. Due to uncongenial edapho- climatic conditions in the hot arid ecosystem, the development of nurseries in the areas is still insufficient to meet the demand of the growers, which necessitates the development of chain of hi-tech nurseries. Even today, farmers are using seedling plantations of several fruits like aonla, bael, jamun, lasora, khejri, mulberry, karonda, datepalm, *etc.* resulting in high variability and poor quality fruits. The vegetative propagation has been standardised recently for lasora, khejri, ker, mulberry, fig, *etc.* but availability of planting materials is a constraint. There are some fruits like ber, bael, aonla, pomegranate, sapota, where vegetative propagation techniques have already been standardized under sub-tropical conditions and if these techniques are adopted under arid and semi-arid conditions the results are not promising. Therefore, there is need of some refinement related to adaptation under changed set up of agro-climatic situations.

The time and technique for vegetative propagation of some of the arid fruit crops have been developed. There is need to work out some mechanism of propagation to overcome the problem of low rate of sucker multiplication in datepalm. The biotechnological approach like tissue culture has been attempted but even after 15 years of it, effect ahs not been converted into a commercial venture. This needs critical examination to address the gap to make it success. Joint efforts of all the palm scientists (coconut, arecanut, oil palm and datepalm) may bring the devidents.

Orchard Establishment

Establishment of fruit plants under arid and semi-arid ecosystem is very difficult due to poor soil and uncongenial atmospheric conditions. The poor quality planting materials brought from distant places have high mortality when planted under arid-agro climatic conditions. For better survival, planting materials are produced under same agro-climatic conditions. Pre-hardened, having fibrous and well developed root system, healthy and uniform size based on plant standard should be used for planting depending upon the topography, fruit species and soil type. On sloppy lands, fruit trees are planted on contours, staggered trenches, bounds and half moon terraces. In micro-catchments, tree should be planted at the point of maximum runoff concentration. It is challenging to establish orchards and further rearing the plants in the light soils and harsh climate of arid zone. *In-situ* establishment of orchard of fruit crops particularly ber, aonla and bael has been very successful under arid and semi-arid conditions and also in degraded lands. Initial care, watering, use of pitcher system of irrigation, use of bentonite/clay layer in the pits, mulching of basin, intercultural operations, adoptions of plant protection measures, *etc.* may be helpful in orchard establishment under resource poor situations. There is need to work out complete orchard management system for arid ecosystems separately for light desert soils and undulating heavy but sloppy soils in the arid region.

Agro-techniques

Success of arid horticulture is only possible through proper management of available resources. However, it is difficult to suggest the common package of practices for various horticultural crops, agro-techniques are highly site specific and vary for different plant species. Mulching with organic materials like straw, hay, dry leaves, local weeds *etc.* has been found beneficial in reducing evaporation loss. It also suppresses weed growth, prevents erosion and adds organic matter to the soil. Black polythene mulch is very effective in ber orchards in Western India. Although local organic mulch material are cheaper than polythene mulch but they require proper care and effective cover thickness. Sugarcane thrash mulch was found effective in pomegranate, fig and custard apple in Maharashtra. The local dry bush of bui (*Avera tomentosa*) is identified as suitable mulch material for datepalm.

Fruit thinning in date palm (one-third strands from centre of bunch) and retaining of 50-60 fruits per tree in pomegranate cv. Muskat and pre-harvest spray in ber (750 ppm Ethephon) and date palm (1000 ppm Ethephon) have ben recommended for improvement of fruit quality. For better yield and quality in date palm, 1:6 bunch and leaf ratio was found optimum under Bikaner conditions. In crop regulation studies, ambia bahar was found better in both Ganesh and Muskat cultivars under Rahuri conditions. NDUAT (Faizabad) has reported the phenomenon of self incompatibility in most of the commercial varieties of aonla and need of proportionate planting of pollinizer varieties. The variety NA-6 was found to be the best pollinizer for NA-7. At Banglore, hand pollination with Squamosa (Custard apple) pollen is recommended to achieve not only higher productivity, but also large size symmetrical Arka Sahan fruits. In case of pomegranate, fruit cracking is minimized significantly by spraying of borax (0.2 per cent). Spraying of sulphuric acid can minimize the adverse effect of frost in ber in Jobner. The productivity of fruit crops is low in the arid ecosystem, in addition to several factors responsible, occurrence of seedling plants of different fruit trees like; ber, bael, aonla, custard apple, tamarind, karonda, lasora, *etc.* in the orchards are growing naturally having poor quality fruits. The productivity of such plants can be improved through rejuvenation into productive conditions by top working.

High Density Orcharding

High density planting technology has been standardised for many crops and adopted by many fruit growers in India. It has become a success story in banana, pinapple, citrus, papaya, mango, cashew and few arid fruit crops. High density orchards not only have provided higher yield and net economic returns per unit area in the initial years, but also facilitated more efficient use of inputs. In high density planting, closer spacing (5 x 2 m) has given two and a half times more yield than normal spacing (5 x 5 m) In Tamil Nadu planting of ber at 8 x 3 m and 8 x 4 m spacing, respectively for Banarasi and Kaithali varieties were found suitable for high density plantation. Technologies for high density planting, canopy management and rejuvenation of old and senile orchards of been developed and successfully demonstrated for many fruit crops. Plant density for optimum yield under local conditions of the centres is required to standardized to harness solar energy efficiently.

Plant Architectural Management

In order to accomplish successful orcharding, plant architectural management assumes great importance. Training at initial stage of growth gives strong framework to the trees. Under the arid ecosystem to regulate the fruiting behaviour of trees, pruning is important to make a balance between vegetative and reproductive phase. Both time and severity of pruning for optimum yield varies at locations and with climatic conditions, have been worked out for different crops. It is suggested that the bushy pomegranate plant should be trained keeping 3-4 stems from the ground level and also retaining of 3-4 stems in Poona fig under Rahuri conditions has been found useful while in most of the other fruits single stem training keeping 3-5 main branches above 1-1.5m from ground level in various directions are preferred. The ber is pruned during Jan-feb in Tamil Nadu, by the end of April in Maharashtra and end of April to end of May in North India. The main shoot of previous season is cut by retaining 15-25 node or 50 per cent of the extension growth, depending upon location cultivar age and vigour of the tree. All the secondary shoots are completely removed. As a result of light pruning for several years, long non-flowering shoots develop. To eliminate this, half the number of shoots on the tree should be pruned keeping normal length and remaining half should be pruned keeping 1-2 nodes to induce new growth for fruiting in following year. In case of phalsa pruning is done twice in a year in South India(December and June) and once in North India (January) to invigorate new shoots. In datpalm 1:6 bunch and leaf ratios recorded maximum number of strands per bunch, number of fruits per strands, average bunch weight and yield per plam at Doka stage at Bikaner while at Mundra 9 bunch per palm was found best in respect of fruit quality attributes and yield. There is need to intensify the research to device plant architectural system for other fruits of arid region.

Organic Fruit Production

Demand for green food is on increase and harnessing the potential of organic farming which address soil health and environmental health is considered of greater significance. In last few years, organic farming has attracted many farmers across the country especially combined with ecotourism and have experimented successfully. India is best known as an exporter of organic Tea and also has niche market for spices, fruits and vegetables. Fresh arid fruits that are having high potential in the world market are Date, Fig, pomegranate, aonla and bael. Large scale investment in the sector will overcome the marketing constraints of these produce and hence will be a highly remunerative venture in the days to come.

Integrated Nutrient Management

The quality of fruits and vegetables is excellent in arid ecosystem having better taste, good appearance and low pathogenic presence but less emphasis has been given on integrated nutrient management. In arid fruits to improve their productivity on sustained basis in high value horticultural crops like pomegranate, aonla *etc.*, there are several reports of over fertilization. There are many arid fruit tress like ker, phalsa, karonda, bael, jamun, mulberry,pilu, khejri, bordi *etc.* where fertilizers are hardly applied, rather they are either growing naturally or some amount of FYM or compost is added in the basins during rainy season. In true sense, arid fruit crops are well prone for organic farming. Use of local waste and vermi-composting has special significance in integrated nutrient management of arid fruits as it improves the soil physico-chemical properties of the growing site also. The use of vermi-compost along with inorganic fertilizers in improving the vegetative vigour and yield of pomegranate has been found successful under irrigated ecosystem of north-western Rajasthan.

There are several considerations in deciding the nutritional requirement of the fruit trees, as they are perennial in nature. They also recycled some amount of nutrients on growing site of every year in form of leaf litter. The tissue analysis, based on leaf sampling standard is one of the criteria while assessing the nutritional requirement of the fruit trees and leaf sampling stage for such assessment has been worked out on ber, bael, aonla, pomegranate, phalsa and custard apple. The quantity, time and method of application of fertilizers are very important for achieving better response. The fertilizers should be applied at the site of active root zone after ensuring proper moisture conditions in the field, which is varied not only with fruit types but also with growing conditions. Micro-nutrients are often found deficient in semi-arid and arid soils. Foliar feeding of nutrients such as nitrogen (0.5-2 per cent urea), zinc (0.05- 1 per cent zinc sulphate) and boron (0.05- 1 per cent borax) has given beneficial results in several arid fruits. The nutritional requirement of some arid fruits for different agro-climatic zones has been worked out and successfully adopted by farmers. But the focus is required for the use of biofertilizers, VAM fungi, biological N fixers and other beneficial microbial agents under the typical condition of arid zone for effective nutrient use efficiency.

Water Management

Water is precious commodity in arid ecosystem; hence each and every drop of water should be utilized judiciously. Most part of the arid region depends upon rainwater from little rainfall, therefore horticulture production under these conditions is only possible through proper management of rainwater. The modern concept of watershed approach has been used by mankind since antiquity. The cultivation of crops and trees in Khadins in Jaisalmer district of Rajasthan is one of the classical example of *in situ* water harvesting. The hilly, rocky and slopping lands generate runoff, which can be collected in ponds and recycled for establishment of new plantation and irrigation of fruit trees and crops at critical stages of plant growth and development. The size of optimum micro-catchment and plant population for different arid fruit crops such as ber, pomegranate has been recommended.

Most of the horticultural production in arid areas comes from the small pockets having assured irrigation facilities either by tube well or canal command covering less than 4 per cent of these lands. The underground water in most of the areas is brackish in nature and thus not ideally suited for irrigation of horticultural crops. Therefore, use of suitable crops coupled with proper irrigation system like drip and sprinkler system of irrigation have better impact on improving productivity of horticultural crops. By these systems about 25 per cent water can be saved over conventional method of flood irrigation. Though, the installation of drip system requires additional cost but the system avoids the development of water logging conditions and salinity build up as common in case of flood irrigation and the quantity of water saved can be used for bringing of additional area under cropping. The system can also be utilized for fertigation purpose. The work conducted at various centres have revealed saving of 25-35 per cent of water if drip irrigation system is used in pomegranate ad aonla. Provision of windbreaks also modify microclimatic situations and thus reduce transpiration loss. The other means of enhancing water productivity can be development of varieties with drought tolerance or requiring less water and further supported by efficient water management technology. Keeping in view the resources available, water management system is required to be worked out for different regions for arid fruit crops.

Cropping Systems in Arid Region

Horticulture is the best option for crop diversification and maximizing system productivity. Need based integration of various components like tree, crops, grasses and animal on the same land management unit is a long tradition in the arid ecosystem. In hot arid region of western Rajasthan, growing of more than one crop as mixed cropping particularly during kharif season coupled with animal husbandry and retaining of perennial trees like khejri (*Prosopsis cineraria*) in an around farm bunds is being practiced since long back in order to minimize the risk of crop failure and to get multiple outputs for fulfilling the household needs. Farmers have adopted growing of fruit trees in their farm land mainly ber, aonla and pomegranate as inter-plantation and karonda, lasora, phalsa, bael, jamun, mulberry, *etc.* as boundary plantation in order to generate income and employment on sustained basis from the beginning by intensification of various compatible components. Various fruit based cropping systems have given appreciable response and aonla/ber-okra/ brinjal/cowpea system have been adopted largely by the farming communities of tribal area of semi-arid region of Gujarat. In ber based cropping system ground story combination with Indian aloe gave the highest economic return followed by cluster bean in arid conditions of Western Rajasthan. In aonla based cropping system aonla-ber-brinjal-moth bean-fenugreek and aonla-khejri-suaeda-moth bean-mustard are found promising combinations in hot arid region.

Besides these, grass such as *Cenchrus ciliaris* and pasture legume such as *Stylosanthes hamata* can be grown without adverse effect on the fruit trees. In arid areas, the system could have combinations such as khejri + ber + grasses such as *Cenchrus ciliaris, C. setigerus and Lasiurus sindicus.* In low rainfall zone (300-500 mm) of asrid region, the combination of khejri or ber +ber or drumstic + vegetables such as legumes and cucurbits can be used. However in a date palm based cropping system, all the inter plantations have reduced the plant vigour in shade as compared to open. Various multi-storey systems with ber, aonla and Khejri for arid region of western Rajasthan have been standardized. They have also advocated for the adoption of back yard poultry, rearing of sheep and goat, small- scale postharvest units like dehydration of sangria (tender pods of khejri) and kachri, pickling of ker and lasora, preparation of aonla shreds, *etc.* Aonla based diversified cropping models in arid ecosystem performed well without affecting the growth and development of main and component crops. Under semi-arid conditions aonla performed well in double hedge row system followed by cluster and hedge row system. There is need to work out such viable cropping systems for the arid region which could cover the risk and ensure returns to farmers by making sustainable use of farm resources and environment on a long term basis.

Hi-tech Horticulture and Precision Farming

Hi-tech horticulture is the development of modern technology which is capital intensive, less environment dependent, having capacity to improve the productivity and quality of produce. On the other hand, Precision farming involves the application of technologies and principles to manage special and temporal variability associated with all the aspects of horticultural production for improving crop performance and environment quality. Precision farming calls for efficient management of resources through location-specific hi-tech interventions. Hi-tech horticulture encompasses a variety of interventions such as micro-irrigation, fertigation, protected/green house cultivation, soil and leaf nutrient based fertilizer management, mulching for *in situ* moisture conservation, micro-propagation, biotechnology for germplasm, genetically modified crops, use of biofertilizers, vermiculture, high density planting, high-tech mechanization, green food, soil less culture, biological control *etc.* The Site Specific Nutrient Management (SSNM) approach compared to farmers practice has good response. A fertilizer calculation chart for a particular yield goal based on yield in nutrient omission trials need to be conducted. About 17 Precision Farming Development Centers (PFDC) have been established in different agro-climatic regions under NHM. Activities like green house contraction, mulching, shade net and plastic tunnels are also being promoted and 195.2 ha have been already brought under protected cultivation. Utilization of these interventions orchestrated together having aim of achieving higher output in given time period leads to precision farming, which is largely a knowledge driven.

It has been recognized that hi-tech horticulture would play a major role in horticulture sector in the coming years to improve production and productivity of horticultural crops. Therefore, adoption of hi-tech horticulture and precision farming assume greater significance. The crops where some of the components of precision farming have been practiced are banana, grape, pomegranate, capsicum, tomato, cashew and selected flowers. Many of these hi-tech needs to be tested in arid fruits.

Integrated Pest and Disease Management

Besides wild animals, rodents and birds there are several pests and diseases causing severe losses to arid horticultural crops. However, the damage caused by pests and diseases in arid region is comparatively less than those of humid regions. Among different pests, termites cause considerable damage particularly in low rainfall areas. Fruit fly (*Carpomyia vesuviana*) and fruit borer (*Meridarchis scyrodes*) in ber, butterfly (*Virachola isocrates*) bark eating caterpillar in pomegaranate and leaf gall midge in aonla are major pests causing serious damage to fruits. The chemical control measures for various pests and diseases have been worked out at various centres, but there is need for eco-friendly practices. The approach suggested to bring the loss below economical level is adoption of integrated pest management (IPM) strategy. In fact, the basic consideration of IPM is to reduce the use of chemicals particularly those having high residual toxicity and harmful influence to the flora and fauna of the ecosystem and to promote practically feasible, economically viable eco-friendly practices like cultural practices and use of bio-control measures.

Biological Control

Biological control has been the most significant approach to plant health management. It promises through modern biotechnology to be even more significant in the twenty-first century. Although the number of bio control products is increasing, these products still represent only about 1 per cent of agricultural chemical sales. Yet these are important contributions because bio control agents offer disease management alternatives with different mechanisms of action than chemical pesticides. Trends in research include the increased use of bio rational screening process to identify microorganisms with potential for biocontrol, increased testing under semi commercial and commercial production conditions, increased emphasis on combining biocontrol strains with each other and with the other control methods, integrating biocontrol into an overall system. Plant health management in horticultural crops involves not only pre-harvest but also postharvest health management strategies such as production of pest and disease-free planting materials, use of bio-inoculants and other growth enhancing soil amendments, indexing of major pathogens and certification of planting materials, seed plot technique and mother garden technique and other such measures. Several biocontrol agents have been identified for various fruit crops but under arid ecosystem, the new biocontrol agents from the native zone are required to be identified and could only prove effective under the harsh climate of arid ecosystem.

PCR-based diagnostic protocol need to be developed for rapid detection of diseases. Forecasting based disease management (for downy mildew and powdery mildew) could result in considerable saving of sprays. A comprehensive approach involves cultural, biological and chemical measures are required to be evolved for the integrated management of the important pests. Similarly, IPM methods with emphasis on biological control or use of plant based pesticides need to be developed for the management of pests and diseases affecting arid horticultural crops for the different regions. Further thrusts should include identifying and cloning resistant genes and improving the efficiency of conventional and molecular breeding, studies on mode of parasitism to develop resistant lines. Effective control of the disease can be achieved by judiciously following the *Orchard Health Management Schedule* formulated on the basis of research outcome.

Postharvest Technology

The postharvest losses of horticultural produce are to the extent of 25-30 per cent. The cost of saving these losses is less than the cost involved for their production. Therefore, postharvest losses must be controlled by adopting suitable strategy; integrating right time of harvesting and adopting appropriate method of harvesting, grading as per international and domestic standard, proper packing by using desired containers, safe transportation, development of storage facilities, development of cooperative making system, processing into delicious products *etc.* In case of arid fruits and vegetables, there are no much-processed products commercialized so far except aonla preserve, bael syrup and guava jam. However, there is good scope of natural dehydration due to high solar ration available in arid ecosystem. The practice of natural dehydration of *sangri*, ker, kachri *etc.* Is common in arid region of Rajasthan but neither stage of harvesting nor method of dehydration is proper and thus quality of product is not up to the mark, which needs development of suitable postharvest techniques for processing of various value added horticultural products.

In a preliminary study, an experiment was initiated to observe the effect of blanching treatment on re-hydration quality of *sangri.* From this investigation, it was found that 5 minute blanching in 2 per cent salt solution had given light green colour after re-hydration and was preferred after organoleptic scoring as compared to either un-blanched or prolonged blanching. Similarly, some attempts were also made to prepare cookies from dried pods (*khokha*) or khejri. It was found that on organoleptic scoring 10 per cent replacement of wheat flour by *khokha* powder the taste of biscuits was bitter. The work on dehydration and making of kachri powder, preparation of aonla shreds as mouth freshener, aonla candy, powder of bordi, utilization of cracked fruits of pomegranate, *etc.* are also in progress at CIAH, Bikaner.

So far there are no gluts in the market due to over production of arid fruits. Sometimes, prices come down during the main season due to unorganized supply in the markets of major growing areas in case of pomegranate, kinnow, aonla and ber. There is good scope in postharvest technology for researcher particularly in dehydration, standardization of grading, packaging and recipes of various value added products. Development of new products like juice punches, fruit wines, dehydrated products from grape, pomegranate, bael, ayurvedic products particularly dehydrated powder of rind of pomegranate, bael, aonla have demand. Setting up of small scale industries in rural areas and participation of woman and rural youth can play key role in this context. For preventing the postharvest losses proper storage, cold preservation, packaging and transport methods with Hazard analysis Critical Control Point (HACCP) norms have to be given more thrust.

Mechanization in Horticulture

India is having largest number of tractors in the world, (about 3.2 million) producing about 0.25 million per year. Still most of the horticultural operations in India are done manually or with animal power. Wherever, the farming operations are mechanized the crop productivity is high. The machinery developed by different SAU's and agricultural research institutes need to be popularized. The tractor drawn implements have found wide acceptability in potato cultivation, from sowing to harvesting. The suitable varieties of fruits, vegetables, spices and plantation crops are to be selected and the cultivation practices are to be standardized for mechanization. In fruit crops, the tractor operated pit hole digger and bucket excavators (JCB) have been developed by different firms and they need adoption for planting fruit trees. The fruit nurseries need the adoption of media siever, media mixer and plastic bag filler. The automatic grafting machines developed in advanced countries are yet to be adopted. The commercially available roto slasher is to be adopted for weeding. The electrically/pneumatically/hydraulically operated secateurs, loppers need adoption for pruning. The self propelled pruner for tree crops and grapes orchard need adoption. The harvesting aids for guava, sapota, orange, pomegranate developed by different institutes need adoption by farmers.

Critical Gap Analysis

Past achievement in horticulture is laudable in terms of growth but the opportunities are immense in the emerging scenario. How to harness the potentiality and face the challenges are the issues which need to be addressed. In this context it is pertinent to analyse the critical gaps which can be addressed in systematic manner. Critical gap needing attention are low productivity and poor quality of the product, inadequacy of infra-structural facilities for postharvest management and marketing, inadequate efforts for product diversification and consumption, inadequacy of quality seed and planting material, inadequacy of human resource in horticulture, lack of appropriate database for effective planning, inadequacy of trained manpower and infrastructure in the states, poor delivery system, credit support and price support and slow pace in adoption of improved technology. Horticultural development has to be seen as integrated approach, addressing important gaps, in harnessing the potential for horticultural development.

Thrust and Strategies for Development

The thrust areas identified for the development of horticulture are improving production through productivity enhancement and utilization of wasteland/surplus land, improve productivity through the use of technology, reduce cost of production by efficient utilization of resources, improve quality of products by adoption of good agricultural practices, add value to the produce through efficient postharvest management and processing, promote marketing and export, develop market coupled with price stabilization, reduce the risk through insurance, strengthen the institutional support system to address the needs of human resource development and address relevant policy issues. These thrust areas have to be achieved through production and productivity enhancement along with policy initiative.

Acceptance of the fruits in the high quality conscious era require the produce to be of high standard free from toxins and other elements of health hazards, which could be achieved through, popularization of improved agro-technique, emphasis on cultivation of specific varieties for table, processing and export, emphasis on harvesting horticultural produce at optimum maturity, grading, packaging and transport, setting up of quality control laboratories, wide dissemination of international quality standards for various commodities and, ensuring pesticide residues within permissible limits. With the opening up of the economy and emergence of free trade there shall be increasing competition to capture the markets, which is equally true for horticultural products. The strategies, to overcome the implication on Indian Horticulture, would include emphasis on creating awareness about implication of WTO regime among horticultural entrepreneurs, identification of horticultural cropping zones, harmonization sanitary and phyto-sanitary standards and strengthening of mechanism to meet the requirement and, strengthening of quality assurance.

Future Thrust and New Initiatives

Fruit production has been one of the success stories of the last decade, and to continue to build on success, sector has to face challenges. Therefore, there is need to prioritize the action outlining the research, development and extension, to make this sector a key driver in rural and regional economic development. Demand for high value produce is growing both in domestic and overseas market at the same time, competition is also increasing. New changes in retailing participation of corporate sector means that retailing will depend upon strategic alliance and supply chain management. Strengthen research on impact assessment of climate change on horticultural crops using controlled environmental facilities and simulation models, analysis of past weather data and integration with productivity changes (including extreme events). Production, demand and supply of commodities, economic and trade, sensitive stages and process during crop development, diversity of dynamics of major insects, microbes and pathogens, intensification of studies on pest, disease and weather relationship *etc.* Therefore, sustainability will depend upon improving competitiveness, reducing impact on environment, quality assurance and food safety and communities engaged in this sector to manage change. The new initiatives to be taken to strengthen the research activities are as follows:

1. Biotechnological tools must be used in conjunction with conventional breeding to tag genes of interest and in marker assisted selection. There is need to develop varieties with durable resistance to multiply disease and pests; heat, drought and salt tolerant varieties; and varieties with efficient nutrient and water use efficiency.

2. Eco-region specific technology generation based on maximum productivity of available natural resources like climatic condition, soil fertility and water. Develop system for productive use of water to get enhanced water productivity by increasing the water and nutrient use efficiency. The technology packages to be developed for various fruit crops should be an integral component of multifunctional agriculture of specific zone. IT based enabling mechanism for technology transfer such as decision support system needed to developed.
3. Holistic approach for water, nutrient, pest and disease management, adoption of recent agro-techniques *etc.* In order to get optimum return in minimum investment, emphasis on use of locally available inputs should be given. Promotion of organic farming is also desired.
4. Understand social needs of communities and build the capabilities for practice change, and for effective utilization of resources and adoption of technology. And also to facilitate accelerated adoption of improved technologies and best practices and respond to needs including bio-security threats.
5. Integrated management of emerging diseases and pests. Emphasis should be on identification of new and effective bio-molecules for management of biotic stresses be encouraged for eco-friendly and sustainable management of diseases and pests. Development of new innovative diagnostic technique for rapid, accurate and cost effective detection of high impact pests and diseases.
6. Develop production and postharvest technologies to improve product quality and minimize environmental impacts. Increasing the value of production by reducing variability in yield, quality, reducing crop losses and increasing marketability. Develop the production systems that minimize wastes and maximize recycling.
7. Plenty of wind and solar energy is available in arid region, which can be utilized in mechanization, such as running of small equipments and dehydration of fruit and fruit based products. Comparison to other regions, arid region having better potential of processing. Therefore, establishment of processing units, standardization of recipes for various products, certification and marketing network *etc.* should be given priority.
8. Sustainable path for development to meet challenges in fruit production through proper technological innovations and interventions specifically in areas like value chain management to improve global competitiveness.
9. Inter institutional mechanisms to network and review the ongoing programme of biotechnology, cost effective production technologies, postharvest technology, farm mechanization, transfer of technology and organic farming.
10. Improve the understanding of interactions between native ecosystems and production systems and develop best practices to conserve biodiversity.

Chapter 5

Food Safety and Export Standards of Important Fruit Crops

B.B. Dhakare, S.D. Patil and R.V. Patil

Department of Horticulture,
Mahatma Phule Krishi Vidyapeeth, Rahuri – 413 722, Dist. Ahmednagar, M.S.

India's diverse climate ensures availability of all varieties of fresh fruits and vegetables. It ranks second in fruits and vegetables production in the world, after China. As per National Horticulture Database published by National Horticulture Board, during 2016-17 India produced 92.85 million metric tonnes of fruits and 170.00 million metric tonnes of vegetables. The area under cultivation of fruits stood at 6.48 million hectares while vegetables were cultivated at 10.30 million hectares. India is the largest producer of Ginger and Okra amongst vegetables and ranks second in production of potatoes, onions, cauliflowers, brinjal, Cabbages, *etc.* Amongst fruits, the country ranks first in production of Bananas (22.94 per cent), Papayas (44.03 per cent) and Mangoes (including mangosteens, and guavas) (37.57 per cent). The vast production base offers India tremendous opportunities for export. During 2015-16, India exported fruits and vegetables worth Rs. 8,391.41 crores which comprised of fruits worth Rs. 3,524.50 crores and vegetables worth Rs. 4,866.91 crores.

Mangoes, Walnuts, Grapes, Bananas, Pomegranates account for larger portion of fruits exported from the country. The major destinations for Indian fruits are UAE, Bangladesh, Malaysia, Netherland, Sri Lanka, Nepal, UK, Saudi Arabia, Pakistan and Qatar.

Though India's share in the global market is still nearly 1 per cent only, there is increasing acceptance of horticulture produce from the country. This has occurred due to concurrent developments in the areas of state-of-the-art cold chain infrastructure and quality assurance measures. Apart from large investment pumped in by the private sector, public sector has also taken initiatives and with APEDA's assistance several Centers for Perishable Cargoes and integrated postharvest handling facilities have been set up in the country. Capacity building initiatives at the farmers, processors and exporters' levels has also contributed towards this effort.

1. What is Food Quality?

In the marketing and consumer economics literature, two main approaches are taken to define food quality. The holistic approach includes within the concept of food quality "all the desirable characteristics a product is perceived to have". By contrast, the excellence approach views food quality as referring only to characteristics that pertain to a higher, more restrictive or "superior" specification of the product. The holistic approach has been adopted in this report to cover the whole range of quality assurance schemes currently existing.

The holistic approach leaves wide scope for interpretation: quality can mean conforming to standards (including standards pertaining to the environment, local specialities, organic production, ethics, and even taste and smell) and it can refer to subjectively perceived quality attributes.

Quality is also a factor that involves the entire production process, from raw materials, processing and packaging up to consumption of the product. As the idea of quality is continuously evolving, any attempts to classify it are immediately overtaken by events, as new meanings are added to existing ideas without replacing them. It is helpful to simplify the different quality requirements into two categories - "musts" and "wants":

1. *musts* have to be present in order for the product or service to be assessed as acceptable;
2. *wants* depend on the wishes or expectations that influence choices.

2. What are Quality Assurance and Certification Schemes (QAS)?

Definition

Quality Assurance and Certification Schemes (QAS) enable stakeholders of the food chain to guarantee that their products or processes fulfill defined quality requirements.

QAS can generally be defined as any code of practice, standard or set of requisites, which enables stakeholders of the food supply chain to guarantee compliance with what is declared and to signal this to the end or next user, underlying this statement there is some independent verification process that adds authority to the stakeholders' statement.

For the purpose of this report, QAS are schemes implying a voluntary participation and enabling stakeholders to involved in the food chain to claim that products or processes fulfill defined quality requirements. This definition of QAS includes two categories of schemes. The first is a category of schemes known as "quality management systems" or "within-chain standards", of which the consumer may hardly be aware at the point of final food purchase. The second category of schemes covers those food QAS that explicitly aim to segment the final product market by differentiating the product(s) covered under the scheme, using labeling or branding to signal specific quality attributes to consumers.

Efficient quality control and food safety is most important in improving the export potential in India. These efforts will help in achieving the quality standards of foreign countries comprising of European and other developed countries.

1. Importance of Food Safety and Quality Control

Safe food is one which has minimum pesticide residues, metals and other contaminants and is practically safe in the hands of consumer without any harmful effects. There have been many instances when commodities exported by India have been rejected by importing countries. Recent examples are concerns of EU regarding high pesticide residue levels in grapes during 2004-05. Thus, strict monitoring from production to transport, processing, handling, packing, *etc.* is absolutely necessary encompassing the entire food chain.

2. International Framework for Food R egulations

In the past, institutional mechanism used to be lightly enforced in trading. However in last 10 years, multilateral rules have become much stronger on the development and use of standards. Importing countries have become very rigid on this issue and do not accept the consignments even with minor variation. Therefore, adherence to the food regulations of various countries is essential.

2.1 WTO Sanitary and Phytosanitary Measures Agreement

WTO provided an International framework by adopting Sanitary and Phytosanitary Measures (SPS). Sanitary and Phytosanitary standards, as covered under the SPS agreement include health, hygiene standards or regulations to avoid the spread of animal and plant diseases and epidemics. These have been adopted by the Codex Alimentarious Commission (CAC) of Food, Food and Agriculture Organization (F.A.O.) and World Health Organization (W.H.O.) WTO member countries are encouraged to use International standards by framing their own laws, regulations, *etc.* India being member of W.T.O, is also adopting these regulations.

2.2 Development of Food Safety and Quality Standards

The consumption of fresh fruits and vegetables has increased substantially over the last two decades, so also the global trade in fresh fruits and vegetables. However, recent reports of food borne illness associated with fresh fruits and vegetables have raised concerns from public health agencies and consumers about the safety of these products. This has led to development of various guidelines for cultivating fruits and vegetables and also commodity standards by Codex Alimentarious Commission. Similarly such guidelines for cultivation of commodities have emanated from European Union under the EUREPGAP renamed as GLOBALGAP. Some of the features of these standards are described below:

(A) GLOBALGAP (EUREPGAP)

GLOBALGAP (EUREPGAP) is an internationally used management system for Good Agricultural Practices (GAP) for fruits and vegetable trade. This system has become quite popular and acceptable. EUREPGAP (now GLOBALGAP) was initiated in 1997 with the aim to provide safe food products to consumers. GLOBALGAP is being followed by many leading European food retailers like Tesco, Sainsbury's, Marks and Spencers, Metro, Kesko, *etc.* for doing business in the food sector. The requirements of GLOBALGAP for fruits and vegetables can be categorized into two group's *i.e.*

(i) General requirements for all farm based produce
- ☆ Record keeping and internal self assessment/inspection
- ☆ Site history and site management
- ☆ Workers health, safety and welfare
- ☆ Waste and pollution management
- ☆ Environment and conservation
- ☆ Traceability

(ii) Product specific for fruits and vegetables
- ☆ Propagation material
- ☆ Soil and sustainable management
- ☆ Irrigation and fertigation
- ☆ Harvesting
- ☆ Produce handling

GLOBALGAP certification is very important as it not only imparts increased acceptability of the product in foreign markets, but also leads to reduced cost of production due to lesser use of pesticides, chemicals, *etc.* GLOBALGAP has developed a checklist for self assessment purposes. Producers receive their GLOBALGAP approval by verification by an independent approved body like A.B. Cert, EUROcert, and SGS Agro control. Once GLOBALGAP criteria are met, a certificate is issued by GLOBALGAP approved certification body.

(B) Codex Hygienic Practices for Fresh Fruits and Vegetables

Codex Alimentarius Commission has also come forward with guidelines about Good Agricultural Practices (GAP's) and Good Manufacturing Practices (GMP's) that help to control microbial, chemical and physical hazards associated with all the stages of the production of fresh fruits and vegetables from primary production to packing. It is difficult to implement these guidelines where primary production is taken up in small holdings. Thus these guidelines are flexible to allow for different systems of control and prevention of contamination of different groups of commodities. All aspects like quality of water for irrigation, washing, contamination from unsterlized manures, biosolids, pesticide residues, safety of agricultural workers, personnel hygiene and sanitary facilities, cleanliness and sanitation of post-handling equipment, machinery, *etc.* have been taken into consideration while finalizing the standards.

(C) Codex Standards for different Fruits and Vegetables

Codex Alimentarius Commission has prescribed codex standards for various fruits, vegetables and other commodities which have been universally accepted. The codex standards define minimum requirements in form of quality classification details like provisions concerning sizing and tolerance for sizes and quality, details about marking, labeling and guidance about contaminants, pesticide residues and hygiene.

(D) HACCP (Hazard Analysis and Critical Control Point)

Guidelines HACCP guidelines are a systematic approach to identifying, evaluating and containing food safety hazards, which can be biological, chemical or physical agents. Such agents are likely to cause injury in the absence of their control. HACCP system is a preventive system of hazard control rather than a reactive one and is designed to prevent the occurrence of potential food safety problems. Essentially HACCP is a system that identifies and monitors specific food borne hazards that can adversely affect the safety of the food product. Critical Control Points identify those points in the process that must be controlled to ensure the safety of food. Monitoring and verification are important steps in HACCP to ensure control of potential hazards. Principles have been developed which provide guidance for an effective HACCP plan.

(E) Food Safety Management System Certification

International Organization for Standardization (ISO) has published Food Safety Management System and is classified as ISO 22000. This system promotes FARM TO FORK APPROACH and is the latest and comprehensive food safety and quality management system. It demonstrates the ability to control food safety hazards and communicates food safety information throughout the food chain (through traceability) by emphasizing on pre - requisite programmes *i.e.* Good Agricultural Practices, Good Manufacturing Practices and Good Hygienic Practices *etc.* ISO 22000 ensures compliance to sanitary requirements assesses and meets consumer requirements and certifies that food is safe at the time of consumption. Moreover, Food Safety Management is fully compatible with quality management system as per ISO 9001-2000.

Traceability is the ability to track the movement of food through all the stages of production, processing and distribution. With this system one can verify inputs used, Lab tests performed, SPS measures adopted, grading and packing details *etc.* Many developed countries have built traceability in their food standards.

GS 1 – Global Traceability standard is the most recent standard which is in pipeline for coming in to effect from Jan, 2009. This standard mainly emphasizes on critical issues of seamless traceability and may also reach into the packaging area as is done in HACCP or FSMS (ISO – 22000).

GS 1 is being rapidly adopted by retail chains in Europe. The GS 1 Traceability Standard is a high level description of the traceability process enabling and promoting supply chain collaboration. At the same time, it allows each company to design its own traceability system in terms of scope, precision and automation in accordance to its own commercial objectives.

(F) Other Trends

In European Market, GLOBALGAP certified supplies of fresh fruits and vegetables of conventional (non-organic) type are coming more and closer to organic produce, because of ever reducing MRL requirements by the authorities.

3. Status of Food Quality Standards in India

India being a member of W.T.O, has adopted sanitary and phytosanitary measures agreement. With our experience of exports of fruits and vegetables especially to EU countries, India has realized that to stay in the competitive export market, our exports will have to learn to grapple with existing food standards or else our farmers run the risk of missing the market access in other countries that guarantees them competitive prices. Thus, food safety and quality is inevitable for export market. Few years back, our export of apple to SAARC countries dwindled mainly because of supply of better quality apples (properly graded, shining and having attractive appearance) made available by Australia, USA and New Zealand. Moreover, Indian consumers too deserve safe quality food. WTO's National Treatment Principle relent us from enforcing standards on imported goods in the absence of similar domestic standards and because of this consumers and farmers run the risk of being dumped with cheap and sub-standard food products. Indian consumers are also getting quality conscious and demand for quality food is increasing. If domestic standards are not improved, the upcoming retail chains like Reliance, Spencer's, *etc.* may prefer to source their products through imports to meet the growing demand of quality food and as a result of which Indian farmer would suffer not being competitive. India has started making sincere efforts to improve upon quality standards of farm produce for boosting exports. The measures taken in this direction are given below:

(A) Adoption of Quality Standards

India has accepted the importance of quality standards for different commodities among fresh fruits and vegetables and even Govt. of India has been participating in the meetings of Codex Alimentarius Commission for consideration of draft proposals of different commodities, so that various features and required parameters pertaining to Indian products are incorporated in the final version of quality standards. Recently, various standards applicable for exports of fresh fruits and vegetables have been notified under AGMARK Act and have been named under title of GRADE DESIGNATION AND QUALITY. These standards under Agmark Act have been prepared keeping Codex Standards as a base. These updated standards are available with www.agmarknet.nic.in to exporters for improving quality of produce for exports.

(B) Adoption of GLOBALGAP

GLOBALGAP–Good Agricultural Practices are being widely adopted by the growers; GAP (Good Agricultural Practices) eliminates food safety hazards from indiscriminate use of farm inputs, besides ensuring compliance with hygienic measures, personnel hygiene, environment protection and pest control.

Even many well known exporters in India like M/s Fresh Trop Fruits Limited, Nashik and Field Fresh of M/s Bharti Enterprises Limited, Ludhiana are encouraging farmers (who supply fruits and vegetables) to have GLOBALGAP certification. Many farmers in the ambit of Field Fresh and Fresh Trop have received GLOBALGAP certification especially who supply grapes and pomegranate for export from Maharashtra and also who supply fresh vegetables for export from Punjab.

Indian farmers and exporters were initially apprehensive about EUREPGAP certification but compliance to it has significantly increased India's grape export and improved the unit price realization.

Further, GLOBALGAP certification has been so much appreciated at all levels that a draft IndiaGAP programme designed in consultation with various stakeholders is under consideration. The IndiaGAP programme is dovetailed to Indian conditions and seeks to encourage farmers to constantly upgrade their infrastructure and agricultural practices. In the long run, IndiaGAP too would need to be harmonized with globally accepted norms as that of GLOBALGAP to avoid multiple audits.

(C) Adoption of Traceability

India has realized the importance of traceability, after EU raised concerns over high pesticide residue levels in grapes and adoption of traceability was initiated in Indian grapes in 2006. This has improved the confidence of importing countries in our agricultural products. It also helps to do away with repeat testing of products in importing countries. Because of above facts traceability is quickly percolating in other fruits and vegetables.

(D) Adoption of Food Safety Management System certification

FSMS certification scheme has already been accepted by Bureau of Indian Standards. In due course it will be put into place at levels like exporters, packers, *etc.*

4. Initiative from APEDA

APEDA has also been taking various steps for promotion of quality and its control. APEDA provides assistance in the form of subsidy up to 25 per cent for setting up laboratories for analyzing quality parameters like pesticide residue, metal content, *etc.* APEDA also promotes quality control by giving 50 per cent subsidy for installing quality management, quality assurance and quality control systems such as ISO series, HACCP and GAP Certification.

5. Certification Facilities

With the awareness on importance of quality control especially pesticide residue, more and more farmers are registering with GLOBALGAP. Facilities for

GLOBALGAP Certification are available at a number of locations (Annexure–III). This has resulted in registration of more and more grape and pomegranate farmers in Maharashtra. Similarly, registration of contract farmers (supplying vegetables to Field Fresh) is rapidly increasing and becoming quite popular in Ludhiana/ Ladowal region of Punjab. Also facilities for testing for pesticide residues are now available in almost all states.

APEDA

The Agricultural and Processed Food Products Export Development Authority (APEDA) was established by the Government of India under the Agricultural and Processed Food Products Export Development Authority Act passed by the Parliament in December, 1985. The Act (2 of 1986) came into effect from 13th February, 1986 by a notification issued in the Gazette of India: Extraordinary: Part-II [Sec. 3(ii): 13.2.1986). The Authority replaced the Processed Food Export Promotion Council (PFEPC).

In accordance with the Agricultural and Processed Food Products Export Development Authority Act, 1985, (2 of 1986) the following functions have been assigned to the Authority.

Assigned Functions of APEDA

- ✰ Development of industries relating to the scheduled products for export by way of providing financial assistance or otherwise for undertaking surveys and feasibility studies, participation in enquiry capital through joint ventures and other reliefs and subsidy schemes;
- ✰ Registration of persons as exporters of the scheduled products on payment of such fees as may be prescribed
- ✰ Fixing of standards and specifications for the scheduled products for the purpose of exports;
- ✰ Carrying out inspection of meat and meat products in slaughter houses, processing plants, storage premises, conveyances or other places where such products are kept or handled for the purpose of ensuring the quality of such products;
- ✰ Improving of packaging of the scheduled products;
- ✰ Improving of marketing of the scheduled products outside India;
- ✰ Promotion of export oriented production and development of the scheduled products;
- ✰ Collection of statistics from the owners of factories or establishments engaged in the production, processing, packaging, marketing or export of the scheduled products or from such other persons as may be prescribed on any matter relating to the scheduled products and publication of the statistics so collected or of any portions thereof or extracts there from;
- ✰ Training in various aspects of the industries connected with the scheduled products;
- ✰ Such other matters as may be prescribed.

Export Standard and Export of Important Fruit Crops

Mango

Indian mangoes come in various shapes, sizes and colours with a wide variety of flavour, aroma and taste. The Indian mango is the special product that substantiates the high standards of quality and bountiful of nutrients packed in it. A single mango can provide up to 40 per cent of the daily dietary fibre needs – a potent protector against heart disease, cancer and cholesterol build – up.In addition, this luscious fruit is a warehouse of potassium, beta - carotene and antioxidants.

For maintaining highest quality standards, State-of-the-art packhouses have been set up in major production zones. Keeping in view the different country requirements, internationally recognized treatment facilities like Hot water treatment, Vapour heat treatment and Irradiation facilities have also been set up at various locations across the production belt.

Unique product identification system, compliant to the traceability networking and Residue Monitoring Plan has been developed for the consumer safety and readiness to product recall in case of any emergency.

Mango Varieties Exported

- Alphonso
- Banganapalli
- Dashehari
- Langra
- Kesar
- Chausa

Potential Exportable Varieties

- Ratoul
- Mallika
- Himsagar

Strength of India in Mango Export

- Climatic variability - allows to harvest mango for a longer period
- Availability of different type of variety, which suits the different groups of consumers
- Research and R&D system in place
- Developed infrastructure for export through air and sea.

Areas of Cultivation

The major mango-growing states are Andhra Pradesh, Uttar Pradesh, Karnataka, Bihar, Gujarat and Tamil Nadu. Andhra Pradesh ranks first in mango production with a share of 24.48 per cent and highest productivity.

India Facts and Figures

India is also a prominent exporter of fresh mangoes to the world. The country has exported 36329.01 MT of fresh mangoes to the world for the worth of Rs. 317.10 crores during the year 2015-16.

Major Export Destinations (2015-16)

United Arab Emirates, United Kingdom, Saudi Arabia, Nepal and Kuwait.

Consumer Preference of Importing Countries

In most of the importing countries

- ✰ Appearance of fruit (peel colour -yellow/red colour)
- ✰ Fruit quality
- ✰ Fleshy with attractive pulp colour
- ✰ Fiber less pulp
- ✰ Firmness
- ✰ Sugar and Acid blend
- ✰ Free from any damage

Minimum Quality Standards of Mango

Fruit shall be:

- ✰ Whole, firm, sound and fresh in appearance;
- ✰ Clean, practically free of any visible foreign matter;
- ✰ Free of marked bruising,
- ✰ Free from damage caused by low/or high temperature;
- ✰ Free of any foreign smell and/or taste;
- ✰ Free of damage caused by pests;
- ✰ Sufficiently developed and display satisfactory ripeness;

Table 5.1: Export of Fresh Mango (Quantity: MT, Value: Rs. in Lac)

Country	*2013-14*		*2014-15*		*2015-16*	
	Quantity	*Value*	*Quantity*	*Value*	*Quantity*	*Value*
United Arab Emirates	23046.65	17231.11	29231.90	21497.85	19,973.60	19,199.34
Nepal	1106.44	322.92	3574.93	695.42	8,273.99	1,393.53
Bangladesh	2899.85	411.01	2475.33	473.42	46.30	7.01
Saudi Arabia	1721.91	1218.95	2171.49	1428.59	1,399.08	1,675.19
Qatar	770.08	655.08	998.10	810.81	1,016.25	1,023.29
Kuwait	4601.44	823.55	787.28	1238.18	748.35	1,298.33
Total	41280.01	28542.84	42998.31	30253.65	36329.00	**31710.04**

☆ When a peduncle is present, it shall not be no longer than 1.0 cm.

☆ Mangoes shall comply with the residue levels of heavy metals, pesticides and other food safety parameters as laid down by the Codex Alimentarius Commission for exports.

Grapes

Grade Designation and Quality of Table Grapes

1. Table Grapes shall be fruits obtained from varieties (cultivars) of *Vitis vinifera* L.
2. Minimum requirements:
   i) Bunches and berries of table grapes shall be:
      a. Clean, sound, free of any visible foreign matter;
      b. Free of pests, affecting the general appearance of the produce;
      c. Free of damage caused by pests and diseases;
      d. Free of abnormal external moisture;
      e. Free of any foreign smell and/or taste;
      f. Free of all visible traces of moulds;
   ii) Berries shall be intact, well formed and normally developed,
   iii) Table grapes shall comply with the residue levels of heavy metals, pesticides and other food safety parameters as laid down by the Codex Alimentarius Commission for exports.
   iv) Table grapes shall have minimum soluble solids of 16° Brix.
   v) Table grapes shall have minimum sugar/acid ratio of 20: 1.

Table 5.2: Criteria for Grade Designation

Grade Designation	*Grade Requirements*	*Provision Concerning Sizing*	*Grade Tolerances*
1	*2*	*3*	*4*
Extra class	Grapes must be of superior quality. The bunches must be typical of variety in shape, development and coloring and have no defects. Berries must be firm, firmly attached to the stalk, evenly spaced along the stalk and have their bloom virtually intact.	As per table 'A'	5 per cent by weight of bunches not satisfying the requirements of the grade, but meeting those of class I grade or exceptionally coming within the tolerances of that grade.
Class I	Grapes must be of good quality. Bunches must be typical of variety in shape, development and coloring. Berries must be firm, firmly attached to the stalk and, as far as possible, have their gloom intact. They may, however, be less evenly spaced along the stalk than in the extra class.	-do-	10 per cent by weight of bunches not satisfying the requirements of the grade, but meeting those of class II grade or exceptionally coming within the tolerance of that grade.

Grade Designation	*Grade Requirements*	*Provision Concerning Sizing*	*Grade Tolerances*
1	*2*	*3*	*4*
	Following slight defects may be there, providing these do not affect the general appearance of the produce and keeping quality of the package. - a slight defect in shape. - a slight defect in coloring		
Class II	Bunches may show defects in shape, development and coloring provided these do not impair the essential characteristics of the variety. The berries must be sufficiently firm and sufficiently attached. They may be less evenly spaced along the stalk than Class I grade. Following defects may be there, provided these do not affect the general appearance of the produce and keeping quality of the package. - defects in shape - defects in coloring - slight sun scorch affecting the skin only, - slight bruising, - slight skin defects	-do-	10 per cent by weight of bunches not satisfying the requirements of the grade, but meeting the minimum requirements.

Table 'A': Provision Concerning Sizing

Size is determined by the weight of bunches (in gms). The following minimum (in gms) requirements per bunch are laid down for large and small berries grapes.

Grade	*Large Berries*	*Small Berries*
Extra class	200	150
Class I	150	100
Class II	100	75

Size Tolerance

Extra Class, Class I, Class II: 10 per cent by weight of bunches not satisfying the size requirements for the grade, but meeting the size requirements for the grade immediately below.

Areas of Cultivation

Major grape-growing states are Maharashtra, Karnataka, Andhra Pradesh, Tamil Nadu, and the north-western region covering Punjab, Haryana, western Uttar Pradesh, Rajasthan and Madhya Pradesh. Maharashtra ranks first in terms of production accounting for more than 82.56 per cent of total production and highest productivity in the country.

India Facts and Figures

Grape is one of the important fruit covering an area of 116 thousand hectares occupying 1.70 per cent of the total area. The country is also a major exporter of fresh Grapes to the world. The country has exported 1,56,218.34 MT of Grapes to the world for the worth of Rs. 1,551.32 crores during the year 2015-16.

Major Export Destinations (2015-16)

Netherland, United Kingdom, Russia, United Arab Emirates and Saudi Arabia.

Table 5.3: Export of Fresh Grapes (Quantity: MT, Value: Rs. in Lac)

Country	*2013-14*		*2014-15*		*2015-16*	
	Qty	*Value*	*Qty*	*Value*	*Qty*	*Value*
Netherland	46,136.91	52,705.69	32,859.42	36,167.92	50,702.44	58,814.61
United Kingdom	17,149.00	20,870.33	13,475.81	16,028.53	18,014.68	20,877.91
Russia	24,503.00	22,663.55	12,625.78	11,372.30	13,804.66	13,673.07
United Arab Emirates	17,212.25	12,962.09	11,860.53	11,354.52	13,075.43	11,661.03
Saudi Arabia	9,387.85	5,029.44	5,319.73	4,689.93	8,140.47	6,940.44
Germany	3,158.24	3,201.57	2,939.01	3,236.63	5,143.99	5,491.35
Thailand	3,807.94	5,020.10	2,260.11	2,873.85	3,193.77	4,041.99
Sri Lanka	3,498.69	1,976.15	2,072.47	2,064.37	3,139.82	3,211.27
Ukraine	4,453.97	3,629.14	1,374.00	1,145.05	3,358.39	2,234.35
Finland	901.77	1,282.23	831.08	995.66	1,574.50	2,013.04
Bangladesh	31,188.36	7,575.89	3,562.02	1,312.34	7,425.93	1,975.10
Lithuania	1,395.24	1,346.94	737.79	760.62	1,905.07	1,710.13
Malaysia	1,010.87	1,082.37	830.99	1,041.01	1,375.74	1,580.93
Sweden	2,787.41	3,984.95	940.08	1,041.39	1,282.16	1,485.11
Norway	1,131.95	1,493.79	943.26	1,173.96	1,894.00	1,453.60
Hong Kong	1,644.71	2,409.27	1,141.35	1,655.60	1,056.44	1,424.69
Taiwan	935.54	1,309.96	711.55	944.20	1,079.69	1,396.39
Mexico	1,275.00	956.17	742.00	581.60	2,423.00	1,338.07
Poland	1,043.00	839.54	270.00	232.57	1,703.73	1,233.59
TOTAL	**1,92,616.92**	**1,66,647.40**	**1,07,257.85**	**1,08,649.02**	**1,56,218.34**	**1,55,131.67**

Pomegranate

Grading of Fruits

Grading based on size and colour.

1. Super size - Fruits are free from spots and individual fruit weight is more than 750grams.

2. King size - Fruits are attractive and individual fruit weight is 500-700 grams.
3. Queen size - Fruits are attractive, red and individual fruit weight is 400-500 grams.
4. Prince size - Fruits are attractive, red and individual fruit weight is 300-400 grams.

The desirable fruit characters of fresh pomegranate for export purpose

- Dark rose pink colour of the fruit.
- Fruit weight around 500 gms.
- Round shape of the fruit.
- Uniform size and shape of the fruit in a pack or box.
- Dark rose pink arils.
- Softness of the seeds.
- Higher sugars near about 16-17° Brix.
- Free from scars, russetting, disease spots, insect injury, scratches, *etc.*
- Smooth cutting at the stem end.
- Bracts/calyx without any damage and having freshness.
- Pleasant flavour and aroma.

Marketing and Export Potential

- No organized marketing systems for pomegranate.
- Domestic marketing for pomegranate is fairly balanced as of now.
- Farmers normally dispose their produce to contractors who take the responsibility of transport to far off markets.
- Some progressive farmers who have developed contacts in urban areas despatch them on their own and are able to get a reasonable price for their produce
- Changes in varietal selection and in cultivation practices.
- Pest and disease incidence, fall in demand and price - effect on the cultivation of pomegranate in the state.
- Cultivation for export purpose - improved due to exploring of market potentialities.
- Farmers of Bijapur and Bagalkot districts - exporting pomegranate through their associations to Gulf and European Union for the last 5 years.

Export Potential

- The pomegranate fruit is considered an exotic in European and Middle East countries.

- ☆ There is scope for exporting Indian pomegranates to Bangladesh, Bahrain, Canada, Germany, United Kingdom, Japan, Kuwait, Sri Lanka, Omen, Pakistan, Qatar, Saudi Arabia, Singapore, Switzerland, U.A.E. and U.S.A.
- ☆ The other potential markets are, Belgaum, Spain, France, Italy, Netherlands, Philippines, South Africa, Brazil and Mexico.

Table 5.4: Export of Fresh Pomegranates (HS Code: 08109010) (Quantity: MT, Value: Rs. in Lac)

Countries	*2009-10*		*20010-11*		*2011-12*		*Per cent Share in Value*
	Qty.	*Value*	*Qty.*	*Value*	*Qty.*	*Value*	
U A E	17,350.00	5,170.01	11,229.57	3,884.42	15,899.74	6,416.44	43.57
Bangladesh	4,783.71	468.16	1,964.84	252.55	4,593.46	2,383.70	16.18
Netherland	2,567.71	1,985.31	389.07	330.17	732.69	1,028.17	6.98
U K	973.20	1,051.47	531.27	825.45	890.85	975.31	6.62
Saudi Arab	1,789.35	627.77	1,182.83	368.44	2,196.63	870.39	5.91
Russia	470.63	521.48	186.46	279.87	398.08	573.32	3.89
Thailand	280.26	195.99	104.94	135.50	298.07	310.61	2.11
Nepal	1,447.64	157.11	819.18	111.07	1,438.26	246.85	1.68
Kuwait	505.77	222.43	204.40	102.48	489.65	236.30	1.60
Oman	415.58	151.55	310.92	85.85	391.87	198.95	1.35
Other Countries	2,785.08	1,391.53	1,288.18	719.44	2,832.97	1,487.80	10.10
Total	**33368.93**	**11942.81**	**18211.66**	**7095.24**	**30162.27**	**14727.84**	**100.00**

Table 5.5: Export of Fresh Apple (Quantity: MT, Value: Rs. in Lac)

Country	*2012-13*		*2013-14*		*2014-15*	
	Quantity	*Value*	*Quantity*	*Value*	*Quantity*	*Value*
Bangaladesh	20549.57	3787.37	30701.91	6375.98	13801.86	2968.66
Nepal	3243.55	1359.12	4914.31	1863.07	6172.21	2088.45
China, Peoples Repub	0.00	0.00	0.00	0.00	289.59	139.18
United Arab Emirates	0.27	0.11	30.61	34.17	42.75	32.43
Saudi Arabia	0.00	0.00	0.42	0.15	40.00	31.44
Singapore	5.48	1.39	0.00	0.00	22.00	32.28
Total	**23805.89**	**5151.39**	**36098.07**	**8623.78**	**20374.70**	**5292.73**

Table 5.6: Export of Fresh Banana (Quantity: MT, Value: Rs. in Lac)

Country	*2012-13*		*2013-14*		*2014-15*	
	Quantity	*Value*	*Quantity*	*Value*	*Quantity*	*Value*
United Arab Emirates	0.00	0.00	13897.45	6711.88	19.010.88	8755.03
Saudi Arabia	0.00	0.00	4744.24	2482.71	9314.43	4557.57
Nepal	0.00	0.00	5054.46	899.19	8637.73	955.12
Pakistan	0.00	0.00	21.68	6.45	7243.35	1472.21
Oman	0.00	0.00	3493.86	1500.49	4926.01	2004.43
Kuwait	0.00	0.00	2142.75	1227.07	4684.43	2241.77
Iran	0.00	0.00	885.95	297.40	3377.60	1246.50
Qatar	0.00	0.00	19.33.50	1023.65	3210.70	1671.64
Total	**0.00**	**0.00**	**34830.81**	**15384.93**	**63274.40**	**24194.77**

Table 5.7: Export of Fresh Oranges (Fresh/Dried) (Quantity: MT, Value: Rs. in Lac)

Country	*2012-13*		*2013-14*		*2014-15*	
	Quantity	*Value*	*Quantity*	*Value*	*Quantity*	*Value*
Bangladesh	32205.07	3161.96	25849.11	4879.95	14210.73	2456.01
Nepal	1060.47	263.90	2305.53	651.03	2845.36	661.37
Bhutan	0.00	0.00	0.00	0.00	82.05	12.78
Saudi Arabia	0.03	0.01	1.32	0.26	42.77	34.47
United Arab Emirates	30.58	26.25	1.33	2.32	32.05	16.04
Oman	57.35	13.03	52.03	13.39	5.60	1.73
Total	**36628.13**	**3528.26**	**28236.55**	**5558.31**	**17231.45**	**3194.00**

Table 5.8: Export of Fresh Papaya (Quantity: MT, Value: Rs. in Lac)

Country	*2012-13*		*2013-14*		*2014-15*	
	Quantity	*Value*	*Quantity*	*Value*	*Quantity*	*Value*
United Arab Emirates	5548.06	1031.37	4299.41	1328.40	4378.81	1402.14
Saudi Arabia	3179.13	753.77	1940.02	764.89	2406.25	870.57
Nepal	1882.09	222.58	1113.14	165.81	1768.08	163.48
Bahrain	1223.48	177.21	699.22	239.06	744.95	246.69
Netherland	1177.24	470.14	561.71	395.11	691.79	473.85
Qatar	1400.00	170.91	375.20	172.17	520.10	232.41
Total	**16491.23**	**3329.21**	**9921.76**	**3510.86**	**11484.82**	**3826.41**

Table 5.9: Export of Walnut (Quantity: MT, Value: Rs. in Lac)

Country	*2013-14*		*2014-15*		*2015-16*	
	Qty	*Value*	*Qty*	*Value*	*Qty*	*Value*
Germany	404.59	1,977.14	108.57	570.89	511.56	2,040.31
United Kingdom	639.75	2,931.38	392.50	1,941.69	452.01	1,693.91
Netherland	685.50	3,592.52	254.66	1,419.47	424.50	1,310.50
Egypt Arab Repub.	581.00	3,634.74	241.00	1,631.15	252.60	1,267.69
United States	360.64	2,153.10	230.01	1,735.60	178.69	1,138.84

Chapter 6

Nutritional and Therapeutic Importance of Dryland Fruits

A.M. Musmade, V.P. Kad and J.K. Dhemre

Department of Horticulture,
Mahatma Phule Krishi Vidyapeeth, Rahuri – 413 722, Dist. Ahmednagar, M.S.

Introduction

Fruits are amongst the first food items known to human beings. In fact, they have been in existence since their inception of mankind (even Adam and Eve ate apple-the forbidden fruit). In many religions, fruits have been given the status of food of gods. Fruits whether fresh or dried, have always formed a part of human beings. The reason for this is that they are rich in nutrients and provide some of the essential minerals, vitamins and the alike, to our body. Apart from that, they also help in curing a number of diseases. Fruits along with vegetables come under the class of protective foods on account of presence of vitamins and minerals which help protect the human body from nutritional deficiency diseases and also to build resistance against infectious diseases.

Fruits and vegetables are highly perishable products. Water loss and postharvest decay account for most of the losses. These have been estimated to be more than 20-30 per cent in the tropics and subtropics. The problems of postharvest technology of fruits are critical in the battle to minimize wastage and to extract the maximum potential from the harvested crops. Wastage of fruits is so high in some instances that between the field and the consumer, bountiful amounts of highly nutrition's crops are reduced to heaps refuse. Reduction of high postharvest losses of fruits and vegetables includes proper harvest handling, low temperature storage and environmental control (controlled/modified and hypobaric storage), irradiation and use of chemicals and fungicides as well as packaging techniques, and processing of fruits and vegetables into suitable products.

A) Nutritional Importance

Fruits and vegetables play an important role in human nutrition particularly in supplying certain constituents in which other food materials are deficient. Fruits and vegetables contribute vitamins and minerals and the bulk of dietary fiber to the diets of human beings. They neutralize the acid substances produced in the course of digestion of foods such as meat, cheese *etc.* The nutritional composition of fruits is presented in Tables 6.1–6.3.

Table 6.1: Proximate Composition of Major Fruits Grown in India/100 g Edible Portion

Fruit	*Water (Per cent)*	*Energy (KCal)*	*Proteins (Per cent)*	*Fats (Per cent)*	*Carbohydrates (Per cent)*	*Minerals (Per cent)*
Apple	84.4	58	0.2	0.6	14.5	0.3
Banana	75.7	85	1.1	0.2	12.6	0.6
Ber	85.9	55	0.8	0.3	12.6	—
Grape	81.6	67	1.3'	1.0	15.7	0.4
Guava	83.0	—	1.1	0.4	12.7	0.7
Mango	83.4	59	0.5	0.2	15.4	0.4
Orange	86.0	49	1.0	0.2	12.2	0.6
Papaya	90.7	32	0.5	0.1	8.3	0.4
Pineapple	85.4	52	0.4	0.2	13.7	0.3
Pomegranate	78.0	58	1.6	0.1	14.6	0.7
Sapota	74.0	98	0.7	1.1	21.4	0.5
Strawberry	89.9	37	0.7	0.5	8.4	0.5

Fruits, fresh or dried have been natural staple diet of human beings since ancient times. Replete with minerals, vitamins, enzymes, they are easily digestible. Fruits are not only good source as food; they can serve as medicine and treat ailments. Fruits, eaten raw or consumed as fresh juice are an excellent way to retain and balance moisture level in the body. The low level of sodium in fruits plays an important role for people who avail of salt-free diet. The pregnant ladies eat dry fruits like apricots, raisins and dates are a store house of calcium, iron, and essential for strengthening of bones and good blood. Custard apple is an excellent calcium source. One or two fruit a day cleans the digestive tract and aids easy bowl action. Guavas, custard apples, lemons and oranges are effective sources of vitamin C. Papaya is an excellent source of Vitamin C and carotene. Nutrients in apple, date and mangoes sharpen memory, prevents exhaustion, hysteria, insomnia and mental tension. All forms of berries are rich in iron, phosphorus and sodium that are beneficial for blood building and nerve strengthening. Lemons are good for liver ailments, indigestion and rheumatism. Watermelons are good kidney cleansers. Pineapple, pomegranates help in soothing inflammation of nose, throat, hay fever and other chronic or bronchial ailments. Grapefruit juice is an effective aid for common cold. Fresh and ripe fruits like grapes, apples, bananas and figs are good for all brain deficiencies. Eating liberal amount of fruits daily ensures a healthy energetic life and active life lifetime.

Table 6.2: Vitamin Content of Major Fruits Grown in India/100 g Edible Portion

Fruit	*Vitamin A (IU)*	*Thiamine (mg)*	*Riboflavin (mg)*	*Niacin (mg)*	*Ascorbic Acid (mg)*
Apple	90	0.03	0.02	0.1	4
Banana	190	0.05	0.06	0.7	10
Ber	70	0.03	0.04	0.5	80
Grape	100	0.05	0.03	0.3	4
Guava	250	0.05	0.03	1.2	337
Mango	630	0.05	0.06	0.4	53
Orange	200	0.10	0.04	0.4	50
Papaya	2020	0.03	0.04	0.3	46
Pineapple	15	0.08	0.04	0.2	61
Pomegranate	—	0.06	0.10	0.3	16
Sapota	—	0.02	0.03	0.2	6
Strawberry	60	0.03	0.07	0.6	59

Table 6.3: Mineral Content of Major Fruits Grown in India/100 g Edible Portion

Fruit	*Calcium (mg)*	*Phosphorus (mg)*	*Iron (mg)*	*Magnesium (mg)*
Apple	7	10	0.3	8
Banana	8	26	0.7	33
Ber	30	30	0.8	—
Grape	16	12	0.4	13
Guava	17	28	1.8	—
Mango	12	12	0.8	—
Orange	41	20	0.4	11
Papaya	20	13	0.4	—
Pineapple	18	8	0.5	—
Pomegranate	10	70	0.3	12
Sapota	28	27	2.0	—
Strawberry	21	21	1.0	12

Ascorbic acid is the principal vitamin supplied by fruits and vegetables in the diet. About 90 per cent of a Pearson's dietary requirement is obtained from fruits and vegetables. In addition, fruits and vegetable are rich sources of (ß-carotene thiamine, niacin and folic acid, minerals such as Ca, Mg, K and sugars. The nutritive constituents of fruits that have a positive impact on human health and their sources are mentioned. Nutritional contributing of fruits and vegetables compared to per cent of total food supply.

Table 6.4: Nutritional Constituents

Constituent	*Sources*	*Established Effect on Human-Wellness*
Ascorbic acid (Vitamin C)	Citrus fruits, guava kiwifruit, pineapple, strawberry, anola, watermelons	Prevents scurvy, aids wound healing, healthy immune-system, cardiovascular disease.
Carotenoids (Vitamin A)	Pumpkin, mango, orange, papaya, pineapple *etc.*	Night blindness prevention.
Fibers	Most fresh fruits	Diabetes, heart disease
Calcium	Papaya, raisins, oranges, pumpkin	Osteoporosis muscular/skeletal teeth, blood pressure.
Magnesium	Banana, Cashews	Osteoporosis, nervous system, teeth, immune system.
Potassium Banana	Banana	Hypertension (Blood pressure) stroke arteriosclerosis

Other Constituents of Fruits

Fruits are particularly rich in pectin and gums. Pectin has been shown to delay gastric emptying which may lead to favorable changes in glycemic response to foods. Pectin may also induce a satiety effect, it acts as a general intestinal regulator and detoxifying agent. It is used as a standard additive in commercial baby food formulations. Pectin has been shown to have antibacterial, antiviral, wound-healing and metal-binding properties. It is shown to be effective in reducing cholesterol levels. A substantial proposition of carbohydrates in fruits and vegetables is present as dietary fiber in the form of cellulose, hemicelluloses, peptic substances and lignin. The dietary fiber relives constipation and can be the panacea to cure several human diseases, especially in the affluent societies. Fruits contain some flavour compounds. These include sugars, amino acids and organic acids, volatiles such as aromatic hydrocarbons, aldehydes, acetals, ketones, alcohols, esters and sulphur compounds, when used along with other food items. These flavorings make food more palatable.

Therapeutic Importance

Fruits have highly beneficial effect on the human system. The main physiological actions of fruits are;

Hydrating Effect

Fruits whether taken as a whole or in the form of juice, have hydrating effect. The water in the form of fruit juices has additional benefits of providing sugars and essential minerals.

Diuretic Effect

Most of the fruits are rich in K, Mg and Na. These nutrients act as diuretic and dieresis, increasing the frequency of passing urine. This high frequency results in a lower urine density, which ultimately speeds up the process of elimination of nitrogenous waste and chlorides from the body.

Alkalinizing Effect

The salts present in fruits have organic acids, which when transformed in the body, produce alkaline carbonates. These leads to alkalization of the fluids and in turn, promotion of intestinal elimination. In other words, regular consumption of fruits helps the body stay free of toxic wastes.

Mineralizing Effect

Fruits are mostly rich in minerals. Dried fruits such as dates, raisins, apricots are rich in calcium and iron. The custard apple has as much as 800 mg of calcium/ fruit. One fruit can meet almost the entire daily requirement of calcium by the body.

Laxative Effect

The fibrous matter in fruits, cellulose, aids in the smooth passage of the food in the digestive tract and easy bowl emptying. The sugars and organic acids contained in fruits also increase their laxative effect. Hence regular use of fruits prevents constipation.

Tonicaction

Fruits, as dependable source of vitamins, exert a tonic effect in the body. Guava, custard apple, citrus fruits, mango, papaya are rich in ascorbic acid and carotenoids and can meet the requirement of body.

Other Benefits

Fruits like apple, lemon, orange and pomegranate aid in proper functioning of the heart. Nutrients in apple, date, mangoes sharpens memory, prevents exhaustion, hysteria, insomnia and mental tension. All forms of berries are rich in iron, P and Na that are beneficial for blood building and nerve strengthening. Lemons are good for liver ailments indigestion and rheumatism. Watermelons are good kidney cleansers. Pineapple, pomegranates help in soothing inflammation of nose throat, hay fever and other chronic or bronchial ailments. Grapefruit juice is effective aid for common cold. Fresh and ripe fruits like grapes, apples, banana and figs are good for all brain deficiencies. Eating liberal amounts of fruits daily ensures a healthy energetic life and active life lifetime.

Non-nutritive Plant Constituents that may be Beneficial to Human Health

Many therapeutic drugs in use in modern medicine originated as plant extracts. Certain fruits and vegetables compounds exit pharmacological or therapeutic effects. Limonin and nomilin (and other limonoids) are present in citrus fruits such as orange, lemon, lime and grapefruit. These compounds (Table 6.5) are believed to have a role in inhibiting the development of certain form of cancer, and recent research has indicated that the antioxidant properties of ß-carotene may also play a role in the prevention of some forms of cancer. Fruits and vegetables are very well

Table 6.5: Non-Nutritive Plant Constituents

Constituent	*Compound*	*Sources*	*Established effect on human wellness*
Phenolic compounds			
Proanthocyanins	Tannins	Apple, grape, cranberry, pomegranate	Cancer
Anthocyanidins	Cyanidin, malvidin delpinidin peonidin petunidin	Red, blue and purple fruits such as apple, cranberry, pomegranate, strawberry	Heart disease, caner initiation, diabetes, cataracts, blood pressure, allergies
Flavan-3-ols	Epicatechin, epigallocatechin, catechin, gllocatechin	Apples, apricots	Platelet aggregation, cancer
Flavanones	Hesperitin, naringenin, eriodictyol	Citrus (orange, grapefruit, lemon, limes tangerine)	Cancer
Flavones	Luteolin, apigenin	Guava	Cancer, allergies, heart disease
Flavonols	Quercetin, Kaempferol, myricetin, rutin	Cranberry	Heart disease, cancer initiation, capillary protectant
Phenolic acids	Caffeic acid, chlorogenic acid, coumaric acid, ellagic acid	Black berry, raspberry, strawberry, apple	Cancer, cholesterol
Carotenoids			
Lycopene	-	Water melons, papaya, guava, red grapefruit	Cancer, heart disease, male infertility
a-carotene	-	Pumpkin, mango, papaya, kiwifruit	Tumor growth
p-carotene	-	Mango, papaya, pumpkin	Cancer
Xynthophylls	Lutin, zeaxanthin 0-cryptoxanthin	Vegetables	Muscular degeneration
Monoterpenes	Limonin	Citrus (Grape fruits, tangerine)	Cancer
Sulphur compounds			
	Glucosinolates, isothiocyanates, indoles, allicin, diallyl isosulphide	Vegetable	Cancer, cholesterol, blood pressure, diabetes

known sources of different types of flavonoids which include flavonones, flavones, flavonols, anthocyanins, catechins and biflavans. Apart from fat soluble tocopherols, flavonodis are commonest and most active antioxidant compounds in human food, being active in hydrophilic as well as in lipophillic systems. It protects foodstuff from oxidative deterioration, prolongs their shelf life and keeping quantity and improves taste, acceptability and wholesomeness of mixed dishes by inhibiting oxidation of accompanying lipids. Flavonoids increase or stabilize biological activity of ascorbic acid. All fruits rich in flavonoids are marked by the unusual stability of their ascorbic acid. The ascorbic acid stabilizing effect of flavonoids results in increased production of ascorbic acid-2-sulfates which improves membrane stability by syntheses of sulfated mucopolysaccharides and enhances cholesterol excretion. Ellagic acid is a naturally occurring phenolic constituent present in fruits, especially straw berries and other berries and has been shown to be effective as antimutagen and anti carcinogen and a potential inhibitor of common inducers of cancer. Fruits are processed into several products such as juices, jams, jellies wines, dried fruits, and canned products employing different method of food preservation. There are some losses of nutrients during processing. Once processed, the nutrient content of the product remain relatively stable. Hence, processing condition should be such that processed products will suffer minimum loss of nutrients compared to raw material considering the nutrition and pharmacological significance of fruits and their products, they have a major role in nutrition and health of human population.

The Importance of Fruits Juices in Nutrition

Fruit juices have been scientifically proven to give certain health benefits, provided that they are taken in moderation. Nutritionists also suggest that fruit juices should be taken in their pure state and people.

Other fruit juices that are rich in antioxidants, vitamins and anthocynanins (water soluble pigments found in plants which act as antioxidants) include grape juice (purple), pomegranate juice and cranberry juice. It is believed that juices that are made from fruits with rich colors such as grapes, pomegranates, cranberries and blueberries are rich in vitamins and antioxidants. The studies have claimed that the antioxidants found in most fruits juices can help lower a person's risks of developing Alzheimer's disease. Antioxidants are naturally occurring substances found in most plants and have the potential to help combat heart diseases and fight cancer. Freshly squeezed juices from fruits are excellent sources of minerals and vitamins which catalyze chemical reactions occurring in the body. Another helpful benefit of fruits and fruit juices is their ability to promote detoxification in the human body. Fruits help to cleanse the body, especially those with high acid levels.

Food Value of some Dryland Fruit Crops

1. Pomegranate

Pomegranate consists of antioxidant, antiviral and anti-tumour properties. It is said to be a good source of vitamins as it includes vitamin A, C and E as well as folic acid.

Pomegranates are known mostly for curing the problems related to heart and for maintaining effective and healthy blood circulation. Other health benefits include cure of stomach disorders, cancer, dental care, osteoarthritis, anemia and diabetes. Pomegranates are being traditionally used for clearing the skin and reducing inflammation. Pomegranate's juice can participate in treatment of a sore throat. Pomegranate syrups are useful in treating the cartilage damage and protest the babies from the brain damage during their birth. Pomegranate helps to reduce the illness caused in various forms, like atherosclerosis and osteoarthritis. The damages that are caused due to the thickening and hardening of the arterial walls and in cartilage and joints can be cured by eating this fruit. Also, pomegranate is capable of preventing the creation of enzymes that are liable for breaking down the connective tissues. Pomegranate reduces the possibility of having premature babies and it is also beneficial for the expected mothers to avoid having low weight babies during birth. Pomegranate also reduces the possibility of developing Alzheimer's disease among the older generation. It helps to control aging problems like wrinkles and thus, facilitates youthful and glowing skin. Other than this, it helps a woman to overcome from her depression period, especially form the menopause period.

2. Aonla

Aonla is mainly recognised due to its high nutritive and medicinal properties. It contains high amount of vitamin (500-700mg/100g), it is also rich in pectin, iron, calcium and phosphorous. Aonla probably is the only fruit to fill the gap of astringent food recommended by Ayurvedic system of medicine for balanced diet and sound health. It is acidic, cooling, refrigerant, diuretic and laxative. Aonla is the main ingredients in Chavanprash and Triphala used in the treatment of headache, constipation and enlarged lier.

3. Custard Apple

Custard apple is a storehouse of Vitamin C, which is an anti-oxidant and helps in neutralizing free radicals. Vitamin A present in the fruit is good for hair, eyes and healthy skin. Custard apple contains magnesium, which plays vital role in relaxing muscles and protecting heart against diseases. Potassium and Vitamin B6 are also present in custard apple. Copper is yet another useful ingredient of custard apple. It is a rich source of dietary fiber, which helps in digestion. As it contains low fat levels, it is good for maintaining optimum health. The paste of the flesh of the fruit can be used for local application on ulcers, abscesses and boils. The fruit, in its unripe form, can be dried, crushed and used for treating diarrhea and dysentery. Custard apple serves as an expectorant, stimulant, coolant and haematinic and is even useful in treating anemia. The seeds of the fruit have insecticidal and abortifacient properties

4. Mango

The mango is used as food in all stages of its development. Green or unripe mango contains a large portion of starch which gradually changes into glucose, sucrose and maltose as the fruit begins to ripe. It disappears completely when the fruit is fully ripe. Green mango is a rich source of pectin which gradually diminishes after the formation of the stone. Unripe mango is sour in taste because of the presence of oxalic, citric, malice and succinic acids.

Food Value	*Minerals and Vitamins*
Moisture - 81.0 per cent	Calcium - 14 mg
Protein - 0.6 per cent	Phosphorus - 16 mg
Fat - 0.4 per cent	Iron - 1.3 mg
Minerals - 0.4 per cent	Vitamin C - 16 mg
Carbohydrates - 16.9 per cent	Small amount of Vitamin B Complex
Fiber - 0.7 per cent	
Value per 100 g edible portion	Calorific Value - 74

The raw mango is a valuable source of vitamin C. It contains more vitamin C than half ripe or fully ripe mangoes; it is also a good source of vitamin Bl and B2 and contains sufficient quantity of niacin. These vitamins differ in concentration in various varieties during the stages of maturity and environmental conditions. The ripe fruit is very wholesome and nourishing. The chief food ingredient of mango is sugar. The acids contained in the fruit are tartaric acid and malic acid, besides a trace of citric acid. These acids are utilized by the body and they help to maintain the alkali reserve of the body.

Natural Benefits and Curative Properties of Unripe Mangoes

The mango is well-known for its medicinal properties both in unripe and ripe states. The unripe fruit is acidic, astringent and anti scorbutic. The skin of the unripe fruit is astringent and stimulant tonic. The bark is also astringent and has a marked action on mucous membranes. Mango pickles preserved in oil and salted solution is used throughout India. However, these pickles, if extremely sour, spicy and oily are not good for health and should be specially avoided by those suffering from arthritis, rheumatism,.sinusitis, sore throat and hyperacidity. The ripe mango is anti scorbutic, diuretic, laxative, invigorating, fattening and astringent. It tones up the heart muscle, improves complexion and stimulates appetite. It increases the seven body nutrients, called *'dhatus'* in *Ayurveda.* The fruit is beneficial in liver disorders, loss of weight and other physical disturbances. The natural benefits and curative properties of mangoes are; Heat Stroke: The unripe mango protects men from the adverse effects of hot, scorching winds. A drink, prepared from the unripe mango by cooking it in hot ashes and mixing the pith with sugar and water, is an effective remedy for heat exhaustion and heat stroke. Eating raw mango with salt quenches thirst and prevents the excessive loss of sodium chloride and iron during summer due to excessive sweating.

Gastro-Intestinal Disorders

Unripe green mangoes are beneficial in the treatment of gastro-intestinal disorders. Eating one or two small tender mangoes in which the seed is not fully formed with salt and honey is found to be very effective medicine for summer diarrhea, dysentery, piles, morning sickness, chronic dyspepsia, indigestion and constipation.

Bilious Disorders

Unripe mangoes are an excellent fruit remedy for bilious disorders. The acids contained in the green mango increase the secretion of bile and act as intestinal antiseptic. Therefore,eating green mango daily with honey and pepper cures biliousness, food putrefaction *i.e.* when proteins are decomposed by bacteria; urticaria and jaundice. It tones up the liver and keeps it healthy.

Blood Disorders

The green mango is valuable in blood disorders because of its high vitamin C content. It increases the elasticity of the blood vessels and helps the formation of new blood cells. It aids the absorption of food-iron and prevents bleeding tendencies. It increases body resistance against tuberculosis, anemia, cholera and dysentery.

Scurvy

The *amchur*, a popular article of diet in Indian houses, consists of green mangoes skinned, stoned, cut into pieces and dried in the sun. 15 g of it is believed to be equivalent to 30 g. of good lime on account of its citric content. It is valuable in the treatment of scurvy.

Natural Benefits and Curative Properties of Ripe Mango

Eye Disorders

Ripe mangoes are highly beneficial in the treatment of night blindness caused by vitamin A deficiency. It is very common among children who are victim of malnutrition due to poverty. Liberal use of mangoes during the season will be very effective in such conditions. It will also prevent many other eye diseases which may ultimately cause total blindness. Eating mangoes liberally will also prevent development of refractive errors, dryness of the eyes, softening of the cornea, itching and burning in the eyes.

Infections

All bacterial invasions are due to poor epithelium the tissue that covers the external surface of the body. Liberal use of mangoes during the season contributes towards formation of healthy epithelium, thereby preventing frequent attacks of common infections such as colds rhinitis and sinusitis. This is attributable to high concentration of vitamin A in mangoes.

Loss of Weight

The mango-milk cure is an ideal treatment for loss of weight. They should be taken thrice a day-morning, noon and evening. The mangoes should be taken first and then followed by milk. The mango is rich in sugar but deficient in protein. On the other hand, milk is rich in protein but deficient in sugar. The deficiency of the one is made up by the other. Mango thus combines very well with milk and exclusive mango milk diet taken for at least one month, will lead to improvement in health, vigor and gain in weight.

Diabetes

The tender leaves of the mango tree are considered useful in diabetes. An infusion is prepared from fresh leaves by soaking them overnight in water and squeezing them well in water before filtering it in the morning. This infused water should be taken every morning to control early diabetes.

Diarrhea

The mango seeds are valuable in diarrhea. The seeds should be collected during the mango season, dried in the shade and powdered and stored for use as medicine. It should be given in doses of about one and a half gram to two grams with or without honey. Female Disorders: Mango seeds are considered useful in certain disorders connected with women reproductive organs. Juice of the fresh mango bark is also valuable in heavy bleeding during menstruation, *i.e.* menorrhagia, leucorrhoea, mucus and pus discharges from the uterus and bleeding or hemorrhages from uterus.

Throat Disorders

The mango bark is very efficacious in the treatment of diphtheria and other throat diseases. Its fluid is locally applied and also used as a gargle. The gargle is prepared by mixing 10 ml of the fluid extract with 125 ml of water.

5. Bael

Bael fruit is rich in mineral and vitamin contents. The sherbet made out of this fruit has all the important nutrients and health growing ingredients. It should be thick and syrupy enough to be taken with spoon and it should be thoroughly masticated. If taken hurriedly, it may produce heaviness in the stomach. The bael fruit should also not be taken in excess at a time as excessive intake of bael may produce a sensation of heaviness in the stomach and may cause gastric discomfort.

Food Value	*Minerals and vitamins*
Moisture -61.5 per cent	Calcium - 85 mg
Protein - 1.8 per cent	Phosphorus - 50 mg
Fat - 0.3 per cent	Iron - 0.6 mg
Minerals - 1.7 per cent	Vitamin C - 8 mg
Carbohydrates-31.8 per cent	Small amount of Vitamin B Complex
Fibre - 2.9 per cent	
Values per 100 g edible portion	Calorific Value - 137

Natural Benefits and Curative Properties

Bael tree is one of the most useful medicinal plants of India. Its medicinal properties have been described in the ancient medical treatise in Sanskrit, Charaka Samhita. All the parts of this tree including stem, bark, root, leaves and fruit at all stages of maturity has medicinal virtues and has been used as traditional medicine for a long time. The fruit is of considerable medicinal value when it just begins to ripen. The ripe fruit is aromatic, astringent which helps construction of skin, coolant and laxative The unripe or half-ripe fruit is astringent, digestive stomachic which improves appetite and anti scorbutic, *i.e.* which helps to fight scurvy caused due to vitamin C deficiency.

Constipation

Ripe bael fruit is regarded as best of all laxatives. It cleans and tones up the intestines. Its regular use for two or three months helps evacuate even the old accumulated faecal matter from the bowels.

Diarrhea and Dysentery

The unripe or half ripe fruit is perhaps, the most effective food remedy for chronic diarrhea and dysentery where there is no fever. Best results are obtained by the use of dried bael or its powder. It is also a valuable remedy for chronic dysenteric conditions characterized by alternate diarrhea and constipation.

Peptic Ulcer

An infusion of bael leaves is regarded as an effective food remedy for peptic ulcer. The leaves are soaked overnight in water. This water is strained and taken as a drink in the morning. The pain and discomfort are relieved when this treatment is continued for a few weeks. Bael leaves are rich in tannins which reduce inflammation and help healing of ulcers. The bael fruit taken in the form of beverage has also great eating properties on account of its mucilage, *i.e.* sticky or viscous content. This substance forms a coating on the stomach mucous and thus helps in the healing of ulcers.

Respiratory Affections

Medicated oil prepared from bael leaves gives relief from recurrent colds and respiratory affections. The juice extracted from bael leaves is mixed with equal quantity of sesame oil and heated thoroughly. A few seeds of black pepper and half a teaspoonful of black cumin are added to the hot oil. It is then removed from the fire and stored for use when necessary. A teaspoonful of this oil should be massaged into the scalp before a head bath. Its regular use builds up resistance against colds and coughs.

6. Jambul

Food Values	*Minerals and Vitamins*
Moisture-83.7 per cent	Calcium - 15 mg
Protein - 0.7 per cent	Phosphorus - 15 mg
Fat - 0.3 per cent	Iron - 1.2 mg
Minerals - 0.4 per cent	Vitamin C - 18mg
Carbohydrates - 14.0 per cent	Small amount of Vitamin B Complex
Fibre - 0.9 per cent	
Value per 100 g edible portion	Calorific Value - 62

The jambul fruit is regarded in traditional medicine as a specific against diabetes because of its effect on the pancreas. The fruit as such, the seeds and fruit juice are all useful in the treatment of this disease. Polyuria: The powder of the seeds is valuable in polyuria or production of excess urine. Diarrhea and Dysentery: Powder of the seed is an effective remedy for diarrhea and dysentery. An infusion of the tender leaves, which contain a high concentration of gallic and tannic acid is also given as a medicine in diarrhea and dysentery. A decoction of the bark taken with honey is also useful medicine for chronic diarrhea and dysentery. Piles: The jambul fruit is an effective food remedy for bleeding piles. Fresh jambul fruit taken with honey is also an effective medicine for bleeding piles. Liver Disorders: Natural acids in the jambul fruit play an important role in the secretion of digestive enzymes and stimulate the liver functions. Female sterility: An infusion of the fresh tender leaves of jambul fruit, taken with honey or butter-milk, is an effective remedy for sterility and miscarriage due to ovarian or endometrium functional disorder.

Chapter 7

Postharvest Physiology and Biochemical Changes in Dryland Fruit Crops

U.D. Chavan

Head, Department of Food Science and Technology
Mahatma Phule Krishi Vidyapeeth, Rahuri – 413 722, M.S.

The number of plant species which contribute to man's diet is probably between 1,000 and 2,000. In addition, there are often many sub-species and cultivars for a given species. Of this large number of crops, about 10 per cent are of major importance in world trade and about 1 per cent provides the bulk of crops important in the world. Edible plants may be classified or categorized in a variety of ways *e.g.* according to botanical systematics (leguminosae, gramineae, *etc.*), by plant organ or part (root, leaf, *etc.*), or on the basis of usage and economic considerations (vegetables, spices, extractives, *etc.*). Most aspects of chemical change as they relate to secondary metabolism in postharvest crops are at best incompletely understood. Even events of fairly general importance and which represent major changes, such as starch-sugar transformations, are not f d y understood at the biochemical level.

The quality components of edible plants are summarized in Table 7.1. The relative importance of each of these elements varies with the commodity with the intended use, and with the intended consumer. Hence a green peel color in orange fruit may be unacceptable in one market and yet be highly valued in another market place; or potato tubers which contain a relatively high content of reducing sugars may be considered desirable for consumption after boiling but unacceptable for consumption after frying due to the occurrence of excessive Maillard browning during the latter method of preparation. Each chemical component is influenced by both genetic and environmental factors before as well as after harvest. For example, most cultivars of Irish potato accumulate reducing sugars in response to exposure to temperatures below 5°C; however, the degree to which "cold sweetening" occurs is quite variable among cultivars of potato.

Table 7.1: Quality Components of Fresh Fruits and Vegetables

Factor	*Components*
Appearance	Size: dimensions. weight, volume Shape and form: diameter/depth ratio, smoothness Color: uniformity, intensity Gloss: wax Defects: external, internal
Texture	Firmness; hardness, softness Crispness; Succulence, juiciness; Mealiness, grittiness Toughness, fibrousness
Flavour	Sweetness; Sourness; Astringency; Bitterness Aroma; Off flavors
Nutrition	Energy; Protein; Vitamins; Minerals; Fiber
Safety	Naturally occurring toxicants Environmental contaminants Mycotoxins Microbial commendation

In addition to quality definition and quality change, postharvest physiology also affects the storability of crops. Estimation of postharvest losses is difficult to document and for given commodity losses may range from 0 to 100 per cent. Losses of crops may occur at various levels of postharvest distribution and can de substantial even in locations which have the benefit of our best technology and

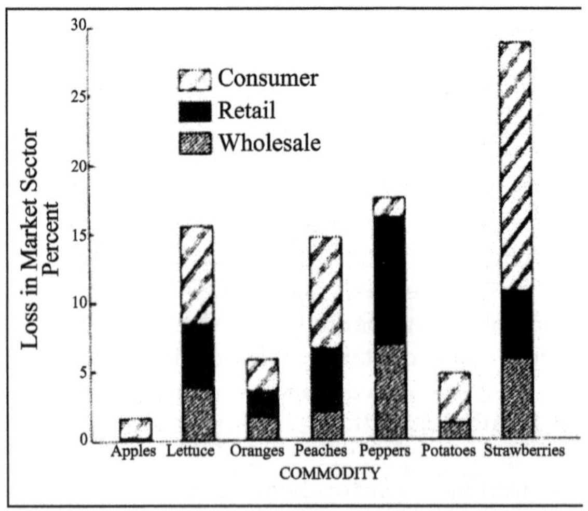

Figure 7.1: Postharvest Losses at Wholesale, Retail and Consumer Levels for Selected Commodities.

know-how for extending storage life (Figure 7.1). Estimates of postharvest losses in less developed countries are generally in the range of 25 to 50 per cent. Such losses represent a weak link in the world's food provision system.

Factors responsible for postharvest losses are varied and a systems approach is necessary to deal adequately with the problem (Table 7.2). In certain situations the weak link in the system may be the choice of genetic stock at planting time or the inability to control environmental conditions during cultivation of the crop. In other instances, the weak link may he non-technical in nature and be the result of socio-economic factors.

Table 7.2: Factors Determining Postharvest Losses

Stage/Phase	*Factors*
Pre-harvest	Genetic Agronomic practice Environmental conditions
Harvest	Maturity Physical damage
Immediate postharvest	Pre-cooling Sorting Curing Washing Chemical treatments
Postharvest storage	Temperature Modified atmosphere Ventilation Physical protection
Transportation and handling	Packaging Temperature Atmosphere
Non-Technical	Managerial skills Administrative skills Inadequate extension service Shortage of capital Shortage of foreign exchange

The storability of postharvest crops varies considerably, and some of the salient features of durable and perishable crops are outlined in Table 7.3. Here we will focus on two key elements which are central to chemical transformations in the postharvest cell. These two factors respiration and genetic control appear to play a pivotal role in the myriad of events which lead to desirable and undesirable changes in the quality of edible plants after harvest. First it will be instructive to consider some examples of the diversity and dynamic nature of the chemical components of postharvest crops.

Table 7.3: Some Characteristics of Durable and Perishable Crops

Durables	*Perishables*
Low moisture (10-15 per cent)	High moisture (50-95 per cent)
Low respiration rate	High respiration rate
Low heat production	High heat production
Hard texture	Prone to physical damage
Dormant phase of ontogeny	Senescent phase of ontogeny

Chemical Composition

Major Components

While it is more or less apparent that minor constituents such as pigments and compounds responsible for odor and taste vary considerably, it is perhaps not as recognizable that the major constituents (water, carbohydrate, protein, and lipid) also occur at considerably different levels in fruits and vegetables. However, it is possible to make some generalization with mention of exceptions. Water, the chemical substance in "wet crops," generally represents 70-90 per cent of their fresh weight (Figure 7.2) and represents about 10-20 per cent of the weight of "dry cops" (Figure 7.3). Certain crops, for example tomato or watermelon may contain in excess of 95 per cent water. Carbohydrates, consisting of sugars, starches, and cell wall polysaccharides, account for most of the solid matter of edible plants (about 75 per cent dry weight). Exceptions to this generalization include avocado (Figure 7.2) and walnut (Figure 7.3). For the most part, lipids are confined to cytoplasmic membranes and range from 0.1 to 1 per cent of the fresh weight of fruits and vegetables. However, certain crops accumulate storage lipids (*e.g.* avocado, Figure 7.2; nuts and oil seeds

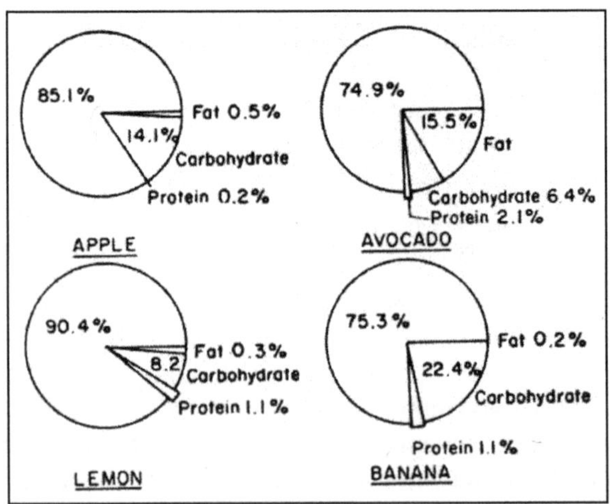

Figure 7.2: Major Chemical Components in Selected Fruits.

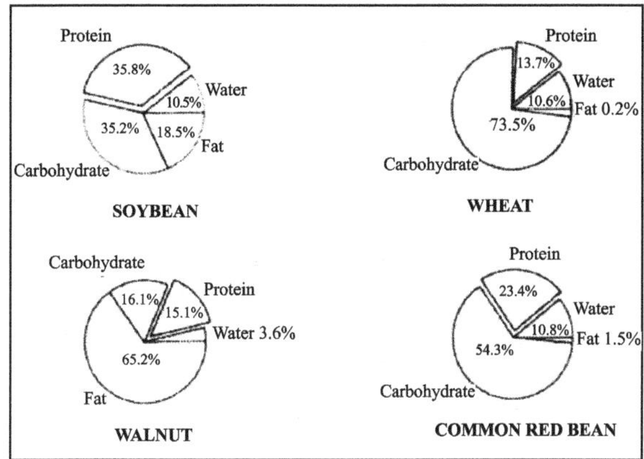

Figure 7.3: Major Chemical Components in Selected Dry Crops.

Figure 7.3). Proteins commonly represent less than 2 per cent of the fresh weight of fruits; while some vegetables in the family Lemmiuosae actually accumulate storage proteins (*e.g.*, lima beans). "Dry crops; by virtue of their low moisture content and in some cases their ability to accumulate storage proteins, may contain as much as 40 per cent protein (Figure 7.3).

Minor Components

Time and lack of complete information do not permit comprehensive discussion of the sundry minor components which add to the special character of different fruits, vegetables, and cereals. A few examples are provided here to illustrate the diversity within and between categories of crops. Edible plants are important sources of vitamins in our diet (Figure 7.4): Fruits, vegetables, and cereals provide an estimated 50 per cent of vitamin A, 60 per cent of thiamine, 30 per cent of riboflavin, 50 per cent of niacin, and almost the entire vitamin C intake in the U.S. diet. The

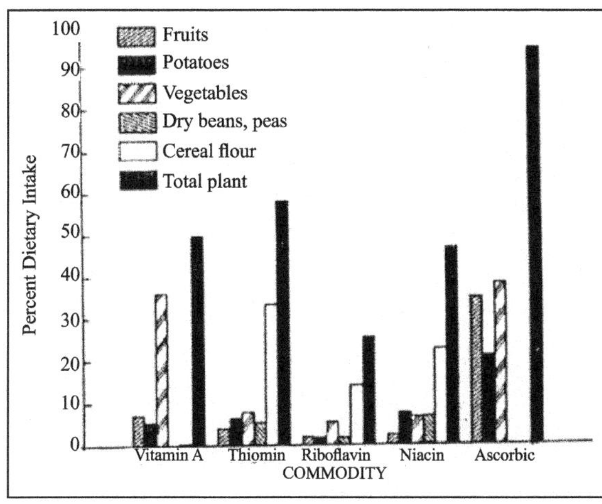

Figure 7.4: Contribution of Plants to Vitamin Intake in the USA.

contribution of plant crops to the intake of vitamins and other nutrients such as high-caloric substances and protein in less developed countries is undoubtedly much greater than it is in developed countries where animal source; of food are more available.

Another category of minor constituents which vary from both a qualitative and a quantitative perspective are the organic acids. These substances range in concentration from less than 1 to more than 50 milli-equivalents per 100 gin fruits and vegetables. The organic acids often are metabolically active in the postharvest tissue and can influence quality attributes in a variety of ways. Examples of carbocyclic organic acids which accumulate in certain crops are illustrated in Figure 7.5. Dihydroxyphenolic substances such as chlorogenic acid can serve as substrate for enzymatic browning and differences in phenol content lead to varietal differences in this browning reaction. Moreover, the fact that the metabolism of phenylpropanoid compounds is involved with lignification and anomalous metabolism resulting in low phenolic content (and hence low tendency brown) may be associated with excessive lignification (*e.g.* "Yuzuhada" disorder in pear fruit).

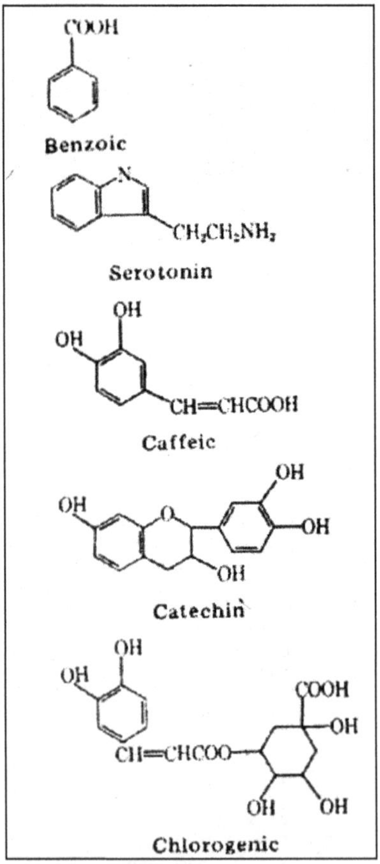

Figure 7.5: Some Examples of Carbocylic Compounds which may Accumulate in Certain Fruits and Vegetables during Development.

Recently it has become apparent that fruits and vegetables may synthesize a multitude of chemical compounds, not present in healthy tissue, as a result of stress following harvest. Many of these compounds detract from quality by virtue of their bitter taste aid possible toxicity. There appears to be a taxonomic basis for the distribution of "stress metabolites" in the plant kingdom. For example, members of the

Leguminosae tend to accumulate isoflavonoids, Solanaceae accumulate diterpenes and steroidal alkaloids, and Umbelliferae accumulate coumarins and furanoteruenoids. The chemical structures of some stress metabolites are shown in Figure 7.6. These substances usually arise as families of compounds having homologous chemical structures. In the white potato more than 24 diterpenes have been identified as stress metabolites. Indeed, the tendency of stress metabolites to accumulate as a result of trauma can depend on the storage history of the commodity. For example, the storage history influences the formation of diterpenes in white potato (Figure 7.7).

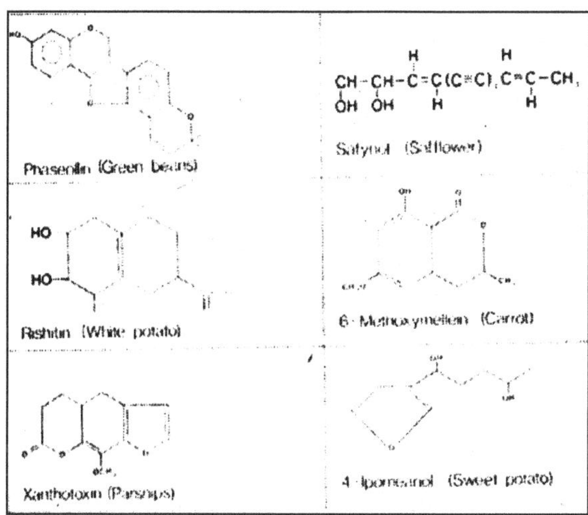

Figure 7.6: Examples of Stress Metabolites which can Accumulate in certain Vegetables after Harvest.

These few examples illustrate the diversity and dynamic nature of minor chemical substances in harvested crops. Because of the enormous numbers of such chemicals, and our incomplete understanding of their relationship to quality, it is perhaps more instructive to dwell on primary metabolic events which appear to relate to coordination and control of secondary metabolism.

Primary Events in Postharvest Fruits and Vegetables

At one time it was believed that the chemical transformations which occur in postharvest fruits and vegetables were the consequence of disorganization and de-compartmentation of the cellular milieu. Further it was assumed by some that chemical transformations were strictly catabolic-a falling apart of the tissue

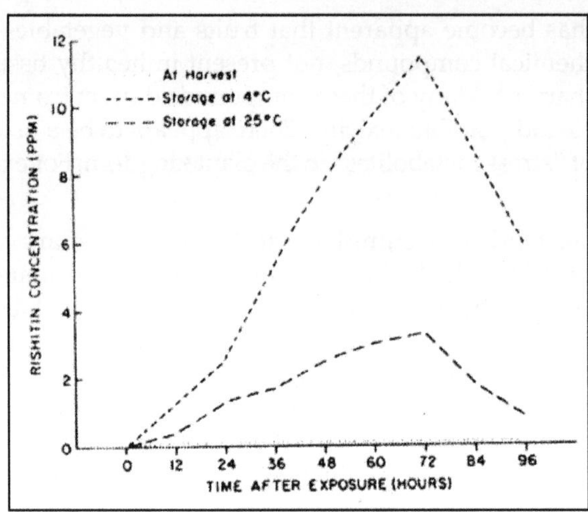

Figure 7.7: Accumulation of a Stress Metabolite Rishitin in Potato Tuber as a Result of Exposure to Ultraviolet Light.

constituents. We now recognize that both anabolic and catabolic reactions play a role in postharvest biochemical transformations (Table 7.4) and that these changes are not the consequence of disorganization but are the consequence of a "biological clock," *i.e.*, are under some form of genetic control. The links between cellular respiration and gene expression and between these primary events and secondary metabolism is by no means clear. However, the above discussion illustrates the importance of these events to chemical transformations and, hence, to quality and postharvest losses.

Table 7.4: Metabolic Changes in Postharvest Fruits and Vegetables

Degradative	*Synthetic*
Destruction of chloroplast	Synthesis of carotenoids and anthocyanins
Breakdown of chlorophyll	Synthesis of flavor volatiles
Starch hydrolysis is	Synthesis at starch
0rgan.c acid catabolism	Synthesis of lignin
Oxidation of substrate	Preservation 01 selective membranes
Inactivation *01* phenol c compounds	Inter conversion of sugars
Hydrolysis of pectin	Protein synthesis
Breakdown of biological membranes	Gene transcription
Cell wall softening	Formation of ethylene biosynthesis pathway

Respiration

It has been recognized for many years that there is an inverse relationship between respiratory rate and storability of postharvest crops (Figure 7.8). Hence, crops which exhibit a relatively high respiration rate tend to deteriorate rapidly;

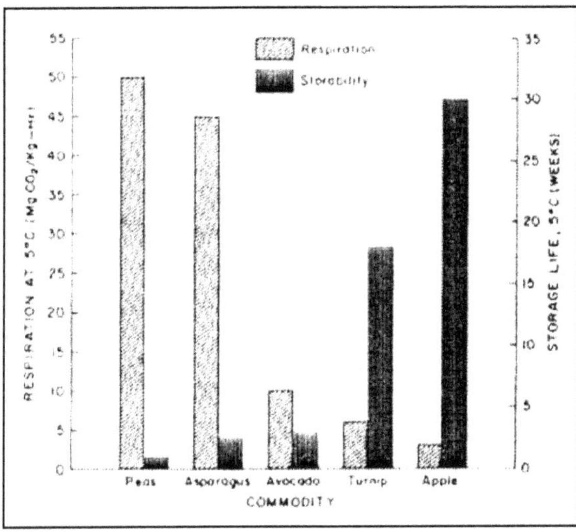

Figure 7.8: Relationship between Respiratory Rate and Storability for Selected Fruits and Vegetables.

whereas, crops which respire slowly can be stored for extended periods of time. In addition, reduction in the respiration rate by physical or chemical techniques is usually accompanied by a corresponding extension in storage life.

One means of reducing respiration is lowering of the environmental temperature. The benefits of lowering storage temperature vary from commodity to commodity for at least two reasons. First, the temperature coefficient of respiration normally about 2.0 can vary considerably and may range from 1 to 7 for different crops; hence, reducing the temperature in a storage facility from 30°C to 20°C may have a more profound influence on both respiration and storability for certain crops than for others. Second, some crops are sensitive to chilling temperatures and the resulting anomalous metabolism can lead to a wide array of physiological disorders. Chilling injury is a result of an imbalance in intermediates arising from respiratory metabolism and other metabolic events.

Similarly, respiration rate and storability respond to modification of the storage atmosphere by reducing oxygen and increasing carbon dioxide partial pressure. Again crops exhibit individuality in their ability to cope with the stress of these changes and for some commodities physiological injury rather than extended storage life results from such treatments (Table 7.5). Although techniques such as temperature and atmospheric control undoubtedly influence processes other than respiration, we will consider here only the effects on respiration.

The respiration rate and storability of crops such as apple fruit is related to the amount of endogenous calcium in the tissue. Studies with various crops have shown that manipulation of tissue calcium by either agronomic practice or past harvest sprays or dips can retard an assortment of physiological injuries and extend storability. The mechanism of calcium action is not clear; however, the concomitant impact on respiration is intriguing.

Table 7.5: Susceptibility of Various Crops to Extremes in CO_2 and O_2 Concentrates

Commodity	*Minimum O_2 Tolerance*	*Maximum CO_2 Tolerance*	*Injury symptoms*
Lettuce	0.5	2 - 4	Brownish-red discoloration of midrib
Broccoli	0.25	20	Tolerant
Cauliflower	-	5	Excessive softening, discolors on cooking
Potatoes	5.0	10	Prevents wound healing

It has long been recognized that ethylene gas can stimulate the respiration of postharvest fruits and vegetables (Figure 7.9). The dependency of respiration on ethylene concentration and the type of effect vary with the commodity. "Climacteric" fruits, which exhibit an autonomous burst in respiration coincident with ripening (Figure 7.10), respond to exogenous ethylene concentrations above certain threshold levels by showing an earlier respiratory climacteric and ripening; that is, ethylene has an all-or-none effect on such crops. Non-climacteric fruits exhibit an ethylene-concentration-dependent respiratory rise, that is, the magnitude of the respiratory burst is greater as ethylene concentration is increased. A recent classification of climacteric and non-climacteric fruits is shown in Table 7.6.

The biochemical basis for the relationship between respiration and storability is not entirely clear. Available evidence indicates that respiration in senescing plant tissues is tightly coupled to the conservation of energy in adenosine triphosphate (ATP). Generally, ATP concentration increases in senescing tissues (Figure 7.11).

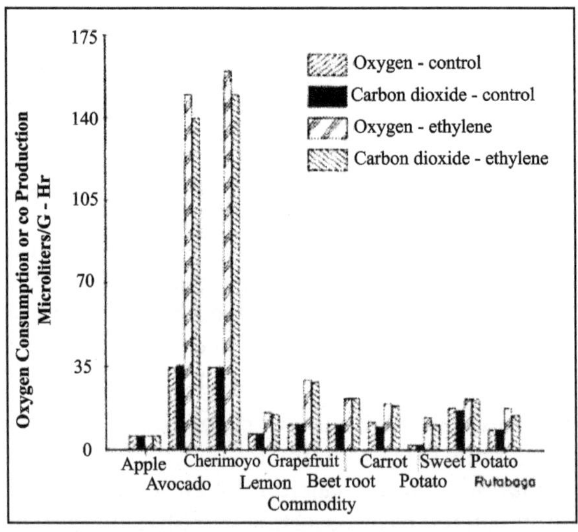

Figure 7.9: Influence of Exogenous Ethylene on the Respiratory Rates of Selected Fruits and Vegetables.

Table 7.6: Classification of Fruit as Climacteric or Non-climacteric

Climacteric	*Non-Climacteric*
Apple	Blueberry
Apricot	Cacao
Avocado	Cashew, apple
Banana	Cherry, sweet
Biriba	Cherry, sour
Breadfruit	Cucumber
Cherimoya	Grape
Chinese gooseberry	Grapefruit
Feijoa	Java plum
Fig	Lemon
Guava	Lychee
Mammee apple	Mountain apple
Muskmelon	Olive
Pawpaw	Orange
Papaya	Pineapple
Passion Fruit	Rose apple
Peach	Strawberry
Pear	Surinam Cherry
Persimmon	Tamarillo
Plum	nor-Tomato
Sapota	rin-Tomato
Soursop	
Tomato	
Watermelon	

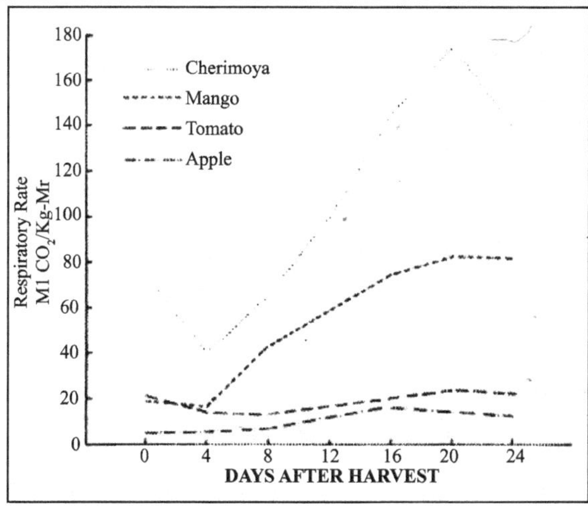

Figure 7.10: Respiratory Patterns of Selected Climacteric Fruits.

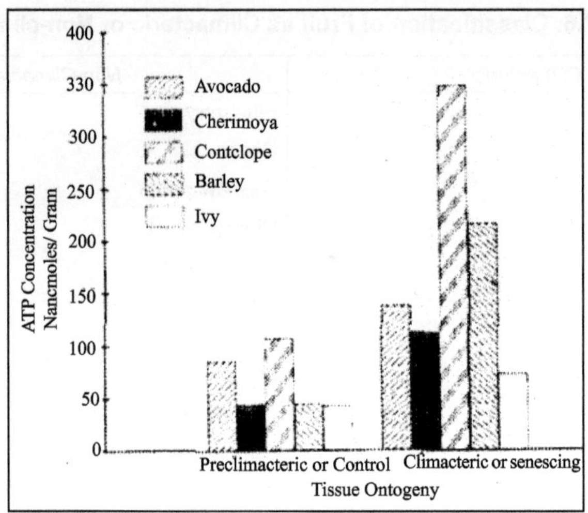

Figure 7.11: Adenosine Triphosphate (ATP) Concentration of Tissue in Selected Fruits and Ivy in the Mature and Senescent Stages of Development.

Although it is known that anabolic reactions which require ATP, such as protein synthesis, occur in senescing plant tissue, it is not easy to account theoretically, using known reactions, for the ATP which is formed during this period. Solomos, assuming that protein turnover in ripening banana is 50 per cent/24 hr (a rather high estimate) and that sucrose formed during this time involves starch hydrolysis via phosphorylase (an ATP-dependent reaction), was unable to account for a large percentage of the A'I'P which presumably is formed by respiration (Figure 7.12). Furthermore, the relationship of respiratory rate and metabolic change in other plant food commodities is very difficult to explain in terms of ATI formation and

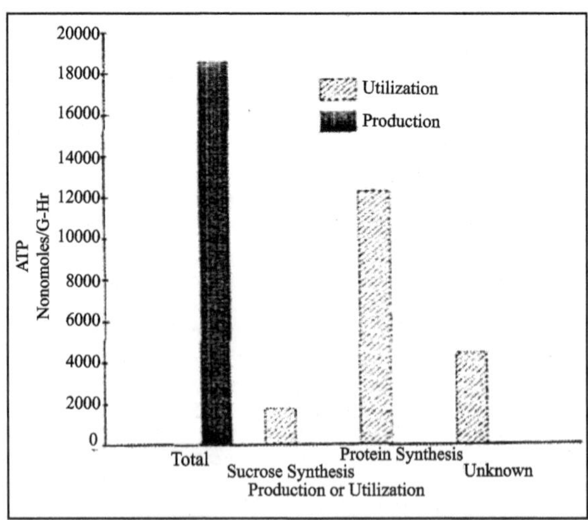

Figure 7.12: Accountability of ATP Formed in Ripening Banana on the Basis of Estimated ATP Utilization.

expenditure. This indicates that we are not yet fully aware of all ATP-dependent processes in postharvest systems. A recently suggested scheme considers protein degradation, in *situ*, to the dependent on activation of proteins by a kinase enzyme. Such energy-dependent activation of protein, preparatory to degradation by proteolytic enzymes, has been demonstrated in microorganisms and animal tissues and has been suggested as a possible mechanism of regulation in senescing plants.

A different approach to rationalizing the accountability of respiration in the postharvest system is to hypothesize that, in *situ*, respiration is partly "uncoupled" from ATP formation. Although the biochemical evidence is not particularly supportive of this thesis, one suggestion relating ethylene to the evocation of alternate chain respiration is especially interesting. Evocation of uncoupled, thermogenic respiration by ethylene has many interesting implications with regard to postharvest metabolism and is worthy of further study.

Genetic Expression

Various lines of evidence now indicate that the metabolic transformations associated with detached plant parts are not simply the result of cellular disarray but are an organized, directed set of events under genetic control. Some of the lines of evidence which have confirmed this view are summarized in Table 7.7. Mutant selection studies have demonstrated that fruit ripening is under the control of nuclear genes. Inhibitors of mRNA synthesis and protein synthesis can block chemical transformations which are associated with characteristic quality changes in ripening and senescing plant tissues. New enzymes, such as the synthase involved with ethylene biogenesis and polygalacturonase in tomato are formed during the period of physiological deterioration. Protein turnover remains high during plant senescence, at least in part due to the formation of new mRNA. In developing tomato fruit, there is a burst in mRNA synthesis just prior to the onset of ripening (Figure 7.13). Further research in this area may lead to a better understanding of the use of genetics and genetic engineering to regulate the storability and quality of fruits and vegetables.

Table 7.7: Possible Control Sites of Gene Expression in Postharvest Commodities

Event	Control	**Evidence**
DNA ↓	Transcription	Mutant studies indicate nuclear genes control ripening. Incorporation of uridine into RNA increases. Ethylene stimulates RNA synthesis.
mRNA protein ↓	Post-transcription Translation	Incorporation of amino acids into protein increases. New enzymes (e.g., polygalacturonase and ACC synthase) appear. Cycloheximide retards or prevents ripening
Active protein amino acid	Post-translation Degradation	Certain proteins are "turned over" rapidly in ripening stags.

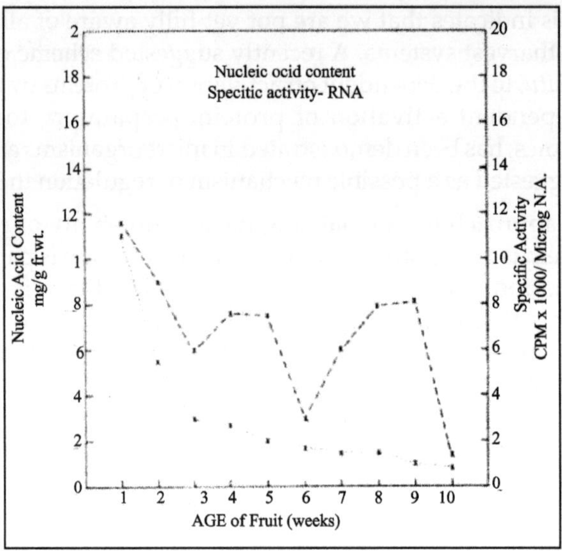

Figure 7.13: Nucleic Acid Content and Synthesis of mRNA in Developing Tomato Fruit.

Control and Co-ordination Hormones

Plant tissues may contain up to five classes of plant hormones (Figure 7.14). The plant hormones appear to play a role in the orchestration of the myriad of chemical transformations in the postharvest cell. The presence of certain hormones (auxins, cytokinins, and gibberellins) appears to maintain the tissue in a "juvenile" state; that is, these hormones appear to prevent ontogenic changes associated with senescence. Two other plant hormones, abscisins and ethylene, appear to promote generally senescence in postharvest crops.

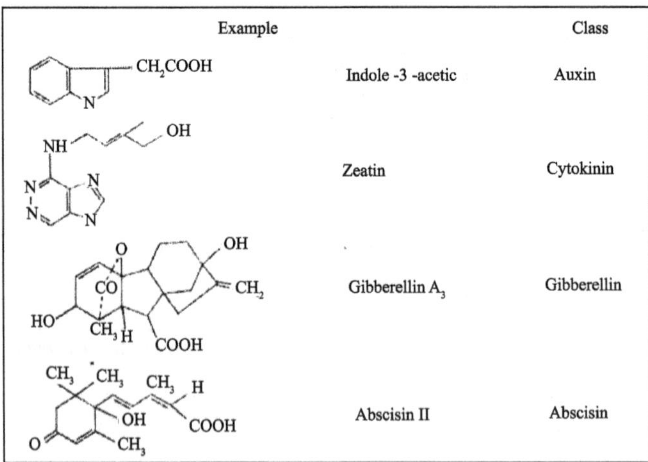

Figure 7.14: Examples of Plant Hormones (Ethylene not shown).

The role of ethylene in postharvest metabolism has been the subject of extensive and varied research. Recently many advances have been made in this Field, although the mechanism and the site ad ethylene action remain obscure. Despite this lack of information, we do know of various ways in which ethylene can be manipulated to the benefit of the postharvest system. For example, knowledge of the internal ethylene concentration of apples can be used to determine the desired market or the storage channel after harvest (Figure 7.15). Apples having an internal ethylene concentration near 1 ppm should be shipped immediately to market by the grower; whereas fruit harvested at a time when internal ethylene is about 0.1 ppm are suitable for long-term storage in refrigerated, controlled atmosphere with little incidence of storage disorders and physiological deterioration.

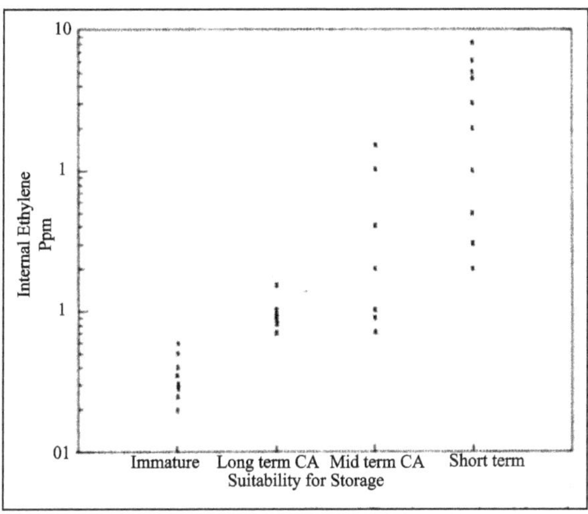

Figure 7.15: Internal Ethylene and Storability of Empire Apples.

The recent discovery of ACC synthetase as the enzyme responsible for the biogenesis of ethylene and its inhibition by compounds such as AVG (Figure 7.16) has broadened the ability of the postharvest physiologist to manage ethylene formation and hence ethylene-mediated changes in the crop. Thus, ethylene-generating compounds, such as ethephon, ethrel, and ACC (1-amino-cyclopropane-l-carhoxylic acid) can be used to trigger certain physiological processes. Inhibitors of ethylene biosynthesis such as AVG (aminoethoxyvinylglycine) can be used to retard other physiological events which are mediated by ethylene (Table 7.8). Blueberries, in which postharvest losses are minimized and quality attributes are enhanced or maintained, provide an example of a benefit arising from the prevention of endogenous ethylene formation.

Exactly how and where ethylene and other plant hormones influence-postharvest metabolism remains somewhat cloudy. Suggested loci of plant hormone action have been outlined. Hormones such as ethylene may act at the level of direct enzyme control, influence energy metabolism and the redox potential of the cell, alter the selective permeability of cellular membranes, or act by controlling gene expression by some direct or indirect mechanism. Hormonal control of gene

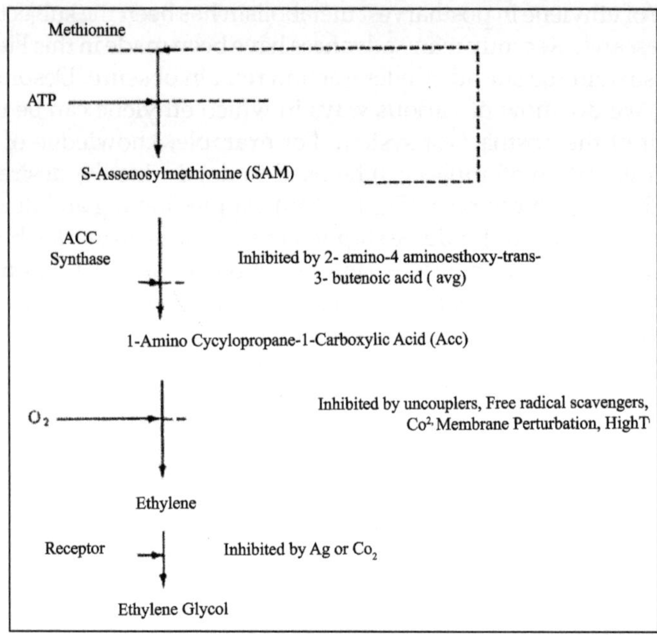

Figure 7.16: Scheme Showing Biosynthesis and Metabolism of Ethylene in Plant Tissue.

expression may be one of the most important mechanisms in senescing plants. We should emphasize that hormones frequently, if not always, act in a concerted way. That is, we must recognize that there is strong interplay of hormones in the evocation of physiological events in plants.

Table 7.8: Scheme Showing Biosynthesis and Metabolism of Ethylene in Plant Tissue

Application	*Effects*
Ethylene-generating compounds:	Desirable
Ethephon (2-chiaoethyl phosphonic acid)	Accelerate ripening
ACC (1-amino-cyclopropane-1-carboxylic)	Uniform ripening for mechanical harvest
	De-green certain fruits or vegetables
	Enhance flavor development
	Undesirable Die-back of parent plant Excessive softening
	Accelerate lignification (e.g. asparagus)
Inhibition of ethylene formation:	Desirable
AVG (aminoethoxyvinylglycine)	Delay senescence
Inhibition of ethylene action:	Extend storage life
$Ag+$	Effect quality indices in desirable way
$C0_2$	Undesirable
	Reduces aroma development in apples

Conclusions

Edible plants exhibit diversity in chemical composition both between and within species. Likewise, chemical transformations in postharvest crops exhibit wide qualitative and quantitative differences which are ultimately reflected in quality change and losses after harvest. Energy metabolism and the ability of the living cells to store and express information via genes are central events which, together with a myriad of secondary catabolic and anabolic events, can be manipulated by hormonal and environmental factors. Our ability to understand these chemical transformations and their control has a considerable hearing on our ability to maximize quality and minimize losses.

References

Beyer, E. M. "Ethylene Action and Metabolism: in "Recent Advances in the Biochemistry of Fruits and Vegetables, (Editors, Friend, J. and Rhodes, M. J. C.I. Academic Press, New York, 1981.p. 107.

Bide. J. B., and Young, R. E. Respiration and Ripening in Fruits-Retrospect and Prospect. *In* "Recent Advance in the Biochemistry of Fruits and Vegetables." Editors: Friend, Lend Rhodes, M. J. C.), Academic Press, New York, 1981.

Bmmlage, W. J., Drake. M. and Baker, J. H., *J Amrr Soe Hort.* Sci. 99, 376 (1974).

Cheema. A. S. and Haard, N. F. Plant Physiol., 13.233 11978).

Composition of Foods: Raw, Processed. Prepared." U.S.D.A. Hand book No. 8, Washington, DC. 1963.

Dekazus, E. D., "Effect of Postharvest Treatments of Growth Regulators on Quality and Longevity of hits and Vegetables. In Postharvest Physiology and Crop Protection. (Editor: Lieberman. M.), Plenum Press, New York, 1983, p. 355.

Dilley, D. Manipulation of postharvest Atmosphere, In Postharvest Physiology and Crop Protection. Editor, Lieberman, M. J. Plenum Press, New York, 1983. P. 391.

Duckworth, R. B., Fruits and Vegetables, Pergamon Pren, New York, 1966, p. 28.

Grierson, D., Control of Ribonucleic Acid and Enzyme Synthesis During Fruit Ria-391.

Haard, N. F., "Edible Plant Tissues: in "Principles of Food Science. Part l-Food Chemistry." (Editor Fennema, 0.). 2nd ed., Marcel Dekker. New York, 1983, chap. 16.

Haard, N. F, Stress Metabolism in Postharvest Physiology and Crop Protection: (Editor: Lieberman, M.), Plenum Press, New York, 1983, p. 299.

Harvey, J. M., Ann. Rev. Phytopathol. l6, 321 (1978).

Hershko, A, Cuechanover, H., Holler, H., Harrd, A. L., and Rose, I. A., Proc. No11 Acod. Sri. USA.73, 1788 (1980).

Jsnich. J., Schm R W., Wds, R. W. and Ruttan, V. M. "Plant Science: W. A. Freeman, San Francisco, 1969, chap. 3.

Kader, A. A, "Postharvest quality maintenance of fruits and vegetables in developing countries in Postharvest Physiology and Crop Protection," (Editor: Lieberman, M.). Plenum Press. NY. 1983. p. 455.

Postharvest Food Losses in Developing Countries." National Academy of sciences. Washington, DC, 1978.

Ransdive, A. S. and Haard. N. F. J. Sri. Fwd Agri, 22, 86 (1911).

So1omm.T. and Latiel, G., Biorhrm. Biophys. Res Comm., 70, 663 (1976).

Solomos, T., Respiration and Energy Metabolism in Senescing Plant Tissues. In "Postharvest Physiology and Crop Protection." (Editor Lieborman, M. L. Plenum Press, New York, 1983, p. 61.

Spurgeon, D., "Hidden Harvest-A Systems Approach to Postharvest Technology. International Development Research Centre, Ottawa. 1976, p. 7.

Woolhouse, H. W., "The General Biology of Plant Senescence: in "Postharvest Physiology and Crop Protection," (Editor: Lieberman, M.). Plenum Press. New York, 1983, p. 1.

Yang, S. F., Biosynthesia of Ethylene and its Regulation: in Recent Advances in the Biochemistry of Fruits and Vegetables, (Editors: Friend. J. and Rhodes. M. J.C.), Academic Press, New Yolk. 1981. p. 89.

Chapter 8

Recent Advanced Techniques in Macro and Micronutrients Use Efficiency for Quality of Dryland Fruit Crops

A.G. Durgude and I.R. Bagwan

*Department of Soil Science and Agril. Chemistry,
Mahatma Phule Krishi Vidyapeeth, Rahuri – 413 722, Dist. Ahmednagar, M.S.*

Soil is important for establishing an orchard. In case of fruit cultivation careful examination of sub soil rather than surface soil is more important. For this, profile observations and layer wise soil sampling for morphological properties and chemical, physical properties are very important in respect of productive span period of fruit crops. For any plantation of dryland fruit crops following physical and chemical properties of soils are very essential before planning of dryland horticultural fruit crops which increased the life of orchards and quality of fruits.

A) Good Physical Condition of Soil for Growing Fruit Crops

1. Porous and well aerated
2. Well soil depth upto 2 m with well drained
3. Fine to coarse loamy textural class
4. Crumb, subangular blocky structure
5. Moderate infiltration rate
6. Medium to high waterholding and nutrient holding capacity
7. No hard pan and kankar pan at depth (within 30 cm) in pedon

B) Good Chemical Properties of Soil for Growing Fruit Crops

1. pH: 6.0 to 8.0 (for better nutrient availability and microbial activity in soil)

2. EC: normal range should be 0.25 to 0.50 dSm^{-1}
3. Organic carbon: > 0.60 per cent
4. $CaCO_3$: 1 to 8 per cent
5. Available N, P, K: Medium to high status
6. Available micronutrients: Sufficient (above the critical limit) *viz.*, Fe (> 4.5 ppm), Zn (> 0.6 ppm), Mn (> 2.0 ppm), Cu (> 0.2 ppm) and B (> 0.5 ppm).

In general, it may be stated that the suitability of soils for fruit growing orchard should be porous, deep and well aerated and should not be waterlogged, marshy, saline, acidic and there should not be hard pan within 30 cm depth.

Table 8.1: Soil Resources of Arid and Rainfed Agro-ecosystems

Soil Orders	*Constraints for the Dominant Soil Order*
Entisols – Aridisols	Very low temperature, shorter growing period, shallow and gravelly soils, and high calcareousness
Entisols – Aridisols	Wind erosion, drought, moisture stress, crusting and secondary salinization
	Rainfed Agro-ecosystem
Aridisols– Inceptisols	Drought, moisture stress, poor drainage, high runoff and secondary salinization
Inceptisols- Vertisols	Erosion, nutrient imbalance, secondary salinization and alkaline ground water
Inceptisols- Entisols	Nutrient depletion
Vertisols– Inceptisols	Massive structure, poor tilth, drought stress and water erosion
Inceptisols– Vertisols	Water erosion, low organic matter and nutrient imbalance
Alfisols – Vertisols	Weak soil structure, crusting, compaction, low AWC and nutrient depletion and water erosion.
Vertisols – Alfisols	Massive structure, poor tilth, drought stress, water logging during early stage and doughtiness at latter, nutrient depletion and water erosion
Inceptisols – Alfisols	Erosion, severe gully erosion in laterites, subsoil gravelliness, P fixation low organic matter content, nutrient imbalance, low PAWC and acidity at places.

Soil Resources	*Major Constraint*
Alfisol	Low O.M. and fertility P deficiency
Aridisols	Physical degradation Accelerated erosion
Inceptisol	Human induced soil degradation
Entisol	Deforestation Over grazing Loss of top soil
Vertisol	Salinity and sodicity Water logging Hard setting soils

Soil Constraints of Nutrient Management for Fruit Crops Production

Physical Constraints

- ☆ Shallow soils
- ☆ Subsurface hardpans (Kankar pans)
- ☆ High water table, poor drainage
- ☆ Excess erosion

Chemical Constraints

- ☆ Inherent nutrient deficiency (Soils in general low in N, P, Zn)
- ☆ Calcareousness
- ☆ Salinity/Alkalinity and acidity

Biological Constraints

- ☆ Soil sickness

Unique Nature of Fruit Crops

- ☆ Need of fruit crops for nutrients differs from annual crops
- ☆ Large root systems exploiting larger soil volume
- ☆ Simultaneous need of nutrients to meet the requirements of both fruit production and vegetative growth which persist for year to year
- ☆ Nutrient supply and uptake has considerable influence on growth and production in subsequent years

Role of Potassium in Development of Fruit Quality

Returns from fruits is largely determined by their quality. Many fruit marketing associations have strict quality control largely determined by size, shape and TSS content. In grape, price to the growers is paid based on the quality of fruit. The quality of fruit crops depend upon the role of various nutrient especially potassium.

Table 8.2: Integrated Nutrient Management in Pomegranate (cv. *Mridula*) (2006-07)

Treatment	*Total Yield (kg/tree)*	*Juice (per cent)*	*T.S.S (per cent)*	*Acidity (per cent)*
T_1 : RDF (625:250:250 NPK g/tree) + 50 kg FYM/tree	30.00	47.11	14.68	0.36
T_2 : RDF (inorganic farming)	17.00	43.23	12.08	0.43
T_3 : RDF (75 per cent N) + FYM (25 per cent N)	27.33	45.90	13.73	0.32
T_4 : RDF (75 per cent N) + Vermicompost (25 per cent N)	20.00	44.73	13.93	0.39
T_5 : RDF (75 per cent N) + Neem cake (25 per cent N)	24.33	43.68	14.26	0.39
T_6 : RDF (50 per cent N) + FYM (25 per cent N) + Neem cake (25 per cent N)	21.66	45.65	13.75	0.35

Treatment	Total Yield (kg/tree)	Juice (per cent)	T.S.S (per cent)	Acidity (per cent)
T_7 : RDF (50 per cent N) + Vermicompost (25 per cent N) + Neem cake (25 per cent N)	18.33	44.78	14.12	0.40
T_8 : RDF (50 per cent N) + FYM (25 per cent N) + Vermicompost (25 per cent N)	18.66	45.78	14.41	0.41
T_9 : RDF (33 per cent N) + Vermicompost (33 per cent N) + Neem cake (33 per cent N)	16.66	43.65	13.16	0.41
T_{10} : FYM + Vermicompost + Neem cake + Jeevaamrit (Total 100 per cent N)	13.00	47.82	14.73	0.34
SE ±	0.64	**0.43**	**0.41**	**0.007**
CD at 5 per cent	1.90	1.28	1.22	**0.021**

Source: Anonymous (2007).

The Table 8.2 revealed that juice content and TSS per cent was found significantly higher in only organic treatment T_{10} (47.82 and 14.73 per cent), respectively. However, it was at par with RDF treatment with FYM. The significant highest yield (30 kg/tree) of pomegranate was recorded in T_1 treatment (RDF + FYM).

Recent Techniques for Increase in Nutrient Use Efficiency for Quality of Dryland Fruit Crops

At present, the nutrient use efficiency of nitrogen is 40 to 50 per cent, phosphorus is 15 to 20 per cent, potassium is 60 to 70 per cent and micronutrients are only 2 to 5 per cent. It is therefore, necessary to give attention for increase in nutrient use efficiency by adopting following techniques.

1. Planting of fruit crops in soil by taking pits as per duration of fruit crop. *e.g.* For Aonla, ber, fig, pomegranate, custard apple required pit size 60 x 60 x 60 cm length x breadth x depth, respectively. Then fill the pits with tank silt having good quality soil chemical properties as mentioned beginning of this text.
2. Use of primary (N, P, K), secondary (Ca, Mg, S) and micronutrients (Fe, Zn, Mn, Cu and B) as per soil test based for balance use of nutrients. Micronutrients and secondary nutrient fertilizers mixed with organic manure as a basal dose for better nutrient use efficiency.
3. Use of organic manures 10 to 50 kg (as per age of orchard) per tree or 2 to 10 kg per tree vermicompost after stress period require for bahar regulation.
4. After planting of fruit crops, first two to three years in between row, cultivation of pulse crops will increased the carbon content and fertility of soil.
5. Application of biofertilizer *e.g.* phosphate solublizing bacteria with organic manure will increased the phosphorus use efficiency.
6. Application of cow dung slurry incubated with micronutrients for one

week (125 kg fresh cow dung + 5 litre urine + 10 kg micronutrient grade I + 500 litre water per hectare) will increase the nutrient use efficiency and quality of different dryland fruit crops.

7. Provision of protective irrigation through construction of farm pond will help for increase in yield and quality of dryland fruit crop.
8. Use of organic mulch *e.g.* sugarcane trash, paddy straw, wheat straw, dry grass, farm waste at the base of tree upto 15-20 cm away from the trunk will reduce the evaporation losses, weed intensity and maintain the soil temperature.
9. Under dryland condition foliar feeding is often a practical method of supplementing nutrient requirement, particularly of nitrogen and micronutrients which increase the fruit set, fruit retention, yield and improve the fruit quality of ber. (Kadlag *et al.*, 2016)
10. Combined sprays of micronutrients (0.03 per cent boric acid and 0.5 per cent $ZnSO_4$) and growth regulator NAA (50 ppm) have been found to improve the fruit quality in ber by increasing total soluble solids, total sugars and ascorbic acid and decreasing acidity. (Kadlag *et al.*, 2016)
11. Pomegranate respond well to foliar application of micronutrient grade II (0.5 per cent) prior to flowering at full bloom and at fruit set for increase in yield and juice content.
12. Application of boron 0.2 per cent was found useful for reducing cracking of fruits in pomegranate and improvement of fruit colour.
13. Foliar spray of 0.6 per cent boric acid thrice in the months of September and October at about 10-15 days interval has also been recommended to control fruit necrosis in aonla which develops due to deficiency of boron.
14. For custard apple in light and gravelly soils, an annual dose of 50:50:25 g $N:P_2O_5:K_2O$ alongwith 10 g FYM per plant was found sufficient. (Kadlag *et al.*, 2016)

References

Anonymous (2007) Annual report of AICRP on pomegranate Solapur, Maharashtra.

Kadlag A.D., Bhosale A.B., Kale S.D. and Todmal S.M. (2016) Dryland fruit crops Daya Publishing House A Division of Astral Int. Pvt. Ltd. New Delhi 110002.

Chapter 9

Recent Advances in Pre-harvest Factors to Enhance the Postharvest Quality of Aonla, Ber and Jamun

V.R. Joshi

Onion Breeder, Department of Horticulture
Mahatma Phule Krishi Vidyapeeth, Rahuri – 413 722, Dist. Ahmednagar, M.S.

The arid regions are spread over about 38.7 million hectares including both the hot arid and cold arid regions mainly in the state of Rajasthan, Gujarat. Andhra Pradesh, Punjab and Haryana. The hot arid region inhabit on an average 61 persons per square km making up the population of nearly 20 million peoples besides high population density of animals. Although. India is the largest producer of fruits in the world and is only second to China in vegetable production; the availability in fruits and vegetables to the people particularly in the arid region is very low. Considering the current trend in population growth, these requirements are expected to reach 2 and 6 million tons respectively during the next two decades. Although the recent scenario indicates an increasing trend in the production of fruits particularly arid fruits like aonla, ber and jamun but the pace has been rather slow. In arid region, the area under fruits increased from few thousand hectares to above lakh hectares during last 8-10 years. This is mainly due to the thrust given by the state agricultural universities from last 30-40 years. Among the different arid zone fruit crops Aonla, Ber and Jamun are three important arid zone fruit crops in nutritional as well as postharvest management point of view.

Aonla

Aonla or Indian gooseberry (*Emblica officinalis*, Gaertn. Syn. *Phyllanthus embli* L.) is one of the important non-traditional fruits indigenous to India. It is hardy, prolific bearer and highly remunerative even without much care and can be successfully grown variable agro-climatic and soil conditions. The tree is much branched and ranges from 9.12 m in height. It is deciduous under North Indian conditions, however, it is considered evergreen in the tropics. The fruit is highly nutritive and is one of the richest source vitamin C. The fruit is also rich in pectin and minerals such as iron, calcium *M* phosphorus. Aonla is the only fruit to fill the gap of astringent food recommended by Ayurvedic system of medicine for a balanced diet and sound health. It is acidic, diuretic and laxative. Aonla is the main ingredients in Chavanprash and Triphla, used in the treatment of headache, constipation and enlarged liver. Aonla will be an important fruit in the near future due to its suitability even in the wastelands of ai and semi-arid areas.

Varieties

In all, there are only 10-12 varieties in India that have been named and described. These varieties are distinguished as early, mid season, late and errect and spreading. The salient features of these varieties are as under:

Early Maturing Group (mid October to mid November)

Krishna (NA-5): The fruits contains 0.39 per cent fibre with 549.20 mg/100 g vitamin C.

Balwant (NA-10): Fruits are attractive medium to large (41.5 per cent), flattened, round, skin rough, yellowish green with pink tinge and the fruit contains 0.52 per cent fibre with 528.0 mg/100 g pulp of vitamin C.

Mid Season Group (Mid November to mid December)

Francis: Fruits are large (45.8 g), flattened, oblong, greenish white in colour. Segments distinct, solid and thick. Ascorbic acid content is 566.5 mg/100 g pulp. Fruits are highly susceptible to necrosis - a physiological disorder, hence, it is not an ideal variety for commercial exploitation.

Neelam (NA-7): It is a selection from Francis. The tree has semi-erect growth habit. Fruits are medium to large (44.0 g), flattened, oblong with conical apex and greenish white in colour with 0.61 fibre content. It is free from necrosis. Juice is highly astringent. It seems to be an ideal cultivar for commercial growing.

Kanchan (NA-4): It is a selection from Chakaiya cultivar. The tree has spreading growth habit and it bears profusely. Fruits are medium sized (30-32 g), yellowish green in colour with ascorbic aid content of 711 mg/100 g. Fruits of this cultivar are highly susceptible to corking.

Amrit (NA-6): This is also a selection from Chakaiya. Fruits are medium to large (35-37g) with bright appearance. Segment 6, distinct and thin. Fibre content is lowest (0.8 per cent). This is an excellent variety for making preserve, candy and jam.

Late Maturing Group (Mid December to mid January)

Chakaiya: The tree has an upright growth habit and bears profusely. Fruits are medium sized (33-35 g), flattened and whitish green in colour. Segment 6 and not distinct. The fibre content of this variety is highest (2 per cent) with ascorbic acid content 789 mg/100 g. The fruits have strong attachment hence premature dropping is not a problem. Due to the upright growth habit this cultivar could achieve a significant place in high density plantations.

Soil and Climate

Aonla has wide adaptability and can be grown under varied conditions of climate. It can be grown on well drained fertile loamy, light as well as heavy soils except sandy soils. It can tolerate salts to certain extent and can be successfully grown even in sodic and saline soils upto 35 ESP and 10 Ece/Ds M , respectively with maximum 9.5 soil pH with slight reduction in yields. However, in pot culture studies, opined that young seedlings are susceptible to salt injury. Seedlings with woody stem could tolerate salinity upto 7.5 dS M^{-1}.

Aonla is a subtropical fruit but its cultivation in tropical climate is quite successful. It responds well to the conditions of dry weather, moderately cool temperature during winter followed by high temperature of summer. It can tolerate freezing temperature as well as temperature as high as 46 °C.

Propagation

Aonla has long been raised by seeds. Seed propagation though easy and cheap, does not ensure true to type plant because of its cross pollinated nature. Besides, such plant take more time to bear fruits. To overcome demerits of seed propagation, vegetative method is desirable. Among the various methods, patch budding in the month of February and June is the most successful. Soft wood grafting before new growth emerges has also been tried in the month of February with good success.

Under dryland conditions, due to vigorous tap root system, raising of such seedling in its permanent place in the field and budding *in situ* has been found successful. For raising rootstock seedlings, seeds of fully mature fruits should extracted. Small size seeds present in the inner hard coat should be removed by carefully breaking of seed coat or simply by sun drying. Seed germination can be improved by soaking the seeds in 500 ppm GA solution. Seeds should be sown in the pits during on set of monsoon. In a trial on survival and growth of aonla seedlings increasing pit size from 30 cm^3 to 90 cm^3 helped the seedlings not only to survive but also growth was vigorous After one year, when these seedlings attain pencil thickness, should be budded with tl scion stick of improved varieties as desired. Only one shoot per tree should be budded and allowed to grow.

The seed of aonla contains only one embryo. Microscopic examination of this young seed also did not reveal the presence of more than one embryos. Hence it is suggested that there is no polyembryony. In aonla, important factor is the proper selection of mother plant which is highly fruitful and the buds should be taken from such a branch which bears good number of female flowers.

Cultivation

Planting

In order to provide space and better medium for development of vigorous tap root system, planting should be done in pits of 60 x 60 x 60 cm. The pits should be dug during summer and filled in by 1 kg single super phosphate at the base and with a mixture of good top soil well decomposed F.Y.M. 100 g 10 per cent BHC or Carbaryl powder should also be mixed in the soil mixture to avoid the problem of white ants. The planting distance depends upon the soil type. It varies from 7 to 9 m (square system) from-shallow to deep soils.

Training

In the initial years, training is very essential to develop the frame work. Plant should be trained to modified leader system. Suckers appearing from rootstock and weak or diseased branches should be removed.

Fertilizer Requirement

As in other fruit trees, aonla also requires regular manure and fertilizers. The dose of manure and fertilizer, recommended by Department of Horticulture, NDUAT, Faizabad depends upon the soil fertility and age of the plant. Ten kg FYM, 125 g nitrogen, 50 g phosphorus and 50 g of potash per plant should be given to one year old plant. This dose should be increased every year in the same proportion upto the age of 10 years (100 g N, 400 g $P_2 0_5$ and 400 g K_20/tree).

The complete FYM and above fertilizers should be given in two equal split doses to mature bearing tree, once during September- October and again during April-May after setting of fruits. The plants need to be irrigated after fertilizer application. In western Maharashtra systematic work on fertilizer recommendation is being initiated. At present ad-hoc recommendation of 50 kg FYM + 500 g N + 250 g $P_2 0_5$ + 250 g K_20 per tree is made.

Irrigation

Aonla can be grown as a rainfed crop and is able to produce fairly good crop without irrigation. However, it responds favourably to irrigation to secure higher fruit set, reduce fruit drop and better development of fruits.

Flowering and Fruiting

Flowers

Flower buds appear as tiny heads by middle of March on the newly emerging determinate shoots which are produced on all the branches. The buds are slightly pink in the beginning but as they grow old they turn creamy-white. The number of flowers/cluster varies from 8-34 and generally there are 18-27 clusters/branches. In the cluster male flower appears first in the axil of leaf lower parts of determinate shoot followed by female flower on the same shoot. All the determinate shoot do not bear female flower. The average number of female flowers/branches ranges from 2 to 10.8.

Anthesis

The anthesis of male flowers takes place both in morning (8.30 - 11.00 a.m.) and afternoon (3.30- 6.30 p.m.) at maximum temperature between 29.7 and 35.6°C and minimum temperature between 10.3 and 19.9°C (Mohammad Allemullah and Sant Ram, 1990). Opening of female flower is a gradual process and it takes about 74 hours to complete their opening from start of anthesis.

Receptivity of Stigma

Fruit set occurs when pollinated between second to fifth day of anthesis. The highest stigmatic receptivity was recorded on the third day of anthesis.

Dehiscence

The dehiscence of anther coincided with the opening of flowers. The anther continued to dehisce upto 7 p.m.

Pollination and Fruit Set

Pollination in aonla is mostly by wind, honey bees and gravity. Honey bees visit the flowers in the evening and early morning. Although actual fertilization could not be traced, zygote was observed within 24-36 hours of pollination. As regards fruit set, open pollination through hand increased initial fruit set and fruit retention. Hence, incorporation of honey bees (pollinator) in aonla appears to be most important for increasing the fruit yield.

Fruit Development

After fruit set the ovary remains in dormant condition upto 3 months and by the end of July and Middle of August, the ovary begins to swell. The increase in diameter of fruit is quite rapid in the month of August and September. The delayed growth of embryo may be because of lack of available nutritive materials which are needed for growth of structural elements of the fruits till middle of August. Afterwards the rapid growth may be due to the enlargement of the cells of the mesocarp, while endocarp cells form the hard stone walls.

Alternate Bearing

Aonla has a tendency to bear a heavy crop in alternate years.

Harvesting

Stage of maturity at which fruits are harvested has great effect on quality, storage life and marketability. Harvesting depends on number of factors which affect the maturity indices of fruits such as location, variety, climate, season, nutrition, soil type and moisture. In general harvesting of aonla should be done when the colour of fruit becomes dull greenish yellow during the month of November- December. It takes 18-20 weeks from fruit set to maturity.

Yield

A full grown grafted tree with good bearing habit may yield between 200-300 kg. Average fruit yield is 200 kg/tree. Under Rahuri conditions, during 1994-95, the yield/tree ranged from 100 to 312 kg for six years old trees.

Ber

The ber (*Zizyphons mauritiana* L.) is one of the important dryland fruit crop. The fruits are nutritious containing high amount of vitamin C, A and B complex. The nutritive value of fruits of ber could be compared to apple fruit and it can be seen that it surpasses Apple fruit in vitamin content (Table 9.1).

Table 9.1: Comparative Nutritive Value of Ber and Apple

Sl.No.	*Character*	*Ber*	*Apple*
1.	Moisture content (per cent)	85.9	85.9
2.	Protein per cent	0.8	0.3
3.	Fat per cent	0.1	0.1
4.	Mineral matter per cent	0.4	0.3
5.	Carbohydrates per cent	12.8	13.4
6.	Phosphorus per cent	0.03	0.02
7.	Iron per cent	0.8	1.7
8.	Calorific value per 100 g	55	56
9.	Carotene (Vit. A) units 100 g	70	Traces
10.	Vit. C (mg/100 g of edible portion)	50-150	

It's drought hardiness, xerophytic nature, tolerance to salt (30 ESP and 8 mm hos ECe) deep tap root system, spiny nature and ability to shed its leaves during harsh summer have helped the plant to adopt to the arid and semi-arid area.

In Maharashtra ber is not an uncommon fruit. However, regular orchards of ber crop were unheard of, till introduction of large fruited commercial varieties from Punjab and Haryana in 1975-76 like Umran, Gola, Kadaka *etc.* at the Central Campus, MPKV, Rahuri.

Realizing the potential and importance of this crop is otherwise scarcity zone in the district of Ahmednagar, Solapur, Osmanabad district, the Directorate of Agriculture, Maharashtra State, Pune collaborated with the State Agricultural University from 1984-85 to train village level workers to transfer, technology of successful cultivation of ber crop as an alternate crop to bajra, kharif jowar *etc.*

These concerted efforts of scientists on one hand, and transfer of technology through the staff of Directorate of Agriculture, Maharashtra State, Pune and last but not the least efforts of popularizing the ber fruit by enterprising and inquisitive growers on the other hand has resulted in increasing the area to 5826 ha (1991-92). Prior to 1985-86 local ber cultivars were grown either in forest or on bunds of the fields. In Maharashtra at present the area under ber is about 71335 ha (Anonymous, 2002).

Climate and Soil

Wild ber is seen in almost in all villages in India except perhaps in those villages where temperature below zero will remain for a longer time or where frost occurs regularly. The ber crop withstands high temperature and aridity by cessation of growth, leaf fall and dormancy phase. In high humidity conditions disease and pest problems increase especially powdery mildew which is a serious disease.

In Maharashtra it was observed that leaf fall is observed around Jan-Feb. New vegetative growth is observed by May (2nd fortnight) or June (1st fortnight). Increase in atmospheric humidity around this period favours new sprout, though rains are not received.

Ber crop can grow in all types of soils ranging from shallow to deep soils. Even it can be grown in class VI or class VII soils. The tap root system, the tree develops, helps survive harsh conditions. Soils with salinity upto 40 ESP and 14 mm hos/cm ECe. But, however, in problematic soils initial establishment will be difficult and later years yields are reduced.

Propogation

Seed propagation was a rule in earlier days. Seminal roots that* develops from embryo helps to establish the plant through tap root system. But the main disadvantage is of heterozygosity and variability in seedling.

For early germination of ber seeds, breaking of endocarp gives quick germination.For economically viable and commercial orchards it is necessary to use identified improved varieties through vegetative means. Though cutting and layering is possible, percentage success is too low to be adopted on large scale. Converting the seedlings through budding (shield bud ring bud and patch bud) are the commercially exploited method. Of these, in Maharashtra, patch budding is most successfully employed.

In northern India though transplant from nursery stock through polythene tul an accepted commercial practice advantage and development of seminal tap root be exploited. Under scarcity condition plants may show symptoms of wilt and in extreme situation may dry up. *In situ* sowing of seedling and converting them to choice varieties through patch budding offers the best choice in Maharashtra.

In studies on raising of rootstock it was observed that *Z.nummularia* is the dwarfest and *Z. rotundifolia* is the vigorous rootstock of ber. By using dwarfest rootstock more number of trees can be accommodated in hectare (Biannual Report. AICRP on AZF, 1993).

Planting

Monsoon season offers good choice, for raising *in-situ* plants in Maharashtra, at a distance of 6x6 m. Such seedlings be made to survive the dry season of summer through watering. Deheading these seedlings to ground level in the month of may helps give new sprout by July. Patch budding these sprouts in July helps convert to choice variety.

Some commercial plantations raised through transplanting have been planted even upto December-January. But these plantations had to be taken care of, by irrigating frequently. In northern India generally planting is done in either February-March or July August at a spacing of 7 to 8 meters.

Training and Pruning

Owing to vine like growth, multiple sprouting of buds to give more than one brittle shoot at the bud, the tree may assume bushy growth, harbouring insect pest, crotch injury and breakage of crotch due to high velocity wind, the training (through pruning) an essential operation in the initial years.

Training

By providing support to the new growth of bud from either insitu or transplanted budling, vertical growth is allowed. Once the growth is of five feet *or five and* and half feet, the terminal growth is pinched, allowing lower buds to sprout to form primary branches. After sufficient growth these primary branches also need to be pinched to develop secondary branches. Thus plants could be kept in manageable size and strong framework. The training process would continue for 2-3 years. After this initial training it is necessary to maintain the framework.

Pruning

In fact pruning where shoots are removed reduces the capacity of the tree. Depending upon.number of buds that are removed the capacity is reduced. But annual pruning - is an essential part by which capacity and vigour need to be balanced and thus every year production is maintained at a certain level.

In a trial on pruning time, it was observed that pruning between April and May, yields were higher. Early or late pruning than this period reduced yield (Anonymous, 1988).

In a trial on pruning intensity, it was observed that 4 primary shoots be maintained per tree and maintaining 4-6 secondary shoots helped to maximize yield. Maintaining less number of primary or secondary shoots reduced yield while maintain more increased bushy nature, disease and pest incidence *etc.*

In a trial on pruning at Kanpur, it was observed that pruning of trees delays flowering. Light pruning *i.e.* heading back of branchlets to 90 cm proved better treatment.

However, Zorasingh (1984) observed that early pruning resulted in early sprout, early flowering and also early maturity.

Bajwa (1986) also has said that light pruning improves fruit quality.

Light pruning (after about 6 secondaries) during summer dormancy period should be done to harvest good yield from ber plants in dry areas.

Spray of 3 per cent thiourea or potassium nitrate 2 days before pruning coupled with pruning of 50 per cent shoots to 6th secondary induced greater number of new bearing shoots.

Irrigation

Irrigation as a rule is not advisable. But if available could increase yield.

Quality of irrigation water and time of application need to be standardized to increase yield. But, however, increase in yield also diminishes quality of the fruit. In general, however, irrigation during developmental stage of the fruit is beneficial.

Irrigation or excess rainfall during the period of flowering may result in flower drop. Irrigation during maturity delays the period of maturity.

Successful ber plantation with runoff concentration can be raised. Mini catchment of compartmentalization helps direct the rain water towards ber trunk. At CAZRI, Jodhpur it has been demonstrated that a catchment area of 54 sq.m. with 5 per cent slope is sufficient to mature cv. SEB (Yadav *et al.*, 1978).

At AICRP on Arid Zone Fruits, MPKV, Rahuri there was successful plantation of total 25 acres of ber of cv. Umran, Chhuhara, *etc.* However, of these cv. Chhuhara is susceptible to water stress condition. While Umran not only withstands water stress b.ut also gives economical yield. The variation in yield is mainly due to rains during flowering especially around 31st Met. Week.

Black polythene is the best mulch material to conserve moisture and to check weed population in ber orchards.

Spray of antitranspirants like 1.5 per cent power oil, 7.5 per cent Kaolin and 10^4 M PMA reduced moisture loss through transpiration and produced higher yield in ber orchard.

Manuring

The response of ber to manure and fertilizer under rainfed situation is poor. However, under irrigated conditions, the response is better.

For the plants of 10 year old ber tree, it is recommended to apply 30 to 50 kg FYM, 250 g N/tree/year (split in two doses), 250 g $P_2O_{5/}$tree and 50 g K2O per tree (Anonymous, 1994).

Application of 250 and 750 g nitrogen/tree tended to increase the productivity of trees at Rahuri and Jobner, respectively.

In Western Rajasthan. application of 10 kg FYM for one year old plantation and then increasing dose by 10 kg for each year, upto fifth year is recommended.

Inrtercropping and Interculturing

Until the ber crop assumes full growth (5-7 years) intercrop can be taken. This helps the farmer to supplement his income.

In a trial on intercrop at Rahuri it was observed that growing of *Stylosanthes hamata* or growing of horse gram *Dolichos hiflorus* has been found to be useful. The adverse effect of intercrop was not observed as both the crops are leguminous, shallow rooted and fair well under drought situation and improve soil fertility.

Leguminous crop like *kulthi* as cover crop improved water holding capacity and organic carbon content of soil in ber orchard.

Among the tree species as an intercrop Karonda, drumstick, curry leaf also fit in well and could be exploited.

As a rule no interculturing operations are recommended except exposing the soil underneath the canopy by stirring or digging immediately after pruning. This will help expose eggs of insect pest to sunlight.

Stirring of soil below the canopy after rains helps close the capillaries to check, evaporation losses.

Use of Growth Regulators

Attempts to increase fruit set, fruit size and early/late maturity through application of growth regulators have been tried by several workers.

Dhillon and Singh (1968) found 2,4-D at 10-20 ppm to be the most effective in reducing fruit drop (from 60-61 per cent in control to 50.8-56 per cent) whereas, 25 ppm GA and 2,4 5-T at 5 and 10 ppm showed little promise in 'Dandan', 'Kaithli' and 'Umran'.

Exogenous application of 10 ppm NAA has been found to increase the fruit size and total soluble solids in 'Banarasi' variety (Singh, 1976).

Ghosh *et al.* (1985) isolated through chromatography factor, a a2, a3 and a4 from young developing fruit and observed similarly to gibberellin A_1, A_3, A_9 and A_4. The activity was highest in early period and decreased as days passed.

Flowering and Fruit Set

Flowers are borne in cymose cluster in the axil of leaf either on mature wood or on current seasons growth (Yamadagni, 1968). Flowers at base of the shoot open first and flowers of the terminal are last. The differences between initial flowering to end of flowering varies as per climate, nutrition and variety.

From the duration of flowering varieties have been classified as early, middle and late.

At Rahuri it was observed between 1985-87 that (Anonymous, 1988), some varieties as:Early, Mid-late, Late. Also it was observed that opening of flowers varied as AM type and PM type. Cross pollination is a rule. Honebee and housefly are the main visitors effecting pollination (Dhaliwal, 1975).Phenomenon of cross incompatibility, self sterility have also been reported (Mehra and Gupta, 1983). In Illaichi variety all the pollens are not viable. Of the total flower only 5-15 per cent flower set fruit.

Avoid use of insecticide during flowering period to allow honeybees to visit the flower and pollinate. Spray of any insecticide reduces the visit of honeybee. Maintaining colony of honeybee during flowering may help increase fruit set.

Harvesting

Harvesting needs to be done when fully mature. Fruits are in colour turning stage when they are harvested. All the fruits on the tree do not ripen at one time. Therefore 4-5 pickings need to be done on the same tree. Fruit requires nearly 120 days to reach maturity. Under irrigated conditions maturity will be further delayed. However, harvesting in Maharashtra of improved varieties will be completed by January end or February first fortnight.

Pre-harvest spray of 750 ppm ethephon about one week of first anticipated harvest induced early ripening and reduced number of pickings.

On an average per tree under dryland conditions 50-60 kg/tree for 10 year old is being harvested. Under irrigated condition yields will be 3-4 times higher.

Quality of the fruit under dryland conditions are far better. Total soluble solids as a measure of quality is around 18-20° Brix under dryland conditions while it is only 11 to 14° Brix under irrigated conditions.

Harvesting in northern Indian conditions is later than that of Maharashtra and will continue upto April.

Research Achievements in Ber

- ☆ Ber cultivar Gola is early maturing, high yielding and vigorous throughout the country and can be grown in soils having pH 8.0 to 8.7 and ESP 26.3 to 33.7. While Umran is a late maturing cultivar with better shelf life.
- ☆ Application of 250 and 750 g nitrogen/tree tended to increase the productivity of trees at Rahuri and Jobner, respectively.
- ☆ Fifth and sixth leaves from secondary shoots representing recently mature leaves should be collected for nutrient analysis.
- ☆ *Zizyphus nummularia* is the dwarfest and *Z. rotundifolia* is the vigorous rootstock of ber.
- ☆ Modified ring method has proved to be the best method for ber propagation.
- ☆ Black polythene is the best mulch material to conserve moisture and to check weed population in ber orchards.
- ☆ Spray of antitranspirants like 1.5 per cent power oil, 7.5 per cent Kaolin and 10^4 M PMA reduced moisture loss through transpiration and produced higher yield in ber orchard.
- ☆ Leguminous crop like *kulthi* as cover crop improved water holding capacity and organic carbon content of soil in ber orchard.
- ☆ High density plantation increased yield in ber on unit area basis.
- ☆ Light pruning (after about 6 secondaries) during summer dormancy period should be done to harvest good yield from ber plants in dry areas.
- ☆ Spray of 3 per cent thiourea or potassium nitrate 2 days before pruning coupled with pruning of 50 per cent shoots to 6th secondary induced greater number of new bearing shoots.
- ☆ Pre-harvest spray of 750 ppm ethephon about one week of first anticipated harvest induced early ripening and reduced number of pickings.

Future Challenges

- ☆ Drought tolerant varieties needs to develop.
- ☆ To develop variety having good yield, quality as well as keeping quality.

- ✰ To breed the varieties having disease resistance to powdery mildew.
- ✰ To develop processing technology.
- ✰ To exploit, local ber genotypes is a need.

Jamun

Jamun *(Syzygium cumini)* belongs to family Myrtaeeae. It is known as black plum, Indian black cheny, Kalajam, Jam and Phalinda. The jamun tree is tall and handsome evergreen generally grown for shade and windbreak on roads and avenues. Jamun fruit is a good source of iron, apart from usual contents like minerals, sugars, protein *etc.* Apart from eating them fresh the fruits can also be used for making delicious beverages jelly, jam, squash, wine, vinegar and pickles. A little quantity of fruit juice is much useful for cure of diabetes. The seed powder of jamun reduces a quantity of sugar in urine very quickly and permanently. It is used in the form of lotion against ringworm. Jamun seeds can be used as a concentrate for animals because it is rich in protein, carbohydrates and calcium.

Jamun is probably originated in India or the East Indies. It is also found in Thailand, Philippines, Madagascar and some other countries. In India, the maximum number of jamun trees are found scattered throughout the tropical and subtropical regions. It also occurs in the lower ranges of Himalayas up to an elevation of 1,300 m and in Kumaon hills up to 1,600 meters.

Soil and Climate

The jamun tree requires deep loamy and well-drained soil for its optimum growth and good fruiting. Its cultivation should be avoided in very heavy or light sandy soils. It is successfully grown under tropical and subtropical climate. It requires dry atmosphere at the time of flowering and fruiting. Early rains are beneficial for proper growth development and ripening of fruits. The young plants are susceptible to cold and drought conditions.

Varieties

At present, there are a number of seedlings strains of jamun in India, which provide a good scope for selection of better germplasm. While making selections, the following points should be kept in mind oval or oblong shape of fruit, deep purple or. bluish black colour, big size of fruit, more pulp, small stone, juiciness, sweetness and earliness. Considering the above parameters Regional fruit Research Station, Vengurla has taken up intensive survey and evaluated 13 seedling strains. Out of which VGLJM-1, VGL-JM-7 VGL-JM-8, VGL-JM-10, and VGL-JM-11 were found superior. Fruit weight of these strains varies from 8.70 g to 13.68 g and T. S. S. ranges from 13 to 19°Brix.

Propagation

Earlier, the most common method of propagation of jamun was by seed. However, at present softwood grafting, like in mango and cashew is commercially followed and it is now a common practice in jamun.

Planting and Aftercare

For jamun 10 x 10 m spacing is adopted. Pits of 1 m3 are dug and refilled with good soil, about 10 kg F. Y. M. and 1.0 kg Single Super Phosphate at May end. At the onset of monsoon one-year-old vigorously growing graft or seeding is planted per pit. When grafts is used, care should be taken not to disturb root ball and keep the graft union above the soil surface. Soil around the grafts/seedling is firmly pressed and a support of split bamboo stick is provided. After planting, plants should be protected from stray catties and sun scorching. Mulching with dry grass may be done around the base of plant to conserve the soil moisture.

Manures and Fertilizers

Experimental evidence on nutritional requirement of jamun is not available. As jamun is perennial crop bearing fruits every year, it is necessary to apply manures and fertilizers for suitable yields.

In the first year, each plant should be applied with 4 kg F. Y.M. + 100 g N + 50 g P_20_5 + 50 g K_20. This dose is increased every year and from fifth year onwards each plant may be given 20 kg F.Y.M. + 500g.N + 250 g P_2Os + 250 K_20. Manures and fertilizers are applied in a ring under the canopy of the tree during the month of August, after the heavy rains are over.

Harvesting and Yield

The fruit ripens in May. It turns deep purple or black in colour when fully ripe. The ripe fruits are immediately picked singly by hand and in all cases should be taken to avoid all possible damage to fruits. The fruits are generally harvested daily and sent to the market on the same day. The jamun fruit highly perishable and can be kept in good condition only for about 2-3 days under normal conditions. Average yield of fruits per, tree is about 802 to 100 kg at full-grown stage of tree.

Pest and Disease

There is no serious pest or diseases known to attack jamun tree. Sometimes the fruits get wormy due to fruit attack: Leaf eating caterpillar and White fly are sometimes observed to attack jamun trees.

As far as diseases are concerned, leaf spot and fruit rot have been reported which may be controlled with sprays of fungicides like Diathane-Z-78 at 0.2 per cent of Bordeaux mixture at 1 per cent concentration.

References

Annual Report. 1986-87. AICRP on AZF, MPKV, Rahuri. Annual Report. 1987-88. AICRP on AZF, MPKV, Rahuri. Annual Report. 1988-89. AICRP on AZF, MPKV, Rahuri.

Annual Report. 1993-94. AICRP on AZF, MPKV, Rahuri. Anonymous. 2002. Director of Horticulture, M.S. Pune.

De Condolle A. 1986. Origin of cultivated plants. Bibliotheque Scientifique Internationale, 43. Paris, P.385.

Dhillon, B.S. and Singh, K. 1968. Effect of some plant regulators on fruit set and fruit drop in *Zizyphus jujube* Linn. *J. Res.*, Ludhiana, 5: 392-4.

Pareek, O.P. 1983. Ber. Book. Pub. ICAR, New Delhi.

Singh, P., Bakshi, J.C. and Singh, R. (1972). *Punjab Hort. J.* 12(2/3): 123-133.

Singh, U.R. 1976. Effect of plant regulators on fruit drop, size and quality of ber (*Zizyphus mauritiana* Lamk) var. Banarasi. Haryana *J. Hort. Sci.* 5: 1-8.

Yadav, R.C., Sharma, K.D., Singh, H.P. and Pareek, O.P. 1978. Proc. Int. Symposium on Arid Zone Research and Development.

Chapter 10

Recent Advances in Pre-harvest Factors to Enhance Postharvest Quality of Custard Apple

S.D. Patil[1] *and S.A. Ranpise*[2] *and B.B. Dhakare*[3]

[1]*Associate Professor,*
College of Agriculture, Dhule – 424 001, M.S.
[2]*Head,* [3]*Professor of Horticulture,*
Department of Horticulture,
Mahatma Phule Krishi Vidyapeeth, Rahuri – 413 722, Dist. Ahmednagar, M.S.

Custard apple is known as "Poor Man's Apple" which belongs to the family Annonaceae. Consist of 40 genera of which Annona is most important which has about 70 species. It is native to the tropical Americas and widely grown in Australia, Brazil, Burma, Chile, Egypt, India, Spain, South Africa, USA, Pakistan and Philippines. *Annona squamasa* (sweet sop), *Annona reticulata* (Ramphal), *Annona cherimola* (Cherimoya), *Annona mericata* (Soursop), *Annona glabra* (Pond apple), and one natural hybrid *Annona Cherimola* X *Annona squamosa* (Atemoya) are the major species available in India. Among these *Annona squamosa* is most popular.

There is great genetic diversity within genus Annona with chromosome 2n = 14 with the exception of *Annona glabra* which is tetraploid. Among the existing species *Annona squamosa* is widely grown under arid and semi arid tropics of India may be due to its appealing sweet test. In India, the custard apple is mainly grouped under two categories *viz.* green fruited and red fruited. The flavonoidal analysis of the species showed two patterns in the cultivars of Annona squamosa. The Annona squamosa is represented by Balanagar group (Balanagar, Mammoth, British Guinea, local custard apple, Barbados, Washington PI 107905), whereas the

'Island Gem' group (Island Gem, Bullock's Heart, Pink Mammoth) seems to be the hybrid between *A. squamosa* X *A. cherimola*.

In India, Andhra Pradesh is the major custard apple growing state followed by Maharashtra, Tamilnadu, Orisa, Assam, Uttar Pradesh, Bihar and Rajasthan. Area covered under this crop was 29,571 ha with the production 47,979 MT. In Maharashtra, it is widely grown in Purandar Tahasil of Pune district, Daulatabad area of Aurangabad, part of Akole and Pathardi Tahasil from Ahmednagar district, scarcity zone of Solapur, Osmanabad, Dhule, *etc.*

Custard apple is mostly consumed as table fruit. It is also used in ice-cream and preserved as jam, jelly or other products. The edible portion of fruit or pulp is creamy, granular with good blend of sweetness and acidity. Custard apple fruit is high in calories and is a good source of iron. The nutritive value of fruit is high. Beside high nutritive value, it has also a high medicinal value. It contain 70.5 g moisture, 23.5 g carbohydrate, 1.6 g protein, 17 mg calcium, 47 mg phosphors, 0.9 g mineral matter and 104 K calories per 100 g edible portion.

Beside high nutritive value, it has also high medicinal value. Root is drastic purgative; leaves are used for unhealthy ulcer and for store grain pests. Seed extract and seed oil have insecticidal properties.

The Balanagar (Sitaphal) is mostly acceptable in Indian markets, however, the Balanagar group though having sweet appealing test, contain more number of seeds per fruit (30-50) with as many of 80 seeds and obviously the less pulp percentage. Fruit size is small with frequent cracking at maturity and poor shelf life. This variation seems to be not only genetically but also environmental. Secondly, fruit contain more number of aeroles with uneven size which is undesirable characteristics from marketing point of view.

Hence, in view of improvement of Annona attempts have already been made at AICRP on Arid Zone Fruits, Department of Horticulture, M.P.K.V. Rahuri since 1990 to combine desirable characteristics of these two groups like less seed number, higher pulp percentage, fruit size and fused carpels of Island Gem Group, reciprocal crosses were made between selected genotype of *Annona squamosa* and Island Gem, pinks Mammoth, Bullocks Heart and Mammoth. The promising selection have been made from the progeny evolutions.

Drought Adaptive Mechanisms of Custard Apple

The Annona species grow well under rainfed conditions because it has a deep tap root system and good sense of timing in its shoot tip abscised by December with the onset of drought which indicates that is an excellent drought evading mechanism, by imposing limitation on its own growth. Infra-petiolar development of axillary buds could be an intrinsic safeguard for the survival of axillary buds under condition of extreme drought and heat. In addition to intrapetiolar buds, production of accessory superposed bud in the season which turns into 'sylleptic' branches appears to be the fair weather conduct of the plants. This helps the plant in maximizing its growth when soil moisture is not limiting.

When the Annona plants undergo moisture stress, they shed their leaves to reduce leaf area and minimize the transpiration at losses of water. The absence of short tip and exposure of axillary bud because of leaf fall, do not initiate new growth, since, a few mature leaves present on a shoot impose 'correlative inhibition' on the axillary bud sprouting. It is only after more than 50 per cent leaf fall, new growth is initiated.

While cultivating custard apple, pre- harvest factors are most important in order to enhance the postharvest quality and due attention contributes quality of custard apple. Selection of soil and climate, propagation, selection of suited variety, manuring, canopy management, water management, bahar management are most important pre-harvest factors which could enhance postharvest quality directly of indirectly. Owever, constrains in custard apple cultivation are:

1. Inadequate supply of genuine and true to type planting material.
2. About 60-70 per cent of custard apple plantations in India are from seedlings which are not true to type.
3. They have become overcrowded and unproductive, thus reducing the productivity drastically and have become uneconomical.
4. Several biotic and a biotic factors cause substantial loss in the field diseases and pests generally mealy bug losses.
5. Highly perishable in nature with very short shelf life and hence not suitable for distant market.
6. The few scientific practices of fruit cultivation are not fully adopted by the farmer thus affecting the productivity and quality of fruit.
7. Spurious plant protection chemicals, which are not effective, are being used extensively.
8. Non-availability of cultivar having longer shelf life.
9. The practice of giving orchards to pre-harvest contractors has been found to lead to poor orchard management, resulting into poor quality of fruits.

Cultivar

1. Balanagar

This is promising selection made from Maheboob nagar district of Andhra Pradesh. Fruit shape is pyramidal or chordate, areoles very rough, pitted usually remains green when ripe.

2. TP-7

It is was recommended in 2000. Fruit weight is 400-500g, 71.75 per cent pulp percentage with 28 per cent TSS. Fruits have attractive cream colour and less portion of seed. It produces 70-100 fruits per tree.

3. Arka Sahan

Hybrid Arka Sahan developed at IIHR, Hessarghatta. It is interspecific hybrid between Island Gem and Mammoth fruits. Skin with smooth surface with better shelf life keeps well for 4-5 days. Skin does not crack after ripening. Pulp 48.6 per cent with pleasant aroma, seeds 8-9 per 100 g, TSS >30°B. yield is 12 t/ha.

4. Red Custard Apple

Fruits are small sized with more seeds attractive red coloured, leaves are faint red. 50-60 fruits per tree.

5. Phule Purandar (JWC-1)

Released from AICRP Fig and Custard apple Research Station, Jadhavwadi. This is selection, high yielding potential, large sized fruits, less seed content, high pulp percentage, more number of white flakes. Recommended for cultivation under Maharashtra conditions.

Soil

The custard apple is not very particular about soil conditions and flourishes in all types of soils like shallow, sandy, marginal but fails to grow if the subsoil is ill drained. It can be grow well in deep black soil provided they are well drained. A little salinity or acidity does not affect it but alkalinity, chlorine, poor drainage or marshy wet lands hamper the growth and fruiting. However for getting a good crop, a fertile well drained sandy loam soil and neutral in chemical reaction (pH 7.0) is preferred.

Climate

All annonas are tropical in origin and grow well in hot and dry climate with varying degrees of difference. Custard apple requires hot and dry climate during flowering and high humidity at fruit setting flowering comes during hot dry climate of May but fruit setting takes place at onset of mansoon. Low humidity is harmful for pollination and fertilization. The custard apple withstands drought conditions cloudy weather and also when the temperature goes below 15°C. In controlled environment, the highest day and night temperatures (32/27°C) resulted into strong vegetative growth and adversely affected floral bud production. Fruit growth and yield at high temperature were further affected bu soil moisture stress. Therefore warm subtropical conditions showed better fruit growth and yield. Annual rainfall of 500-800 mm is optimum, though it can withstand higher rainfall. Custard apple can also tolerate adverse weather conditions, such as extreme temperature, high and low humidity, heavy and scanty rainfall and desiccating winds. In India, inspite of profuse flowering less fruit occurs due to high temperature and low humidity during the flowering period.

Propagation

Annona can be propagated both by sexual and asexual methods. In India, the existing natural population is commonly propagated by seeds. However, seed propagation resulting into variability in plant vigour, prolonged juvenility and inferior fruit quality. Vegetative propagation is therefore, a necessity to maintain the genetic uniformity for obtaining higher yield of better quality fruits. The budding and grafting methods of propagation are nowadays commercially adopted for multiplication. The seedlings of local custard apple have proved a good root stock for many improver varieties and hybrids. Seeds treated with 100 ppm GA3 for 24 hours. Germinate quickly and uniformly. After collection of seed rest of two weeks should be given and thereafter seedcs should be soaked in 500 ppm GA solution or seed may soaked in water for three days and seeds can be sown in polyethylene bags. Budding by shield, patch of chip method and grafting in early spring on the commencement of sap flow using one year old mature scion on more than one year old rootstock give better success. Soft wood grafting followed from second fortnight of February to first fortnight of March under North Maharashtra conditions.

Planting and Season

Planting is done during rainy season. In order to promote growth of the planting material, pits of 60X60X 60 cm at spacing 4X4 m or 5X5 m depending on soil type are dug prior to monsoon and filled with 20 kg good quality FYM, 1 Kg single super phosphate, 2 kg neem or Karanj cake and 100 g carbaryl powder at least a fortnight before planting. The establishment of roots and plant growth is better in *in-situ* planting of custard apple. Hence, it is recommended in Maharashtra. However, it is also desirable that the seedlings be raised in polythene bags and after budding or grafting they are transplanted in the orchard.

Interculturing

For good plant growth, the weeding should be done to keep away the weeds. Intercropping with some legumes, peas, beans and marigold flowers are commonly taken by the growers. Normally, no crop is taken during winter as the plants go under rest.

Pruning

Generally, training and pruning are not in practice in annonaceous fruits. But leaf shedding in winter season, and production of floral buds on new shoots indicate the necessity of some amount of pruning. Hence for obtaining improved growth and better quality fruits a light pruning to break apical dominance *i.e.* removal of unwanted growth and a general thinning of diseased and broken branches are essential to make the orchard a remunerative one. The completion of leaf fall is essential for the initiation of new growth in annonas. Therefore, the dormant pruning in advisable a fortnight before the start of new flush. In an experiment at Mahatma Phule Krishi Vidyapeeth, Rahuri the highest yield, average number of fruits peer tree and average weight of fruit were recorded in the treatment where the custard apple plants were pruned 75 days after previous harvest with 25 per cent pruning intensity.

However, experiment conducted at Dhule condition, pruning at proper time and intensity was found beneficial to improve the growth, yield and quality contributing characters in custard apple. The highest number of flower per shoot (8.31), percentage of fruit set (71.51 per cent), average numbers of fruits per plant (106.04) and maximum yield per plant (25.45 kg) were recorded in treatment D3P2-

i.e. pruning on 1st February with 30 cm of pruning level. Minimum percentage of seed (8.82 per cent) and lowest seed: pulp ratio (0.15) was registered in the treatment D3P2-*i.e.* pruning on 1st February with 30 cm pruning level.

Irrigation

Annonas are considered drought tolerant plants and produce a fair crop even without any irrigation because the fruit set and development occur during the rainy season and thereafter when there is ample humidity in the atmosphere and moisture in the soil. Hence, in India, Custard apple is grown as a rain fed crop, and no irrigation ar given. However, for early and bumper harvest of the crop, irrigation on flowering *i.e.* from May should be given till regular monsoon starts. Two to three irrigations before the onset of monsoon to promote fruit set and one or two irrigations after the monsoon to increase fruit size are beneficial for the higher yields. For better flowering and fruit setting. Mist sprinkling is better over flood or drip system of irrigation as it keeps to lower down the temperatures and to increase in the relative humidity.

Nutrition

Generally, no manures or fertilizers are applied to rain fed crop. However, balanced nutritional schedule is required to be adopted for getting regular higher yield. Since, the custard apple is cultivated mostly on poor soils, the need for fertilizer application becomes all the more essential. For early and bumper harvest with a good keeping quality, the dose recommended is 250g N (550 g Urea) + 125g P_2O_5 (775 g Single supe phosphate) + 125g K_2O (200 g Muriate of Potash) per plant recommended for custard apple under commencement of rainy season. Half nitrogen, full p and k after receiving 1-2 rain showers and half N after one month or or at the time of flowering should be given. The application of optimum quantity of well decomposed manure every year improves the quality of fruits. The application of biofertilizers like Azospirillum and PSB registered production of quality fruits.

Flowering, Fruit Set and Fruit Development

Seedlings as well as vegetatively propagated plants begin flowering within four years of planting. In custard apple the flowering is observed mostly on new shoots as well as on old shoot. Flowering is higher correlated with defoliation and there after emergence of new growth. The flowering period of custard apple is very long commencing from March-April and continuing up to July-August. A floral bud develops into flower in 27 to 35 days.

Pollination as well as fruit set is a major problem in custard apple. Enough flowers are borne on a custard apple plant to give a good crop but the poor fruit set causes low yield. Only 1-8 per cent fruit set has been reported under natural conditions. Fruit set after the onset of monsoon, however late vegetative growth delays flowering and fruit set. However, setting of fruit early in the season is important regarding marketing point of view.

Low fruit set in custard apple is due to poor pollination which has been attributed to both the external and internal factors, such as high temperatures and low humidity prevailing at the time of flowering, soil moisture stress and competition between vegetative and floral growth, hypogyny, dichogamy and poor pollen germination.

For uniform flowering and for early flowering and also to check the flower and fruit drop and for improving fruit size, following growth regulators are used.

1. For defoliation and bringing the plants under uniform rest Ethryl at 1000 ppm is sprayed one month after the harvest of the fruits.
2. 20 ppm NAA is sprayed during the flowering period to improve the fruit set.
3. During fruit development 50ppm GA + 0.5 ppm cppu, foliar spray improves the fruit size and luster of the fruits.
4. For offseason fruiting *i.e.* harvesting before start of monsoon the sowing of bajra between two rows found beneficial as cover crop.

Annona squamosa fruits usually mature in about four months from the time of anthesis, the period is reduced when the soil moisture is optimum. The fruits of other species mature in about six months.

Hand Pollination

Hand pollination is a simple and fast technique that results not only in good fruit set but also produces big size, attractive uniform shape fruits with no loss in edible attributes. Thus fruit yield can be significantly increased by artificial pollination. A tree of Arka Sahan can produce 40-45 kg fruits from the eighth year after planting. Hence in each tree about 150 flowers should be hand pollinated to achieve an expected yield of 25 tone per hectare. Fruits developed from hand pollination command a premium price in the market. As Arka Sahan is an inter-specific hybrid it should be propagated by grafting and not by seeds.

Problems Related to Annona Cultivation

Nutrient Deficiency

Heavy vegetative growth during the fruit, development period competes for nutrients. Such as boron and Ca, is considered a causal factor for these lumps. Applying boron at slight excess can be phytotoxic, especially in sandy soils. A desirable practice is to apply the organic manures, with inorganic fertilizes as a supplement to maintain a balance.

Physiological Disorder

Stone Fruit Formation

With the end of harvesting season, the leaves of custard apple turn yellow and with the onset of winter, leaves fall down. New growth commences in spring after passing a dormant phase in winter. But in certain neglected plants are those under severe moisture or nutritional stress. The dormancy sets well in advance and some fruits turned brown, become quiet hard and without further growth remain on the tree for months even in the following season. Such fruits are referred to as stone fruits.

This disorder can be corrected by practicing clean cultivation, manuring, timely irrigation and application of super phosphate and bone meal.

Harvesting and Yield

If fruits harvested at proper stage it helps to prolong shelf life and ultimately quality. Harvesting of custard apple fruits at proper stage of maturity is very important, because the fruits continuing on the tree after maturity split instead of ripening. Likewise immature fruits after harvest become soft and rot. The Custard apple is a climacteric fruit and harvested at the maturity state when the fruit starts to change colour from green to its varietal colour shade. Immature fruits do not ripe. Swallowing some apical buds – showing inner pulp is also an indication of maturity. A were grown tree yield above 100 fruits weighing 300 to 400 gm. The season of harvest is from August to October. Maturity indices in annonaceous fruits were not yet fixed but the fruits are considered mature on the basis of following indices:

1. The fruit has attained full size.
2. It get light green colour.
3. Yellowish white colour has developed between the segments.
4. The skin between carpel's has initiated cracking.

The fruits are hand picked by determining the maturity of individual fruits on every alternate day during harvesting period.

The yield of custard apples is affected by number of factors, including climate, soil, cultivars, rootstock, age of the plant, pollination and orchard management. In India 2.5 tones fruit yield per acre has been estimated. A good bearing sugar apple tree produce about 60-70 fruits. Each weighing from 113-227 g. African Pride cultivar has given 18.7 tones yield per hectare in the 6th year of planting in Australia.

References

Bose, T.K., S.K. Mitra and B Sanyal.2001Tropical and Subtropical Vol.1. Naya Udyog Publication, Kolkatta Third revised edition.

Chadha, K.L. O.P. Pareekh. 2014 Advances in Horticulture- Fruit crops vol. 1-4. Malhotra Publishing House, New Delhi 11 00 15

Dhakare, B.B, A. J. Kharat, S.D. Patil, C. V. Pujari and P. D. Dalvev 2014. Effect of pruning levels and time of pruning on growth, yield and quality parameters of Custard apple (*Annona squamosa* L.) cv. Balanagar. 6th Indian Horticulture Congress – 2014. November 6-9, 2014, CODISSIA Trade Fair Complex, Coimbatore. IV-22 Pp: 177.

Kumar, N. 2010. Introduction to Horticulture Oxford and IBH Publishing Company Private Limited, new Delhi pp: 286-289.

More, T.A. 2002. Research Recommendations In horticulture. Jointly published by Department of Horticulture and Director of Horticulture, Government of Maharashtra, Pune. MPKV. Res. Pub. 40. Pp: 2, 36 and 90.

Patil, S.D., A.B. Purane, B.B. Dhakare and C.V. Pujari. 2014. Standardization of grafting time in custard apple (*Annona squamosa* L.) cv. Balanagar under shade net condition. 6th Indian Horticulture Congress – 2014. November 6-9, 2014, CODISSIA Trade Fair Complex, Coimbatore. IV-41 Pp: 186.

Ranpise, S.A., B.B. Dhakare, S.D. Patil and D.R. Patil. 2007. Improved production technology of custard apple. MPKV/Ext/525 pp.1-12.

Ranpise, S.A. and T.A. More. 2013. Research Recommendations in Horticulture Part-II (2003-2013) Department of Horticulture, Mahatma Phule Krishi Vidyapeeth, Rahuri. pp-17.

Surve Punam Motiram. 2015. Effect of cover crops on pollen viability and fruit setting in custard apple under North Maharashtra conditions. M.Sc. (Hort) Unpublished Thesis Submitted to Mahatma Phule Krishi Vidyapeeth, Rahuri.

Vora, D.A. 2022. Research Recommendations in horticulture. Jointly published by Department of Horticulture and Director of Horticulture, Government of Maharashtra, Pune. MPKV Res. Pub. No. 40. pp. 36 and 90.

Patil, S.D., A.B. Pujari, B.D. Bhakare and U.S. Bhosale. 2014. "Standardization of grafting time in custard apple (Annona squamosa L.) cv. Balanagar under shade net condition. 6th Indian Horticulture Congress – 2014, November 6-9, 2014. CODISSIA Trade Fair Complex, Coimbatore. IV-31 Pp. 186.

Sangle, S.A., R.S. Dhakare, S.D. Patil and D.K. Patil. 2022. Improved production

Chapter 11

Recent Advances in Pre-harvest Factors to Enhance Postharvest Quality of Fig

S.D. Patil[1] *and S.A. Ranpise*[2] *and B.B. Dhakare*[3]

[1]*Associate Professor,*
College of Agriculture, Dhule – 424 001, M.S.
[2]*Head,* [3]*Professor of Horticulture,*
Department of Horticulture,
Mahatma Phule Krishi Vidyapeeth, Rahuri – 413 722, Dist. Ahmednagar, M.S.

Fig is under cultivation since ancient times. Morphologically it is called as 'syconium' which is a vegetative, fleshy tissue, with tiny true fruits enclosed inside. Fig is a gynodioecious species and some female type needs pollination while other fruit set fruits parthenocarpically. Pollination is affected by a wasp which develops inside the syconium of male fig. this symbiotic relationship is a classical case of co-evolution between a plant and insect.

Fig fruit are often consumed as dried or canned. As a fresh fruit it has a luscious test. Fruit have been known for their medicinal and dietary properties over centuries. In india its cultivation is mostly confirmed to western part of Maharashtra and Gujarat, Utter Pradesh (Lucknow and Saharanpur), Karnataka (Bellary Cnitradurga and srirangapatna) and Tamil Nadu (Coimbatore).

Fig is highly nutritious fruit. It is rich in calories, proteins calcium (higher then milk), iron and highest fiber content. Fig has nutritive index of 11, as against 9, 8 and 6 for apple, raisin and date respectively. The chemical composition and flavor of fig varies with the cultivar. The total sugar content of fresh fig is 6 per cent and of dried is 52 per cent. The edible portion of dried fig (100 g) supplies protein (4 g), carbohydrate (69 g), fat (1g), calcium (200 mg), iron (4 mg), vitamin A (100 IU) and thiamine (0.1 mg). Fig is valued for its laxative properties and is used in the treatment of skin infection. The fruit help to maintain acid alkali balance of the body. Latex is useful for coagulation of milk.

Type of Fig

On the basis of sex of the flower and the method of pollination fig have been grouped into four type depending:

1. Common Fig or Edible Fig

Individual flower are long styled pistillate and fruit develop parthenocarpically. Most popular cultivar includes- Poona, Conardia, Mission Kadota, Brown Turkey.

2. Smyrna

Fruit develops only on pollination by male flowers of caprifig through the fig wasp. Most important variety is 'calimyma'.

3. Caprifig or Wild Fig

Short styled pistillate flower and functional staminate flower. Caprifig are not edible but grown because they harbor fig wasp (blastophega psenes), which is necessary for pollination and setting fruits.

4. San-pedro-Fig

It is an intermediate type where the first crop (known as BREBA) is parthenocarpic while the 2nd crop (main) require pollination like SMYRNA type. Only caprifig produces pollen, while other 3 pistillate types are dependent on pollen of caprifig or develop fruits parthenocarpically. Commercially grown Indian fig belong to 'common' fig group and set fruits parthenocarpically.

Climate and Soil

Fig tree is deciduous, and subtropical. It favours arid or semi arid environment, high summer temperature, plenty of sunshine and moderate winter. Although plants can survive temperature as high as 45°C, the fruit quality deteriorates beyond 39°C. mature trees can withstand temperature up to 4°C, but young ones need protection. In mild climate, plants remain evergreen, lack well defined flowering and fruiting season, and sometimes produce long barren limbs. Climate has been an important bearing on size, shape and color of skin and pulp. A relativity cool climate stimulates production of larger and elongated fruits. Climate condition during fruit development considerably determines fruit quality. Very high temperature (>39°C) induce premature fruit ripening. High humidity results in fruit splitting, while hot breeze during ripening leads to sweet but small fruits.

Medium to heavy, calcareous well drained, deep (about 1m) soil having pH of 7-8 is ideally suited for fig cultivation. Although it does well even on light sandy, shallow soils, deep soils encourage better root establishment. The fruit produced on fertile, light soils are better suited for drying. The crop can be tolerate drought, salts (chlorides and sulphates) but is sensitive to sodium carbonate and boron salts. In general, climate rather than soil is a limiting factor for its cultivation of fig.

Varieties

Nearly 700 verities of fig have been listed in the world. Verities vary for vegetative vigour, pollination requirement, yield, fruit size, shape, skin colour, pulp quality and color. Large sized figs belong to common fig group.

Poona fig is most popular cultivar grown in India. Bangalore, Bellary, Coimbatore, Daulatahad, Dindigual, Ganjam, Hindupur, Lucknow and Saharanpur, have clearly acquired the name from the location in which they are cultivated. Most of them resemble in plant and fruit morphology that of Poona fig. Possibly these are either clones or ecotypes and hardly they warrant varietal status. Black Ischia, Shahi, Maisram and Brown Turkey have not achieved prominence. Dinkar, an improvement over Daultabad for yield and fruit quality, is gaining commercial significance. Phule Rajewadi is the variety released by M.P.K.V., Rahuri for table purpose.

Some well know fig hybrids from Californial have performed well in India in comparison to Poona fig under Bangalore conditions. They produce fruits parthenocarpically. Excel and Conadria fig those are suitable for high density planting. The fruits do not split like Poona and Conadria fig. Conadria, Excel and Deanna are good for drying, canning and table purposes respectively.

Propagation

Although it is possible to propagate fig from seeds, cuttings, layers, grafts and by tissue culture, commercially cuttings are used for multiplication. About 25 cm long cuttings having 3-6 nodes are usually made from wood of previous season and planted in moist sand either in seed pen or in nursery beds. This can be taken up during pruning or just after the onset of monsoon. The cuttings re raised in shade with regular watering. After about 75 days, they are transplanted to polythene bags containing garden soil, sand farmyard manure (1:1:1), and field planted about 4-6 months later.

Higher success can be achieved by: (1) using cuttings, with short internodes, collected from basal portion of the shoot located in the lower part of the crown: (2) storing of cutting in moist sawdust or sphagnum moss for about 4 weeds at room temperature; (3) treating with growth regulators like IBA; (4) pre girdling at the base of canes (removing 2.5 cm width bark) a month prior to taking cuttings; and (5) planting cutting in slightly slanting (80°) position.

Brown Trukey, as root stock imparts vegetative vigour in fig Excel, Conadria and Deanna. Shield or patch budding, cleft or bark grafting enable to top work a desirable genotype on established but inferior tree protocols are now available for micro propagation of fig shoot tip.

Planting

The best time of planting is the onset of the rainy season. The layout for planting can be either square or hexagonal system. The square system is more common and desirable. Spacing depends on variety and soil type. The recommended spacing for Poona fig is 5 m × 5m (400 plants/ha). For light to medium soil it should planted at

4.5 x 3 m. It is 2.5 m × 2.5 m (1,600 plants/ha) for Excel and Conadria. Pits of 60cm × 60cm × 60 cm size are dug and exposed to sun for about 15 days, and then filled with a mixture of compost, top soil and sand (1:1:1); 2 kg of neem cake or castor cake per pit. Planning can preferably be taken up on and overcast days. When grafts are used the graft joint should remain above the ground level. Once the tree is planted the soil around the plant should be trampled firmly. Water is applied immediately after planting.

Training and Pruning

Fig trees are trained initially to single stem to encourage a wide, symmetrical crown with a mechanically strong framework having evenly distributed laterals. The tree is allowed to grow for about a meter and then it is topped, which induces side branches all round the main stem. The interior of the bush should be maintained free of suckers, dry and sick branches.

Pruning in fig is practiced annually to stimulate production of new growth, and bearing fruits. The time and type of pruning vary with location, variety and number of crops harvested annually. The best time to secure a mature crop is hot, dry summer. Therefore, pruning may be done 4-5 months in advance. Generally, a single marketable crop is harvested yearly in our country. Either heavy or light pruning can be adopted in fig. When heavy pruning is practiced, trees are headed back severely every year, leaving about 2 buds on each one year old shoot. If light pruning of terminal shoots 5-6 cm is adopted, lower 2-3 buds sprouts and get new growth. Spray 500 ppm Ethrel which will result in maximum sprouting of buds and new growth within a fortnight.

Notching is practices sometimes in Poona fig for activating dormant buds before the start of vigorous growth. Usually 1-2 buds are selected for notching in the middle portion of about 8 month old canes. Notching involves removing of small slice of bark immediately above the dormant bud just below it to throw out a fruiting shoot. The cut checks the free flow of sap and stimulates the bud just below it to throw out a fruiting shoot. The technique is useful for induction of fruiting laterals on vigorous upright braches and to increase the total bearing area of the plant.

Manuring and Fertilizer

Nutrient requirements vary according to the variety and soil type. For young plants fertilizers can be applied with the onset of monsoon and just after pruning for those which have commenced yielding. Nitrogen is essential for rapid growth of foliage and development of syconia, fruit color and maturation and K for yield and quality. A total recommendation of 900 h N, 250g K_2O and 275g P_2O_5 has been made for an adult tree of 'Poona fig' along with 40-50 kg well rotted FYM.

Time of manuring shall vary with place and accordingly in North India it is done in spring, while in South and West India, September is the preferred period.

Some soils are may be deficient in micronutrients. However, a grower should get the soil tested and do nutrient management accordingly. Application of compost, which is done mostly in the beginning monsoon, also supplies micronutrients to some extent.

Table 11.1: Micronutrients to be Applied for Correcting Deficiencies

Micronutrients	*Soil application*	*Foliar application*
Zinc	30kg Zn SO_4	3-4 sprays of 0.25 per cent ZnSO4 (unneutralized) at 10 day interval
Iron	-	3-4 Sprays of 0.5 per cent FeSO4 at 10 days interval
Boron	12 kg Borax/ha	-
Magnesium	50 kg $MgSO_4$/ha	2-3 sprays of 0.5 per cent Mg SO4 at 10day interval

Either soil or foliar application can be done. Foliar spray is applied when the plants are flushing.

Water Management

July-August is the resting period. After the plants are set in the field, regular watering is essential until they are well established. A basin of 60 cm diameter should be prepared around the plant and is widened as the canopy size increases.

Fig plants can sustain heat and drought. But commercial fig production is possible if plants are timely irrigated. Such plants produce greater shoot growth and higher yields of superior quality fruits. Loose and sandy soils require larger quantities of water than heavy soil. Either drip or flood irrigation can be practiced. The drip irrigation minimizes water requirement and allows fertilizer application through irrigation water.

Flood irrigation may be repeated every 10 days in summer. The frequency may be adjusted depending on the soil type and weather. Excessive irrigation during fruit development causes the terminal buds to initiate growth at the expense of fruit development. Also excessive irrigation or heavy rains during ripening result in fruit cracking and production of insipid fruits. In the absence of adequate and regular irrigation the fruit development is affected, resulting in small and hard fruits. Once harvesting of fruits is completed, irrigation may be reduced regular schedule is resumed after pruning. If drip irrigation is adopted, 15-20 liters of water/day/plant may be supplied.

Weed Management

Basin cleaning is taken up regularly to keep it weed free. The side shoot and suckers should be remover as and when they emerge. Maintenance of weed free orchard is very important. During early years of orchard, raising green manure or intercrop is recommended. Green manure sun hemp suppresses the weed growth and augments the supply of organic matter in the soil. Intercropping vegetables and legumes is beneficial.

Harvesting and Yield

Though the bearing of the fruit starts from the second year, fruits are harvested from the third year onwards. The economic life of the plant may be considered 35 years. The harvesting season in Pune starts in February-March and is over by May-

June. The fruits are harvested once in 2-3 days manually. The average yield of full grown fig per tree produced varying from 18-30 kg.

Bahar Management

Two bahars found in fig fruits produced in rainy season July-August known as *khatta* bahar or off season yielding low volume of fruits but fetches better prices. Less sweet and more acidic. Fruits harvested in summer known *mittha* bahar yielding fruits in March-April. Best quality fruits. Do one light ploughing in in September and give strees for a fortnight. Follow pruning judiciously and apply FYM, phosphorous and potash and halfnitrogen. New growth putforth in October- December and fruit growth starfs. Apply half Nitrogen for fruit growth.

Tips in Bahar Management

1. Consider climatic conditions of particular area befor bahar management.
2. Do soil testing every year.
3. To maintain yield potential and vigour of plant, apply fertilizers balance quality.
4. Apply more organic fertilizers.
5. While applying irrigation for *khatta* bahar, spray 10 ml Hydrogen cynamide 50 per cent per liter of water on stem and branches or rub on all shoots for activation and sprouting of buds.
6. Pest and weed management should be done timely.
7. Follow Integrated Approach for Pest management.
8. Investigation of root should be done every year for detection of nematode infestation.
9. Avoid over watering. Plant windbreaks on North Western side of orchards. Ex. Casuriana, Eucalyptus, Shewari, *etc.*
10. To avoid fruit cracking apply 30-50 g Boron per plant.
11. Stirr soil beneath canopy area to improve fruit size and quality.
12. Inject 20 ml Dichlorovos per liter water through injection in holes of stem borers.
13. Spray 20 g Chlorthalonil + 10 g carbendazim + 25 g Mancozeb per 10 liter water after 15 days of emergence of new shoots after pruning of 20 days.

Physiological Disorder

Apart from the disease and attack by insects, there are some disorders taking place in fig. These are called physiological diseases.

1. Fruit Splitting

When fruits are nearing maturity, a sudden fluctuation in soil and atmospheric humidity causes fruit splitting.

A shower during ripening may also cause severe fruit splitting. These split fruits are attacked by fungi and insects rendering them unfit for consumption. Maintenance of high level of organic matter in the soil can control the situation.

2. Sun Burn

Sum burn is a serious problem of newly planted young trees. As a result of sun burn, the affected parts so exposed gives access to insect borers and the tree does not grow normally. The trunks of young fig trees and bark of older trees suddenly exposed to direct sunshine by heavy pruning are very susceptible to injury. This injury is likely to occur in winter in Northern India when trees are leafless and sum rays strike the bark more directly than in summer. White (lime) painting will guard against the injury.

References

Bose, T.K., S.K. Mitra and B Sanyal.2001Tropical and Subtropical Vol.1. Naya Udyog Publication, Kolkatta Third revised edition.

Chadha, K.L. O.P. Pareekh. 2014 Advances in Horticulture- Fruit crops vol. 1-4. Malhotra Publishing House, New Delhi 11 00 15

Kumar, N. 2010. Introduction to Horticulture Oxford and IBH Publishing Company Private Limited, new Delhi pp: 296-299.

More, T.A. 2002. Research Recommendations In horticulture. Jointly published by Department of Horticulture and Director of Horticulture, Government of Maharashtra, Pune. MPKV. Res. Pub. 40. pp: 2, 36 and 90.

Patil, S.D., Dhakare, B.B., Pujari, C.V. and Dalve P.D. 2013. Improved production technology of Fig. MPKV/Ext/891/2013 pp.1-12.

Ranpise, S.A. and T.A. More. 2013. Research Recommendations in Horticulture Part-II (2003-2013) Department of Horticulture, Mahatma Phule Krishi Vidyapeeth, Rahuri. pp-17.

Chapter 12

Use of Plant Growth Regulators in Dryland Fruit Crops

S.S. Kulkarni

Horticulturist AICRP on Arid Zone Fruits,
Department of Horticulture,
Mahatma Phule Krishi Vidyapeeth, Rahuri – 413 722, Dist. Ahmednagar, M.S.

Growth mainly refers to the quantitative increase in plant body such as increase in length of the stem and root, number of leaves, the fresh weight and dry weight *etc.* On the other hand, germination of seed, formation of flowers, fruits and seeds, emergence of lateral buds, falling of leaves and fruits are qualitative changes, referred to as development.

Growth and development of the plant body are controlled by two sets of internal factors, namely, nutritional and hormonal. Nutritional factors supply the necessary mineral ions and organic substances such as proteins, carbohydrates and others.

These constitute the raw materials required for growth. However, utilization of these substances for proper development of the plant is controlled by certain chemical messengers, called plant growth substances or plant growth regulators, which in minute amounts increase or decrease or modifies the physiological processes in plants.

The term plant growth regulators' is relatively new in use. In earlier literature these were mentioned as Hormones. Hormone is a Greek word derived from hormao which means to stimulate. Now the term phytohormone is used in place of plant hormone.

Plant growth regulators or plant regulators are the organic compounds other than nutrients which modify or regulate physiological processes in an appreciable measure in the plants when used in small concentrations. They are readily absorbed and these chemicals move rapidly through the tissues when applied to different parts of the plant.

Plant hormones or phytohormones are also regulators but produced by the plants in low concentrations and these hormones move from the site of production to the site of action. Therefore, the difference between the plant regulator and plant hormone is in that the former one is synthetic and the latter one is natural from the plant source.

Plant growth substances have key role in different physiological processes related to growth and development of crops. It is obvious that changes in the level of endogenous hormones due to biotic and abiotic stress alter the crop growth and any sort of manipulation including exogenous application of growth substances would help for yield improvement or at least sustenance of the crop. The principles of Conservative Agriculture can be well protected by judicious use of plant growth hormones. Plant growth hormones are organic substances produced naturally in the higher plants, controlling growth or other physiological functions at a site remote from its place of production, and active in minute amounts.

Traditionally five major classes of plant hormones are listed: auxins, cytokinins, gibberellins, abscisic acid and ethylene. However as research progresses, more active molecules are being found and new families of regulators are emerging; one example being polyamines such as putrescine or spermidine. This classification is based partially on the chemical structure and partially on the commonalities of plant physiological effects that certain substances exhibit. Members of one class may not relate from a structural point of view to another.

- ☆ The general functions of different phytohormones are as below.
- ☆ Auxins:Shoot elongation Thin tree fruit, increase rooting and flower formation
- ☆ Gibberellins: Stimulate cell division and elongation Increase stalk length, increase flower and fruit size
- ☆ Cytokinins:Stimulate cell division Prolong storage life of flowers and vegetables and stimulate bud initiation and root growth
- ☆ Ethylene generators:Ripening Induce uniform ripening in fruit and vegetables
- ☆ Growth inhibitors: Stops growth Promote flower production by shortening internodes
- ☆ Growth retardants:Slows growth Retard tobacco sucker growth

Use of Commercial Plant Growth Regulators (PGRs) to

- ☆ Prevent flower abscission.
- ☆ To reduce fruit abscission during fruit set, June drop and pre-harvest drop for late hanging fruit.

- ✰ To increase fruit size.
- ✰ To regulate growth of vegetative shoots and indeterminate floral shoots

PGRs are being used in since long time. However, a short list on usage is narrated below in some important crops.

Aonla

The use of synthetic auxins (naphthalene acetic acid; NAA) and gibberellins (GA3) in combination with thiourea during mid-may and mid-July may provide an effective solution to minimize yield losses in aonla caused by heavy fruit drop. An increase in fruit quality may be the additional gain with the use of these plant growth regulators. Such a strategy may be of particular importance in sodic soils characterized by production constraints such as limited availability of different mineral nutrients for optimum tree growth and yield. (Anshuman and Singh 2015).

The treatment of seeds with GA_3 500 ppm and thiourea were effective in breaking seed coat dormancy in Aonla. (Lilabati and Sahoo 2015).

The soaking of the seeds in GA3 (500 and 750 ppm) for 24 hrs, combined with *Azospirillum* bio-fertilizer was beneficial in increasing seed germination and seedling growth of the aonla. (Supe *et al.*, 2012).

The spray of NAA at 10 ppm was the best to increase fruit retention which consequently resulted in the highest fruit yield. Fruit quality with regard to TSS, total sugar and ascorbic acid content was better in NAA at 10 ppm. (Ghosh *et al.*, 2009).

Pomegranate

Three sprays starting at 50 per cent flowering stage and subsequently at 21 days interval of NAA at 25 ppm gave significantly high fruit set and fruit retention which resulted in highest fruit yield. Fruit weight and quality improved significantly due to growth regulator sprays. (Ghosh *et al.*, 2016).

The importance of synthetic plant growth regulators in achieving higher yield and better quality of fruits.

Application of ethrel 200 ppm has improved the TSS, reducing sugar, non-reducing sugars, total sugars and ascorbic acid content. (Goswami *et al.*, 2013).

The application of 2,4-D at 40 ppm gave significantly high fruit size in length, breadth and volume and higher fruit weight, higher aril percent, maximum number of fruits which resulted in highest fruit yield of Pomegranate. (Reddy and Prasad 2012).

The extent of fruit cracking was reduced significantly with application of 300 ppm pacloputrazol, while grain per cent, fruit juice per cent, TSS per cent and acidity per cent were increased. Yield and fruit weight were increased with applying pacloputrazol.

Application of 40 ppm NAA and 80 ppm GA3 significantly increased fruit length and fruit diameter. Application of 80 ppm GA3 significantly increased total anthocyanin per cent, while tannins per cent was not affected by other two growth regulators (Khalil and Aly 2013).

In cv. Ganesh. NAA (1500ppm) and Ethrel (500ppm) recorded higher thining percentage on percentage of flower thinning as compared to control was in the ratio of 1:1 to 1:5, the NAA spray caused thinning in the ratio of 1:2 to 1:3, whereas ethrel spray caused thinning in the ratio of 1:3 to 1:5, the hand thinning resulted in the ratio of 1:2 to 1:2.5. For ethrel at 250 ppm, the thinning was in order of 1:3. Thus, for one or the other purpose, pomegranate flower thinning by plant growth regulators has effectively been used. The highest percentage increase in fruit yield was of the order of 40.5 per cent over control which was noted for ethrel at 250 ppm, followed by a 15.24 per cent rise in fruit yield for the water spray. (Sheikh 2015).

In cv. G-137, increased fruit size, fruit weight and juice contents were noticed with the treatment of CPPU at 5 ppm. However, total soluble solids, sugar contents, titratable acidity, TSS:acid ratio were significantly increased following the applications of GA3 at 50 ppm and NAA at 20 ppm. CPPU applied at 5 ppm and GA3 at 50 ppm caused a significant reduction in russet formation on fruits. (Sharma and Belasre,2011).

Application of 300 ppm pacloputrazol, could reduced fruit cracking could be reduced significantly and could increase grain per cent, fruit juice per cent, TSS per cent and acidity per cent. Yield and fruit weight were increased with applying pacloputrazol or by using (3 per cent Ca + 0.3 per cent B + 0.3 per cent Zn). Application of 40 ppm NAA and 80 ppm GA3 significantly increased fruit length and fruit diameter. Application of 80 ppm GA3significantly increased total anthocyanin per cent, while tannins per cent was not affected by pacloputrazol and NAA.

In an experiment on cv. Bhagawa with NAA (40, 50, 60 ppm) and GA (40, 50, 60 ppm) it is revealed that NAA 40 ppm was found effective in increasing number of fruits per tree,fruit length, fruit diameter, fruit weight and fruit volume. (Vidya *et al.*, 2016).

Ber

The Maximum fruit retention and minimum fruit drop was recorded in the trees sprayed with 1.5 per cent potassium sulphate and 20 ppm NAA at fruit set stage. Higher fruit size in terms of length and breadth, weight and yield was recorded with 60 ppm NAA sprayed at active growth phase (Singh and Bal 2006).

In a study on a 6-year-old Banarasi Karka cultivar of ber with seven treatments of two growth regulators *viz.*, NAA at 25, 50 and 100 mg/L; GA at 10, 20 and 40 mg/L and control (water spray), it is revealed that application of NAA at 25 mg/L gave significantly highest fruit retention (75 per cent) which resulted in highest fruit yield of 120.5 quintals as against 64.7 quintals/ha in control. No beneficial effect of GA on fruit retention or improving fruit yield was observed. But, fruit quality was significantly improved.(Ghosh *et al*, 2009).

Custard Apple

Significant differences with respect to effect of seed soaking in 50 ppm GA3 for 48 hrs on days required for germination, germination percentage, seedling height, stem diameter and number of leaves per seedling of custard apple were observed. This improvement in seed germination and seedling growth could be due to activation of dormant embryo of seeds with gibberellic acid and also gibberellic acid treatment helps to increase cell division, cell elongation and cell multiplication which might have reflected into maximum seedling growth (Archana *et al.*, 2015).

A trial was conducted to study the influence of wheat straw mulch and different plant growth regulators on fruit set, yield and quality of custard apple. Maximum flowering duration and fruit retention was observed with wheat straw mulch + GA 3 (5 t/ha + 50 ppm) treatments. Highest number of fruits, fruit yield, fruit diameter and fruit pulp were also recorded under same treatments. Wheat straw mulch gave 21 per cent higher fruit yield of custard apple. However, application of 20 ppm NAA was at par with GA 3 (50 ppm). On economic basis, 50 ppm GA 3 + wheat straw mulch followed by 20 ppm NAA + wheat straw mulch gave highest net income. (Patel, N.M., *et al.*, 2010).

Use of PGRs in Custard Apple

1. For defoliation and beginning the plants under the rest Ethryl at 1000 ppm is sprayed one month after the harvest of the fruits.
2. 20 ppm NAA is sprayed during the flowering period to improve the fruit set.
3. During the fruit development 50 ppm GA+ 0.5 ppm cppu, foliar spray improve the fruit size and luster of the fruits.

Phalsa

The application of NAA 200 ppm recorded maximum height of bush and length of shoot An application of NAA 150 ppm, increased number of flowers per shoot, number of fruits per shoot, fruits weight, juice percentage and minimum seed percentage and the maximum yield per plant. The quality of fruits in terms of total soluble solids, reducing sugar, and total sugar were significantly higher in treatment ethrel 1000 ppm. Further, ethrel 1000 ppm also significantly reduced the span of harvesting and number of picking. An application of GA3 150 ppm significantly reduced acidity and increased ascorbic acid content. (Kacha *et al.*, 2012).

Fig

The application of GA3 increased the leaf area, total chlorophyll and percentage of carbohydrate, calcium pictate and firmness. Gibberellins also reduced in the percentage of longitudinal cracking (Abass *et al.*, 2011).

Bael

The role of plant growth regulators on fruit drop in bael consists spray of NAA, GA_3 and etherl. All the growth substances sprayed, proved beneficial in minimizing drop and enhancing quality characters of bael fruits. Maximum fruit set was recorded with NAA 30 ppm, while minimum fruit drop and maximum fruit retention was with NAA 20 ppm. Highest fruit length and breadth was recorded with NAA 30 ppm. The spray of NAA 20 ppm estimated highest fruit weight, volume, pulp content and ascorbic acid but least acidity. Maximum TSS was recorded with GA_3 50ppm (Uniyal and Mishra 2016).

Nodal explants ware transferred MS medium with 0.5 mg BA/l with different concentrations of either kinetin (KN) or gibberellic acid (GA_3) or in combinations has shown healthy shoots with expanded shoot length. Excised shoots (2 cm - 3 cm long with 2 to 3 nodes) when grown on 1/2 MS basal medium with 2.5 mg/l indole-3-butyric acid (IBA) and 0.5 per cent activated char- coal (A.C.)/l has shown rhizogenesis. After excision, in the second passage, the nodal explants also showed bud break when sub cultured on MS basal medium supplemented with 0.5 mg BA/l. These shoots also successfully rooted on the same above said medium.(Puspashree and Shiba,2012).

Jamun

In a study to asses the effect of postharvest treatments on quality of Jamun, fruits were treated with growth regulators GA3 (50 and 100 ppm), chemical CaCl2 (1.0 and 1.5 per cent) and Paraffin wax along with control and kept in with or without perforated polyethylene bag. The treatment $CaCl_2$ 1.5 per cent with perforated polyethylene bag proved to be the best postharvest treatment than the rest of the treatments. The treatment effectively reduced the physiological loss in weight as well as spoilage loss and thereby useful in maintaining good balance between ascorbic acid and sugar content of fruits during storage. The treatment also showed little change in TSS, pH and acidity content and hence, it can be useful in postharvest management of Jamun fruits.(Ayer *et al*,2011).

Jamun (*Syzygium cuminii* Skeels) is an important indigenous minor fruit of commercial value. Foliar application of growth regulators and macro nutrients are known to enhance the growth of seedlings of various fruit crops. The foliar application of GA3 at 300 ppm found to be beneficial to enhance the seedling height (57.66 cm), diameter (6.06 mm), fresh weight (94.66 g) and dry weight (35.86 g) and to reduce the number of days taken to reach graftable size (132.72 days) compared to control (Surakshitha *et al.*, 2014).

References

Anshuman, S. and singh, H. K. 2015. Application of Plant Growth Regulators to improve Fruit Yield and Quality in Indian Gooseberry (*Emblica officinalis* Gaertn.) Journal of Agri Search 2(1): 20-23.

Archana Jadhav, Bhagure and Rajeswari Raundal. 2015. Effect of PGR, chemicals and plant extract on seed germination and seedling growth of custard apple (Annona squamosa). *THE ASIAN JOURNAL OF HORTICULTURE.* Volume 10 Issue 1.184-186.

Ayar, Kiran, Singh, Sanjay and Chavda, J.C. (2011). Effect of postharvest treatments on quality of Jamun (Syzygium cuminii Skeels) fruits during storage, Asian J. Hort., 6 (3): 297- 299.

Ghosh, S.N., B. Bera1, S. Royl, A. Kundu1 and S.K. Dutta Roy. 2009. Effect of nutrients and plant growth regulators on fruit retention, yield and physico-chemical characteristics in aonla cv. NA-10. *J. Hortl. Sci. Vol.* 4 (2): 164-166, 2009.

Kacha, H.L. R.R.Viradia, H.N. leua, Giriraj Jat and A.K.Tank. 2012. Effect of NAA, GA 3 and ethrel on yield and quality of phalsa (Grewia asiatica L.) under South-Saurashtra condition. *The Asian Journal of Horticulture* Volume 7 Issue 2 242-245.

Khalil, Hoda A and Hoda S.H. Aly. 2013. Cracking and Fruit Quality of Pomegranate (Punica granatum L.) As Affected by Pre-Harvest Sprays of Some Growth Regulators and Mineral Nutrients. Journal of Horticultural Science and Ornamental Plants 5 (2): 71-76.

Laishram Lilabati and U. K.Sahoo. 2015. Effect of pre-treatments on seed germination and seedling vigor of emblicaofficinalisgaertn *Global Journal of Advanced Research.*Vol-2, Issue-10 PP. 1520-1526.

M.K. Sheikh 2015 Effect of Growth Regulators and Hand Thinning of Flowers/ Fruits on Size of Pomegranate (Punica granatum L. 'Ganesh') Proc. IIIrd IS on Pomegranate and Minor Mediterranean Fruits Eds.: Zhaohe Yuan *et al.,* Acta Hort. 1089, ISHS 2015.

N. Sharma and C. Belsare 2011.Effect of Plant Bio-Regulators and Nutrients on Fruit Cracking and Quality in Pomegranate (Punica granatum L.) 'G-137' in Himachal Pradesh. Proc. IInd IS on Pomegranate and Minor, including Mediterranean Fruits (ISPMMF - 2009) Eds.: M.K. Sheikh *et al.,* Acta Hort. 890, ISHS 2011.

Patel, N.M., Patel, D.K., Verma, L.R. and Patel, M.M. (2010). Effect of cultural and chemical treatments on fruit set and fruit yield of custard apple (Annona sqamosa Lin.) cv. Sindhan, Asian J. Hort., 5 (2): 498-502.

P. Adi reddya and D. Manohar Prasad.2012. Effect of plant growth regulators on fruit characters and yield of pomegranate (punica granatum l.) Cv.ganesh. International J. of Plant, Animal and Environmental Science. Vol. 2 Issue 2 P-91.

Puspashree Puhan, Shiba Prasad Rath 2012 In vitro propagation of Aegle marmelos (L.) corr., a medicinal plant through axillary bud multiplication.Advances in Bioscience and Biotechnology, 2012, 3, 121-125.

S N Ghosh; B Bera; A Kundu; S Roy 2009. Effect of Plant Growth Regulators on Fruit Retention, Yield and Physico-chemical Characteristics of Fruits in Ber 'Banarasi Karka' Grown in Close Spacing Acta Horticulturae; No.840, Aug, 2009.

Singh, Chaturjeet and J.S. Bal. Effect of nutrients and growth regulators on fruit drop, size and yield of ber (*Zizyphus mauritiana* Lamk.). Internat. J. agric. Sci. Vol.2 No.2. (358-360).

Supe, V.S., Dhanshree Patil, Anju A. Bhagat and Rashmi S. Bhoge. 2012. Seed germination and seedling growth in aonla (*Emblica Officinalis* Gaertn.) *Bioinfolet* 9 (2): 206 - 208, 2012.

Surakshitha, N. C. Dr. Mahadevamma, M. Sharath Kumar, M. 2014 Studies on Application of Plant Growth Regulators And Macro Nutrients on Seedling Growth in Jamun [Syzygium cuminii (L.) Skeels] Indian Journal oOf Applied Research, 4 (12).

Vidya V. Anawal, Narayanaswamy, P. and Suresh. D. Ekabote, 2016. "Effects of plant growth regulators on fruit set and yield of pomegranate International Journal of Current Research, 8, (04).

Chapter 13

Integrated Pest Management of Arid Zone Fruit Crops

A.R. Walunj

Scientist-I (ENT), AICRP on AZF,
Department of Horticulture,
Mahatma Phule Krishi Vidyapeeth, Rahuri – 413 722, Dist. Ahmednagar, M.S.

IPM is holistic approach of crop protection based on the integration of multiple strategies *viz.*, cultural, physical, mechanical, biological, botanicals and chemical. The concept of pest management grew out of the discontent with the purely insecticidal approach to pest control which results in pest resistance, resurgence and residues in fresh fruits. Over the years IPM undergone several changes shifting its focus from damage boundary, economic injury to economic threshold. Currently most stake holders dependent on ETL and tends to apply chemical at the first instance in the event of pest attack. There are certain foods that you should try to buy organic as much as possible. These foods are those fruits and vegetables whose conventionally grown "alternatives" have been found to contain high levels of pesticide residues. This suggests that implementing IPM strategies specifically targeted at avoiding pesticide residues, such as consuming organically-grown produce or avoiding produce known to have large amounts of residues by following postharvest interval and use of eco friendly pesticides.

The area under arid zone fruit crops *viz.* pomegranate, ber, aonla, fig and custard apple is increased due to the their limited requirement of water and sustainability in light–medium soil and government emphasis under dryland horticulture with employment guarantee scheme since last three decades. The rapid increase in area under these fruit crops in the country has resulted in horticulture revolution in the country. Due to change in climate and intensive cultivation of some fruit crops resulted into some extent of problem especially with pomegranate pest and disease throughout the country.

Among these fruit crop, pomegranate has become one of the important fruit crops in the state of Maharashtra. The area under this fruit crop which was barely few hundred acres in 1960-69 has shot up roughly to 1,31,000 ha. in 2014-15. contributing 70 per cent of the total area under pomegranate. Such important fruit crop is attacked by several insect and non-insect pests as well as diseases.

Looking to the constraints in arid fruit crops especially with pomegranate cultivation or management, knowledge of seasonal abundance of pests and various management strategies is must so as to adopt the integrated management effectively. Among the pest complex thrips, fruit borer, fruit sucking moth and root knot nematode are the major threats to to attack pomegranate. Whereas, the fruit borer, pinhole borer, stone weevil, fruit fly, leaf eating caterpillars, chafer beetle, mealy bugs *etc.* are predominately occurs under favorable condition under dryland horticulture fruit crops like ber, custard apple, jamun and aonla.

Efforts have been made to review the information with key pest of some important fruit crop *viz.*(Pomegranate, aonla, custard apple, fig and ber) with recent research based control strategies.

1. Pomegranate

I. Knowledge of Key Pest and their Seasonal Abundance

a. Pests of National Significance

Insect Pests

1. Anar butterfly: *Deudorix isocrates* Fabricus (Lepidoptera: Lycanidae)
2. Stem borer: *Coelosterna spinator* Febricius (Coleoptera: Cerambycidae)
3. Whitefly: *Siphorinus phillyreae* Haliday (Hemiptera: Aleyrodidae)
4. Shot hole borer: *Xyleborus perforans* Wollastan (Coleoptera: Scolytidae)
5. Thrips: *Scirtothrips dorsalis* Hood or *Rhiphiphorothrips cruentatus* Hood (Thysanoptera: Thripidae)
6. Fruit borers: *Conogethes punctiferalis,* (Lepidoptera: Pyralidae)
7. Root knot nematode: *M. incognita*

b. Pests of Regional Significance

Insect Pests

1. Pomegranate aphid: *Aphis punicae* Passerini (Homoptera: Aphididae)
2. Mealy bugs: *Ferrisia virgata* Cockerell (Homoptera: Coccidae)
3. Fruit borer: *Deudorix isocrates* Fabricus and *Spodoptera litura* (Lepidoptera: Noctuidae)
4. Fruit sucking moth: *Otheris fullonica* and *Eudocima* spp. (Lepidoptera: Noctuidae)

5. Scale insect: *parasaissetia nigra*
6. Animal and rodent pests: In some areas, rodents damage the pomegranate crop by making burrows in the pomegranate fields. Sometimes, monkeys, animals also destroy the pomegranate cultivation by grazing or trampling over it.

c. Predator and Parasite (Natural enemies)

1. Coccinelids
2. Predatory mites
3. Predatory birds

d. Key Pest and their Abundance

1. **Aphids** (*Aphis punicae* passerine) More incidence occurs during December to February.
2. **Thrips**: Maximum population build up is found in mrig bahar and severity in July to August whereas, incidence in ambia bahar was noticed in Feb to March.
3. **Whitefly**: More incidence of whitefly occurs during Ambia bahar followed by Hasta bahar and mrig bahar. The whitefly, mealy bugs and scale insect are found abundant during Ambia bahar having significant positive correlation with temp and – ve correlation with humidity.
4. **Shot Hole borer:** Maximum pest intensity found in rainy season because of the congenial condition available for the growth of ambrosia fungus (*Monocrosporus ambrosium*) which is food source for the grub of this pest inside tunnels.
5. **Fruit borer**: Maximum bulid up of this pest has been reported during Mrig bahar (June –October).while the ambia bahar (Feb-June) the infestation is least. The incidence is more when the temp is moderate coupled with humidity.
6. **Stem borer:** Since there is one generation of pest in a year the maximum damage is found in between June to April.
7. **Fruit sucking moth:** The incidence of the FSM is generally maximum August to October and peak period of damage is between 8 to 11 PM in the night. However, In Rajasthan two overlapping generation are observed between july and sept.
8. **Root knot nematode** (*M. incognita*): Maximum activity July to Jan.

e. AESA Methodology

Go to the orchard in groups. Walk across the orchard and choose 10 sites/acre randomly. Observe keenly each of these plants and record your observations: Tree: 5-6 samples per tree (fruits/leaves/inflorescence/stem bark/roots/soil/insects, host plants) should be collected where, one sample from top, four samples from all the four sides (north, south, east, west) and one from bottom/soil, depending upon the requirement of sturdy/observations and if necessary. Pests: Observe and count pests at different places on the tree. Defenders (natural enemies): Observe and count parasitoids and predators. Diseases: Observe leaves and stems and identify any visible disease symptoms and severity.

f. Components of IPM/Integration of Tactics

Cultural, Physical, mechanical methods, Crop diversity and natural enemies of pests, Organic farming, Use of resistant varieties, Biological control of insect, Microbial control, Botanical pesticides and synthetic pesticide-IGR, low use rate outstanding molecules *etc.*

g. Requisites for Decision Making System in IPM

1. Identification of pest and predator parasites
2. Life History and behavior of pest
3. Natural regulation factors
4. Need for control measures
5. Timing of control measure
6. Selection of control measure on the basis of decision making system

(i) FFS to Teach AESA Based IPM Skills

AESA based IPM training for farmers Participatory Practical Regular meetings Learning through field experiments, Problem oriented Learning about crop ecology, Understanding role of beneficial insects

(ii) Pre Management Strategies

1. Avoid planting in heavy soil to avoid wilt
2. Collection and destruction of infested fruit of previous season.
3. To follow recommended plant spacing. Avoid close planting or crowding.
4. Avoid Mrig bahar as per availability of irrigation facilities
5. Follow the crop rotation and avoid cucurbits cultivation
6. Growing/planting of *Agetus* spp.in between the plants.
7. Use of biopesticide, neem based pesticide or 5 per cent Neem seed extract on incidence of pests.
8. Follow the resting period of three months.

A. At the Time of Planting

Pest	*Measures*
Termites White grub	Apply by dusting over pit and mix the 3 per cent Methyl parathion dust with FYM by through mixing around the plant
Aphids/whitefly and Mites	Need base spray of dimethoate @ 1.0 ml/lit water to new seedlings

B. Before Bahar Initiation: At the Resting Period

- ☆ Resting of orchard for 60-70 days after harvest with application of FYM at 25 ton/ha with e *Trichoderma* @ 25 kg/ha. and

Planting of Tagetus Species between Plants

- ☆ Cleaning: Collection and burning of dried/infested fruits to kill overwintering stages of pests. Spraying of Bactinashak @ 250 and 500 ppm with captan 50WP @ 2.5 g/lit as preventative measure for Oily spot on pomegranate
- ☆ Application of recommended dose of N.P.K. (650:250:250) and trace element (Ca,Mg,B,Znso4)
- ☆ Paste application to the stem for controlling hole borer, stem borer and mealy bugs. (Geru 4 kg + 30 ml chloropyrifos + 50 g copper oxychloride/10lit water) (Anon., 1994)

C. After Bahar Initiation: At the Vegetative and Fruiting

1. Split application of Phule *Trichoderma* Plus @ 17g/plant at bahar and 17 g/plant after 90 days after bahar is recommended for the control of root-knot nematode.(Walunj,A.R. and N.L.Mhase, 2015., Anon., 2016)
2. Need base one or two insecticidal application of Imidacloprid 17.8 SL (0.3 ml/lit) Azadiraction 10000 PPM @ 1.5/lit and cyantraniliprole 10.26 per cent OD (1-2 ml/lit) upto flowering stage for the control of aphids thrips, mites and mealy bugs. (Walunj, *et al.*, Anon., 2009)
3. Need base one or two insecticidal application of spinosad 2.5 EC (1.0 ml/ lit followed by neem based (azadirachtin10000 ppm@ 2.0ml/lit), Neem oil, karanj oil each @3 ml/lit up to 30-90 days after flowering stage for the control of thrips.(Anon., 2009)
4. Two spraying of bioagent *Beaveria bassiana, and Metaryza anisopolae* during fruit formatting stage for the control of *Deudorix isocrates* @ 5 g/lit of water at an interval week.
5. Need base application of Emamectin benzoate @ 0.5g/lit, neem seed kernel extract (NSKE) 5 per cent for the control of fruit borer and other pests.
6. One spraying of *Beaveria bassiana* or *verticillium lecani* @ 6 g/lit and acetimipride 0.3 g/lit at an interval of 10 days for the control of mealy bugs/white files. (Anon., 2009)
7. Use of yellow sticky cards (flyfix) for the control of white files.
8. Adjustment of early ambia bahar so as to escape the peak period of fruit seedling moths in September month.
9. Use of repellents like citronella oil at border line of orchards
10. Moth catches during night time (8.00 to 11.00 pm) for the control of fruit sucking moth
11. Use of cover bags as barriers to fruit in specific time of incidence.

Anola

The major bottleneck in its production is the damage caused by various insect pest like fruit borer, shoot gall maker, bark eating caterpillar, aphids, scale insect, mealy bugs, leaf folder, termites and fruit midges. Among these fruit borer thrips and bark eating caterpillar are common occurrence of anola.

Management of Aonla Pests

- ☆ Avoid planting near pomegranate and ber orchard
- ☆ Collection and destruction of infested fruit regularly
- ☆ Keep orchards clean and avoid overcrowding of trees.
- ☆ Insert the Iron spike in larval tunnel to kill the caterpillar
- ☆ Pruning of galled shoots
- ☆ Spraying of systemic insecticide during rainy season
- ☆ The larval parasitoid *Nemertis tectona* Perkins have been reported to be an important natural enemy of the bark eating caterpillar, causing 33.8-45 per cent damage larval parasitisation.
- ☆ Around 100 per cent mortality of the larvae could be obtained by spraying the entomo pathogenic fungus *B. bassiana* at 3.6 X 10^{-4}
- ☆ Raking of soil underneath the trees during summer to destroy the pupae of fruit borer
- ☆ Spot application on infested bark with insecticide like carbaryl 50WDP at 1.0 per cent or Dichlorovos 0.08 per cent or Monocrotophos 0.08 per cent effectively control the pest (Shewale, 2002)
- ☆ Application of fenvalerate 0.01 per cent or quinolphos 0.08 per cent or Monocrotophos 0.1 per cent or Kerosene or petrol @ 5 ml in each larval tunnel through syringe or wash bottle or cotton wool and sealing the holes with mud.
- ☆ The spraying of imidacloprid 17.5 per cent SL @0.5 ml/lit for control of gall maker and 1.0 ml spinosad in 10 lit of water are most effective for the control of *M. scyrodes* (Anon., 2014)

Custard Apple

Mealy Bugs *Ferrisia virigata* (ckll) and *M. hirsutus*

Management

- ☆ Avoid over crowding of the trees in orchard
- ☆ Remove and destruct of affected shoots and fruits
- ☆ Use of natural enemies including parasitoids like *Anagyrus dacylopii* and predators like *cryptolaemus montrouzieri* Muls
- ☆ Use of *Verticillium lecanii* @60 g/10 lit for spraying. (Anon., 2010)
- ☆ Use neem based pesticide or 5 per cent Neem seed extract for spraying.

Fig

1. Fig stem borer: *Batocera rufamaculata*
2. Fruit fly: *Dacous* spp.
3. Fig jassids: *velu caricae*
4. Thrips: *Scirtothrips dorsalis* Hood

Management

- ☆ Remove and destruct of affected shoots and fruits. Spraying of dimehtoate or chlorpyrifos @15 ml/10 lit of water at the time of rainy season @ an interval of 15 days. Or Use neem based pesticide or 5 per cent Neem seed extract for spraying. To avoid fruit fly infestation, fruits should be harvested when these are still firm/unripe.
- ☆ Fallen infested fruits should be collected and destroyed
- ☆ Insert the Iron spike in larval tunnel to kill the caterpillar
- ☆ Application of DDVP or fenvalerate or quinolphos or or Kerosene or petrol @ 5 ml in each larval tunnel through syringe or wash bottle or cotton wool and sealing the holes with mud.
- ☆ Thrips can be controlled by spraying with cypermethrin 10 EC @ 1.0 ml/lit
- ☆ Use of fruit fly: methyl eugenol traps for monitoring and trapping of fruit flies.
- ☆ Two Sprays of thiodicarb 75 WP @1.5g/lit water is recommended fortnightly interval on onset of monsoon for the control of stem borer (Ann., 2012).

Ber

A. Foliage Feeder

1. Cockchafer beetle: *Holotrichia* spp
2. Leaf eating caterpillar: *Thiacidas postica*

B. Borers

1. Ber fruit borer: *Meridarchis scyrodes*
2. Stone weevil: *Aubeus himalayanus*
3. Fruit fly: *Carpomyia vesuviana*

Pest Management

1. Wild varieties should not be allowed to grow near about the ber orchard.
2. Deep raking of soil underneath trees during summer will help in pupal mortality
3. Light traps are also effective for trapping the beetles (Shewale, 2002)

4. Collection and destruction of infested fruits should be a regular feature. Similarly, the shed fruits after the harvest should be destroyed.
5. Application of fenvalerate @1.0 ml or quinalphos 1.5 ml or kerosene or petrol 5 ml in each larval tunnel/hole through syringe or wash bottle or cotton wool and sealing the hole with mud/sticky substance. (Shewale, 2002)
6. Off-season fruits of the various species of ber proved to be an important link in carrying over its population to the main crop-season and hence off-season fruits should not be allowed in or near the ber orcharad.
7. Deep raking of soil underneath trees during summer.
8. Collection and destruction of infested/shed fruits from time to time.
9. Spraying (rotationally) with Fenvalerate, Deltamethrin 1.0 ml, Quinalphos 1.5 ml and Carbaryl 50 WDP at.2 g/lit of water after fruit set at an interval of 30 days (Shevale and Padule, 1992;).
10. First sprays of spinosad 2.5 SC @ 1.0 ml/lit followed by second spray of indoxacarb @ 1.0 ml/lit should be given at 50 per cent flowering and 50 per cent fruit setting for the control of fruit borer and stone weevil. (Anon., 2015)

References

Anonymous, (1994). Research Review committee meeting in Agril Entomology recommendation as pasting for the control of shot hole borer in pomegranate.

Anonymous (2002). Management of Insect pests, disease and physiological disorders of fruit crops Proceeding book by CAS,(Fruit) pp 68-84

Anonymous, (2009). Recommendation report in Research Review Committee report of AICRP on arid zone fruit for the year 2008-09. pp. 1-26

Anonymous,(2012). Evaluation of some insecticide, neem and entomopathogenous fungi against stem borer on fig. Recommendation in Research Review Committee report of AICRP on arid zone fruit for the year 2012-13. pp. 13-16

Anonymous, (2014). Ann. Rept. NRCP, Solapur (M.S.) India, pp.3-4Anonymous,(2014). Chemical control of fruit borer and gall maker (midge) on anola. Recommendation report in X Research Workers Meeting on AICRP on Arid zone fruit at Jobner pp.2-2

Anonymous, (2016). Bio management of root knot nematode, *M. incognita* on pomegranate. Recommendation report in XX Research Workers Meeting on AICRP on Arid zone fruit at Bikaner. pp. 255-260

Anonymous, (2016).Chemical control of ber fruit borer and stone weevil. Recommendation report in XX Research Workers Meeting on AICRP on Arid zone fruit at Bikaner pp. 260-263

Shewale B.S. and Padule,D.N.(1992). Chemical control of ber fruit borer. Mah. J. Horti., 6 (1):17-19

Shewale B.S. (2002). Insect pest management in arid zone fruit crops, Proceeding book on Management of Insect pests, Disease and physiological disorders of fruit crops by CAS,(Fruit). pp. 76-84

Walunj A.R. Supe V. S. and V.R.Joshi (2012). A New Molecule, Cyzapyre10 per cent OD against thrips and whitefly in pomegranate AGRES-An International e-journal ISSSN-2277-9663 pp.84-86.

Walunj, A. R. and N.L.Mhase (2015). Bioagent's for biological management of root knot nematode infesting nematode. Bioinfolet,12 (1A): 30-34

Chapter 14

Disease Management in Important Dryland Fruit Crops

K.S. Raghuwanshi

Associate Professor,
Department of Plant Pathology and Agri. Microbiology
Mahatma Phule Krishi Vidyapeeth, Rahuri – 413 722, Dist. Ahmednagar, M.S.

Dryland fruit crops are mainly grown in tropical and sub-tropical areas where there are less conducive conditions for disease development. Still in such dry climate some diseases are severe on dryland fruit crops and cause heavy losses in fruit production. Therefore, it is necessary to have disease management in dryland fruit crops so as to reduce losses and have better quality of fruits.

Ber (*Zizyphus mauritiana* L.)

Powdery Mildew (*Oidium erysiphoides* f. *ziziphi*)

Symptoms

The disease appears only on fruits in the State of Maharashtra during early September when fruits are bajra grain size/or peanut size. Small whitish spots appear on young fruits which later enlarge and cover entire area of the fruit. These white spots turn brown to dark brown and become cankerous. In severe cases, the whole fruit surface gets covered with powdery masses. The affected fruit either drop off prematurely or become corky, cracked, mis-shapen and under developed. Sometimes whole crop is rendered unmarketable.

Control Measures

This disease is effectively controlled by 80 per cent wettable sulphur. Prophylactic spray (0.2 per cent) of this fungicide should be given at 50 per cent flowering stage and then subsequent four sprays of this fungicide be given at an interval of 20 days. Sulphur dust, Triadimefon (Bayleton) and Dinocap (Karathane) are also effective fungicides for controlling powdery mildew of ber.

Leaf Spots

Two types of leaf spots are mainly occurred in ber. Those are Alternaria leaf spot and Cercospora leaf spot.

Alternaria Leaf Spot (*Alternaria alternata*)

Symptoms

This disease appears with the formation of small irregular pale or dark brown spots on the upper side of the leaves. These spots coalesce to form elongated spot. Under heavy incidence of the disease leaves show blighting symptoms. The affected leaves drop off.

Cercospora Leaf Spot (*Cercospora* sp.)

Symptoms

This disease appears with small, round, grey spots with red border. The spot grow and become visible on the both sides of leaves. Leaves infected by the fungus become dry and fall off.

Control Measures

Both the leaf spots are effectively controlled by spraying Mancozeb (Dithane M-45 or Indofil M-45) @ 0.25 per cent or Zineb (Dithane Z-78) @ 0.25 per cent. First spray should be given immediately after the initiation of the disease. Then subsequent 2-3 sprays of the fungicides should be given at 15 days interval depending upon the disease intensity.

Sooty Mould (*Isariopsis indica*)

Symptoms

Blackish spots appear on the underside of the leaves which later cover the entire lower surface giving a sooty appearance. The leaves look yellow and brown on the upper surface and drop prematurely.

Control Measures

This disease can be effectively controlled by spraying Copper oxychloride @ 0.3 per cent or Mancozeb @ 0.25 per cent or Bordeaux mixture 2:2:50. First spray should be given immediately after the appearance of the disease and subsequent sprays, be given at 15 days interval if required.

Fruit Rot

This disease is caused by number of fungi such as *Phoma* sp., *Colletotrichum* sp., *Alternaria* sp., and *Pestalotia* sp. However, dry fruit rot caused by *Alternaria alternata* is occurred in the Maharashtra on local ber cultivars.

Symptoms

First infection occur at styler end and that portion become brown to dark brown in colour and pointed beak is formed at the styler end. If the infection is occurred at early stages of fruit development, the whole fruit is dried and fall prematurely.

Control Measures

This disease can be effectively controlled by Copper oxychloride @ 0.2 per cent or Mancozeb @ 0.25 per cent. First spray should be given immediately after the initiation of the disease. Subsequent sprays, if required, be given at 15 days interval.

Pomegranate (*Punica granatum* L.)

Up till now, 46 fungi and one bacterium have been recorded on pomegranate in India. These fungi are observed more or less in Mrig, Hasta and Ambia bahar crops of pomegranate. However, their intensity is more in Mrig bahar. Important diseases are as given below:

Leaf and Fruit Spots

Alternaria Leaf and Fruit Spots (*Alternaria alternata*)

Symptoms

Leaf spots caused by this fungus are small, light brown to purple in colour and roundish or irregular in shape. In the advancement of the disease, spots get coalesce and form big spots of dark brown colour. Affected leaves further become pale, yellow and drop down. Small brownish spots are observed on the fruits. They are roundish or irregular. In the advance stage of the disease the black spots are formed and rotting in the interior portion of the fruit begins.

Control Measures

Leaf and fruit spot can be controlled by spraying Mancozeb (Dithane M-45) @ 0.2 per cent or Zineb (Dithane Z-78) @ 0.2 per cent.

Colletotrichum Leaf and Fruit Spots (*Colletotrichum gloeosporioides*)

Symptoms

On leaves small purplish black or black spots are observed. Afterwards yellow hallows are observed around the spots. These spots coalesce to form big spot. Heavily infected leaves drop off. Immature and mature fruits of pomegranate are infected by this fungus. Initially brownish black or blackish, round or irregular deep situated spots are observed. Sometimes these spots form big black spot. If the infected fruits are split open, the rotting of interior portions of the fruit is observed. If the disease is in severe form, the infected fruits drop off. The fungus of this rot enters through wound or holes of the fruits. Infection may take place from stem end or styler end.

Control Measures

This disease can be controlled by Mancozeb (Dithane M-45) or Indofil M-45 (0.25 per cent).

Cercospora Leaf and Fruit Spots (*Cercospora punica*)

Symptoms

On the leaves, small roundish or irregular spots are formed. They are dark black in colour. These spots coalasce to form big spot. Heavily infected leaves drop off. On the fruits, round or irregular big spots having dark black colour are formed. These spots are measured from 1 to 1.5 cm in diameter. Under favourable conditions, large sizes spots are formed.

Control Measures

This disease can be effectively controlled by Copper oxychloride fungicide @ 0.25 per cent or Bordeaux mixture 1 per cent.

Drechslera Fruit Spots (*Drechslera rostrata*)

Symptoms

On fruits, small raised spots are formed. Initially, they are red in colour, afterwards they convert into dark brown in colour. If the infected fruit is taken in hand their presence is easily recognized.

Control Measures

This disease is controlled by Mancozeb (Dithane M-45 or Indofil M-45) @ 0.25 per cent.

Spraying schedule for control of leaf spots and fruit spots of pomegranate caused by *Alternaria, Colletotrichum, Cercospora* and *Drechslera* species of fungi. Leaf and fruit spots caused by these fungi are commonly observed on pomegranate fruit crop in the major pomegranate growing areas of the country. Spraying schedule developed at Mahatma Phule Krishi Vidyapeeth, includes first spray of any one of the fungicides *i.e.* Captan (Captaf) 0.2 per cent, Ziram (Ziride) 0.25 per cent, Mancozeb (Dithane M-45 or Indofil M-45) 0.25 per cent, Copper oxychloride (Blitox-50 or Fytolan @ 0.25 per cent), or Bordeaux mixture 0.4 per cent, be given at the flowering/fruit setting as prophylactic spray. Then, nine sprays of the same fungicide be given alternatively at 10 days interval in Mrig bahar. Similarly, seven sprays of the same fungicide be given at 15 days interval for Hast bahar cropping and five sprays of the same fungicide at 20 days interval for Ambia bahar cropping. For getting better control of diseases, there is a need to add or mix sticker (0.5 per cent) in spray solution in each spray. One fungicides should not spray more than 3-4 times during cropping period.

Fruit Rot (*Aspergillus niger*)

Symptoms

This fungus enters into the fruits through wounds caused by spines. Rotting of peel occurs due to fungal infection and extends it to the interior portion. Infected fruits give rotting or fermented smell. Infected fruits become yellow in colour, pale and gets flattened. In severe infection of disease, whole peel gets rotten and fungal growth is seen on the surface affected fruits drop off.

Control Measures

After fruit setting Carbendazim (Bavistin) @ 0.1 per cent) should be sprayed in which whole fruit surface should be covered. Two to three sprays are sufficient to control this disease.

Declining in Pomegranate

It is a new disease occurring mainly in heavy soils of pomegranate growing area of Maharashtra State. It is a complex disease probably caused by one or more reasons *i.e.* insects, pathogens or both namely shot hole borer, stem borer, root knot nematodes collar rot due to *Rhizoctonia* and root rot due to *Fusarium*. Several patterns of leaf symptoms, and tree deterioration have been first reported by Chaudhari and Desai (1993).

Symptoms

If the pomegranate is cultivated in heavy and ill drained soil, the moisture remains long time around stem and root zone. The main stem is supported by soil mound which favours the infection of *Rhizoctonia* at collar portion. The bark near collar region gets rotten and it leads to wilting from head to trunk.

Due to poor drainage around root zone, the infection of *Fusarium* and *Cerotocystis fimbriata* takes place through infected portion and root hairs and it leads to wilting of the plant. This fungus causes disturbances through xylem tissues which stops the flow of water and food to plant growth. It also produces toxin like fusaric acid. If the main stem is split the blackening of xylem and phloem tissues is seen.

Shot-hole borer, stem borer and root knot nematodes make the site of entries for infection of *Rhizoctonia ceratocystis* and *Fusarium* and also contributes to wilting.

Control Measures

1. Collar rot due to *Rhizoctonia*: First of all rotted bark should be removed by gunny cloth. Apply Bordeaux paste over the exposed area. Bordeaux paste should be prepared @ 1 kg CuSO4 + 1 kg lime + 10 litres of water, having PH 7.0.
2. Wilt due to *Fusarium*: Soil drenching of Carbendazim @ 0.2 per cent (20 g Carbendazim in 10 litres of water). Such type of mixtue @ 10 litres per tree should be drenched near the main stem. It should be followed by use of biological control *i.e.* use of *Trichoderma* one month after Carbendazim drenching. The *Trichoderma* powder @ 25 g + 5 kg well rotted FYM should be mixed thoroughly and such mixture is inoculated in soil. Drenching of Carbendazim followed by use of biological control twice in a season gives better results against pomegranate wilt.

Custard Apple (*Annona squamosa*)

Glomerella Fruit Rot or Anthracnose (*Glomerella cingulata*)

It has been reported from Assam and it is common in Udaipur. It is also observed in Maharashtra State.

Symptoms

The infection begins at blossom end in the form of blackish brown spot or purplish spot which spreads slowly in all the directions and generally covers the entire fruit, transforming into a shrivelled mass. Diseased fruits may cling to the tree or fall down.

Control Measures

Three fortnightly sprays starting from the initiation of symptom with Benomyl (Benlate) @ 0.05 per cent were found to be the best. Mancozeb (Dithane M-45) 0.2 per cent and Carbendazim (Bavistin) @ 0.05 per cent have also proved to be effective to control the disease.

Alternaria Leaf Spot (*Alternaria* sp.)

Symptoms

This disease appears in the form of brownish to dark brown or black spots. Heavy disease incidence results into dropping of the leaves. This disease appears in the month of November and becomes severe at the flag end of harvesting and that results in heavy losses in the production.

Control Measures

This disease can be effectively controlled by spraying of Mancozeb (Dithane M-45 or Indofil M-45) @ 0.25 per cent or Zineb (Dithane Z-78) @ 0.25 per cent. The spraying should be started immediately after the appearance of the disease.

Aonla (*Phyllanthus emblica* L.)

Aonla Rust (*Ravenelia emblicae*)

Symptoms

The leaflet show brown rust pustules. On the fruit black spots appear which sometimes cover the entire surface of the fruits. Affected fruits may drop before maturity.

Control Measures

The disease can be controlled by spraying the trees with Wettable Sulphur @ 0.2 per cent.

Blue Mould Rot (*Penicillium islandicum*)

Symptoms

The disease is characterized by brown patches and water soaked areas on the fruit. As the rot progresses, three distinct colour zones *i.e.* bright yellow, purple brown and bluish-green are seen. There is exudation of drops of yellowish liquid from the fruit. The fruit emits a bad odour. The whole fruit is covered with bluish green pustules.

Control Measures

The rot can be controlled by careful handling of fruits, control of storage hygiene and treatment of fruits with weak Borax @ 3.0 per cent or Sodium chloride solution @ 2.0 per cent.

Karonda (*Carica carandas*)

Anthracnose (*Colletotrichum inamdarri*)

Symptoms

In nature, the infection is occurs on leaves wounded by thorns. However, the fungus may infect in injured leaves under artificial inoculation.

Control Measures

Copper sulphate (0.1 per cent) is recommended for the control of this disease.

Bacterial Canker (*Xanthomonas compestris* pv. *carissa*)

Symptoms

First the spots appear on lower surface of the leaves. These spots look like small and round, water soaked, translucent dots gradually expanding and turning dark brown. A yellowish green halo may surround the spots. The water soaked spots on under surface of leaf enlarge and become raised. The upper leaf surface over the spot may be slightly raised.

Control Measures

The disease can be checked by removing all the diseased leaves from the shrubs pruned to the height of 45 cm from ground level. Spray with Phytomycin or Streptomycin (200 ppm) can also control the disease.

Bael (*Aegle marmelos* (L.) Corr.)

Bacterial Shot Hole and Fruit Canker (*Xanthomonas compestris* pv. *bilvae*)

Symptoms

The symptoms on leaves are characterized by round, water soaked spots (0.5 mm) surrounded by the clear halo. Gradually the spot increases in size (3 - 5 mm) and form brown lesions/with saucer like depression in the centre surrounded by oily and raised margin. The primary localized lesions all over the leaf are always followed by falling out the dead tissues leaving circular or slightly irregular shot holes. The pathogen can also infect the fruit, twigs and thorns.

Control Measures

Spraying with Streptocyclin (200 ppm) can control the disease.

Phalsa (*Grewia asiatica* L.)

Brown Leaf Spot (*Cercospora grewiae*)

Symptoms

The disease causes premature leaf fall, particularly during the rainy season. On the leaves of the affected plants tiny lesions appear on both the sides of the leaf, which afterward cover large area with white mass. The lesions rapidly enlarge and turn blackish in centre. Gradually the black growth becomes like cushion.

Control Measures

It can be controlled by spraying Aureofungin (30 ppm) and Benomyl (Benlate) (0.05 per cent).

Chapter 15

Importance of Pesticide Residues in Fruit Crops

C.S. Patil

Residue Analyst, Department of Entomology,
Mahatma Phule Krishi Vidyapeeth, Rahuri – 413 722, Dist. Ahmednagar, M.S.

India is the second largest producer of the fruits (10 per cent) in the world. India ranks first in the production of banana, mango, papaya, pomegranate, sapota, anola and guava. Productivity of grapes in India is the highest in the world. The horticulture production increased nine fold when compared with 1950-51.The consumption of fresh fruits and vegetables has increased substantially over the last two decades, so also the global trade in fresh fruits and vegetables. Recent reports of food borne illness associated with fresh fruits have raised concern from public health agencies and consumers about the safety of these products.

India as the largest producer of fruits, vegetables and processed products has a very great export potential for this sector. One of the key that determines sustained growth of this sector is maintenance of the quality of the product and increasing the shelf life till it reaches the consumer. India ranks 15th when it comes to exports of these products and share in world trade in fruits has remained dismal at 1.7 per cent.

The global market for horticulture exports is becoming increasingly quality conscious. This has led to the development of certain norms and practices that are mandatory for operating successfully in the global market. These specifications, in turn, lead to certain processes and procedures that need to be incorporated in the pre-harvest and postharvest activities when catering to the export market. These include setting up of pre-cooling facilities, cold storage facilities, undertaking tests for chemical residues, seeking certifications for good agri-practices *etc.*

Importing countries have become very rigid on this issue and do not accept the consignments. WTO has provided an International framework by adopting Sanitary and Phytosanitary Measures (SPS). Sanitary and Phytosanitary standards, as covered under the SPS agreement include health, hygiene standards or regulations to avoid the spread of animal and plant diseases and epidemics. These have been adopted by the Codex Alimentarious Commission (CAC) of Food and Agriculture Organization (F.A.O.) and World Health Organization (W.H.O.). WTO member countries are encouraged to use International standards by framing their own laws, regulations, *etc.* India being member of W.T.O, is also adopting these regulations. Therefore, strict monitoring from production to transport, processing, handling, packing, *etc.* is absolutely necessary encompassing the entire food chain.

Pesticides as Important Inputs

Occurrence of insect pests and diseases are the major constraints in the production of fruits. As a result of insect pests attack, the farmers lose both yield and quality of fruits. To control the damage caused by insect pests, substantial amount of pesticides are being used by the growers throughout the period of growth and sometimes even at the fruiting stage. Fruits receive considerably high quantity of pesticides, and with a cropped area of three per cent, they consume 13 per cent of the total pesticides in the country (Nigam and Murthy, 2000). Indiscriminate use of pesticides particularly at fruiting stage and non -adoption of safe waiting periods leads to accumulation of pesticide residues in consumable fruits.

Pesticides

Pesticide is any substance or a mixture of substances formulated to destroy or control any organism that competes with humans for food, destroys property, or spreads disease. "Pests" include weeds, microorganisms (*e.g.*, fungi or bacteria), insects, and even mammals.

In India, Registration Committee (RC) registers pesticides for their usage whereas, the Ministry of Health and Family Welfare (MOHFW) regulates the safety of food supply by setting tolerance limits, or Maximum Residue Limits (MRLs) (for pesticides on food commodities and in animal feed under Prevention of Food Adulteration Act, 1954 (PFA) The purpose of the tolerance program is to ensure that consumers are not exposed to unsafe levels of pesticide residues in food.

The registration of a specific pesticide for its commercial production and application requires a complete set of analytical data to prove not only its efficacy but also its safety. In general, the tolerance levels in foods range between 0.01 and 10 mg/kg, depending on the commodity and the pesticide used.

Pesticide Residues

Pesticide residue refers to the pesticides or it's residues that may remain on or in food after they are applied to food crops. All chemicals, when applied to crops, animals, soil or water, will leave residues. It also refers to pesticides or their breakdown products remaining in environment following their normal use or accidental contamination.

Pesticide residues may occur in food as a result of the direct application to a crop or a farm animal or as a postharvest treatment of food commodities. Additionally, pesticide residues can occur in food from environmental contamination and spray drift.

Maximum Residue Limit (MRL/Tolerance Limit)

It is the maximum concentration (in ppm) of pesticide residue that is permitted in or on food at a specified stage of harvesting, transport, marketing or preparation of food up to the final point of consumption. MRL makes it possible to ascertain legally whether a commodity containing a pesticide residue is safe and fit for consumption. Produce with residue above MRL is usually not accepted for export and its sale is illegal in domestic markets too. Each country sets its own MRLs and MRLs may vary from country to country. MRLs are the regulatory thresholds set within the context of an Acceptable Daily Intake (ADI) and are set taking into account the Good Agricultural Practice (GAP).

Good Agricultural Practice (GAP)

GAP definition of FAO addresses economic, environmental and social sustainability of agriculture and leads to food safety. GAP in the use of pesticides can be defined as the officially recommended or authorized, safe usage of pesticides under practical condition necessary for effective and reliable pest control. While considering the safe use, the public and occupational health and environmental safety are taken into account. The actual conditions include any stage of production, storage, transport, distribution, sale and use of pesticides *etc.* There is a responsible approach for worker health and safety.

Pre-Harvest Interval (PHI) or Withholding Period (WHP) or Waiting Period

Minimum number of days needed between the last application of a pesticide to a crop and harvest. The WHP is the minimum period which must elapse between last administration or application of a veterinary chemical product/pesticide and the slaughter, collection, harvesting or use of the animal commodity for human consumption.

In order to regulate pesticide residues, number of standards governing the production of fresh fruit, packing and shipping operations is currently being implemented globally (*e.g.* Eurep GAP, Global GAP), These standards primarily focus on food safety and environment protection through GAP. In India, APEDA has prepared India GAP guidelines which has been introduced Ministry of Agriculture. The Food Safety and Standards Bill, 2006 enacted by the parliament on 26.7.2006 is the harmonization of laws relating to food and food safety standards.

The Department Agriculture and Cooperation (DAC) Ministry of Agriculture, New Delhi has initiated a Central Sector Scheme on 'Monitoring of pesticide residues at National Level' in October, 2006 to evaluate the status of pesticide residue levels in fruit and vegetables, cereals, water and animal products. The AINP on Pesticide Residues, Department of Entomology, MPKV, Rahuri is one of the participating laboratories in the Scheme. The validated MRMs (CDFA, Modified Luke method, QuEChERS) are used in residue testing laboratory.

The use of pesticides has become indispensable to control insect pests on fruits. However in order to minimize risk from exposure of pesticides residues, farmers must observe waiting period before harvesting fruits.Pesticides should be used only when pest population crosses Economic threshold level (ETL).Thus reducing pesticide load IPM strategies involving need based application of pesticides should be develop for fruit crop.

Management of Pesticide/Pesticide Residues

1. Education and training in the proper use of pesticides have been very successful in maintaining the residues at minimal levels. The improper use or misuse through lack of understanding creates residue problems.
2. Implement integrated pest management (IPM) in which reduced use of pesticide chemicals is advocated.
3. Observe safe pre-harvest interval (PHI) so that pesticide residue fall within the maximum residue limit.
4. Food processing such as washing, trimming, peeling, juicing *etc.* usually results in large decrease in residues.
5. It is desirable to test the agricultural produce for pesticide residues from the APEDA recognized laboratory before harvest to confirm the residue levels.
6. Always use only recommended pesticides at the specified doses and frequency and at the right time. Never use banned pesticides.
7. Many new insecticides are effective at very low dose rates (10 to 50g a.i./ha) and have less harmful impact on beneficial insects and the environment. Therefore switch to newer pesticides.
8. Always read product labels carefully before applying any pesticide, mix and apply as directed. Do not overdose, do not treat too often. Remember that improper practices can lead to illegal residues even when proper pesticides are used.
9. Keep an inventory of all chemicals. Store all chemicals in their original containers. Never store herbicides with other pesticides.
10. Unused pesticide solution and washings generated by cleaning spray pumps contain pesticide residue. Dispose of them properly to avoid pollution.
11. Follow the correct handling, measuring, mixing and filling procedure as stated on label. Calibrate the sprayers/dusters annually to ensure the accuracy.
12. Maintain healthy soil with compost and mulch so that pest problems are minimal.
13. Select pest resistant or tolerant variety, clean-out diseased plants so disease doesn't spread.

14. Do everything culturally possible to eliminate environment congenial for pest/disease.
15. Reduce spray drift in orchards by using lower pressures, larger nozzles and less volatile pesticides and by spraying when there is little or no wind.

References

Agnihotri, N.P. (1999) Pesticides: Safety Evaluation and Monitoring 173 p

Mittal, S. (2007) Can Horticulture be a Success story for India. Working paper No. 197, ICRIER, New Delhi 70 p

Nigam, G.L. and Murthy, K.S. (2000). An optimum use of pesticides in integrated pest management technology. Pesticides Information 25(2) 145-148.

Sharma, K.K. (2007) Pesticide Residue Analysis Manual, ICAR, New Delhi 251 p

Chapter 16

Postharvest Management of Fruit Crops: A Metabolic and Molecular Perspective

R.M. Naik

Professor and Head,
Department of Biochemistry, PGI,
Mahatma Phule Krishi Vidyapeeth, Rahuri – 413 722, Dist. Ahmednagar, M.S.

Increased recognition of the dietary significance of fruits and vegetables in providing essential vitamins, minerals, fibre and other health-promoting compounds in the human diet, has made their postharvest biology and technology an area of considerable research interest. It is well known that ripening is an irreversible process that confers both positive and negative attributes to the resulting commodity. Although ripening imparts desirable flavour, colour and texture, considerable expense and crop loss result from negative ripening characteristics.

Fruit ripening is a complex developmental process which is orchestrated by the expression of ripening-related genes under the control of a network of signaling pathways. In climacteric fruit components responsible for the production of climacteric ethylene have been identified. Respiration is an essential component in the determination of postharvest shelf-life. The studies have served as abase for the definition of postharvest technologies devoted at reducing respiration such as cold storage, pre-refrigeration and modified, controlled and dynamic atmospheres. In recent years postharvest science has integrated a number of modern methods such as molecular biology, biotechnology and "omics" tools (metabolomics, proteomics and transcriptomics) in order to elucidate the mechanisms of fruit ripening during which fruit develop their sensorial and nutritional quality.

Postharvest losses are relatively higher 20-50 per cent, in developing countries when compared to 5-25 per cent in developed countries. Complete elimination of postharvest losses may be impossible and uneconomical, but to diminish them by

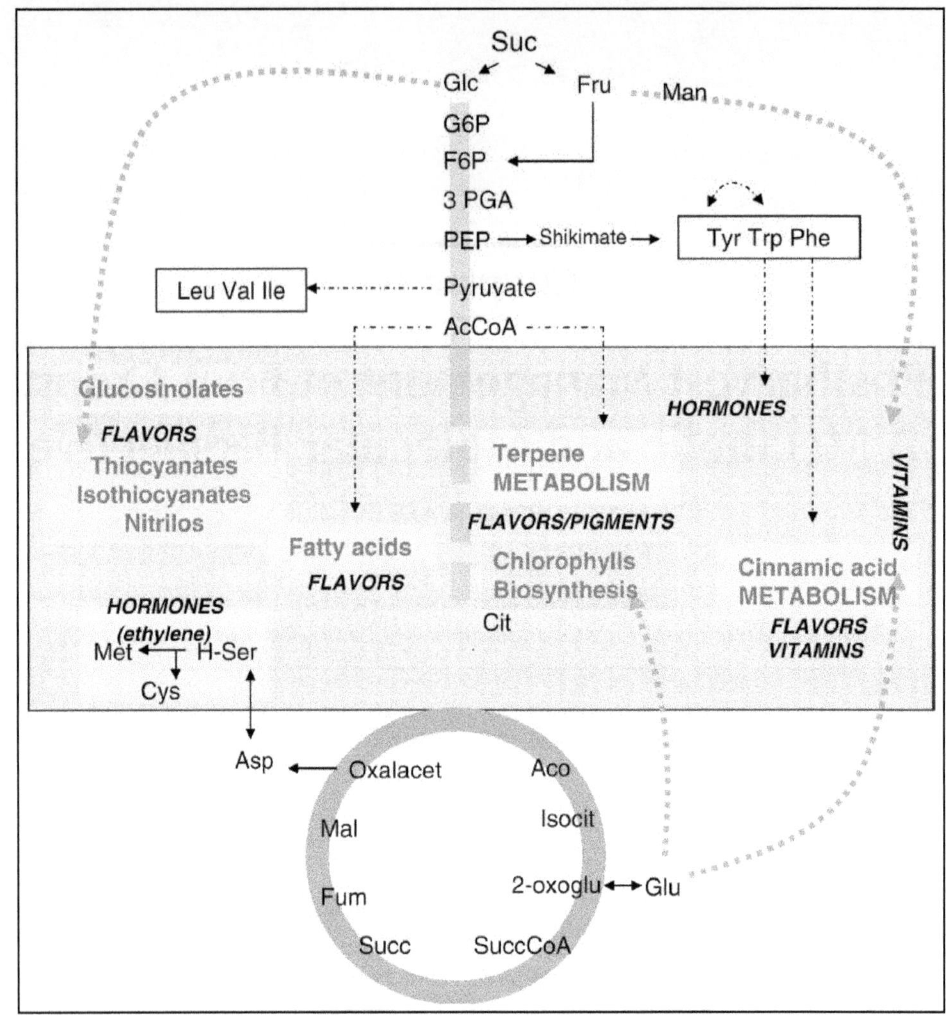

Figure 16.1: Interrelationships of Primary and Secondary Metabolism Pathways Leading to the Biosynthesis of Aroma Volatiles, Hormones, Pigments and Vitamins (Adapted from Carrari and Fernie, 2006).

50 per cent is possible and desirable. To reduce these losses, understanding the metabolic causes of deterioration in fruits and vegetables is the fundamental step, and followed by utilizing appropriate and affordable technological procedures to delay senescence and conserve quality of produce. The main causes of produce deterioration are continuous metabolism and growth, water loss, physiological disorders, mechanical damage and pathological breakdown.

In general the storage life of commodities varies inversely with the rate of respiration. This is because respiration supplies compounds that determine the rate of metabolic processes directly related to quality parameters such as firmness, sugar content, aroma and flavor. Commodities and cultivars with higher rates of respiration tend to have shorter storage life than those with lower rates of respiration.

Factors that slow respiration can slow senescence and maintain quality; however, some respiration must continue or the products will rapidly senesce and die. Cooling the produce can slow many undesirable changes in fruits and vegetables but many fresh commodities are intolerant of low temperatures. Thus understanding the physiology of fresh produce is fundamental to understanding their stability and likely shelf-life. Once severed from the mother plant, the plant organ is deprived of its source of water, nutrients and antisenescent hormones.

Factors Affecting Respiration

Respiration is affected by a wide range of environmental factors that include light, chemical stress (for example fumigants), radiant stress, water stress, growth regulators, and pathogen attack. The most important postharvest factors are temperature, atmospheric composition, and physical stress.

1. Temperature

The most important factor affecting postharvest life is temperature, because temperature has a profound effect on the rates of biological reactions; for example metabolism and requisition. Over the physiological range of most crops. 0 to 30 °C (30 to 36°F) increased temperature cause an exponential rise in respiration. The Van Hoff Rule states that the velocity of a biological reaction increases 2 to 3-fold for every 10°C(18°F) rise in temperature.

2. Atmosphere Composition

Adequate O_2 levels are required to maintain aerobic respiration. The exact level of O_2 that reduces repiration while still permitting aerobic respiration varies among commodities. In most crops, O_2 levels at around 2 per cent to 3 per cent produce beneficial reduction in the rate of respiration and other metabolic reactions. Levels as low as 1 per cent improve the storage life of some crops-for example, apples-but only when the storage temperature is optional. At higher storage temperatures, the demand for adenosine triphosphate (ATP) may outstrip the supply and promote anaerobic respiration.

Increasing the CO_2 level around some commodities reduces respiration, delays senescence, and retards fungal growth. In low O_2 environments, however, increased CO_2 levels can promote fermentative metabolism. Some commodities tolerate brief storage in a pure N_2 atmosphere (for example, a few days at low temperatures) or in very high concentrations of CO_2.

Significance of Respiration

Loss of Substrate

Use of various substrates in respiration can result in loss of food reserves in the issue and loss of taste quality (especially sweetness) and food value to the consumer.

When a hexose sugar (for example, glucose) is the substrate, 180 g of sugar is lost for each 264 g of CO_2.

Biochemistry of Respiration

Measurements of respiration provide an easy, nondestructive means of monitoring the metabolic and physiological state of tissue. For example, events of senescence and ripening are often signaled by abrupt changes in respiration.

Maintaining a supply of high-energy compounds like adenosine triphosphate (ATP). Nicotinamide adenine dinucleotide (NADH), and pyrophosphate (PPi) is a primary function of respiration.

There are three fates for the energy (686 kcal mol^{-1}) released by aerobic respiration. Around 13 kcal is lost due to the increase in entropy (disorder) when the complex glucose molecule is broken down into simpler molecules. Of the remaining 673 kcal that are capable of doing work, around 281 kcal (about 41 per cent of the total energy) is used to produce 38 ATP molecules(38 ATP7.4 kcal ATP^{-1}). The remaining 392 kcal (57 per cent) is lost as heat. In actuality, most energy is lost as heat since energy is lost to heat every time energy is transferred during a metabolic reaction.

Acidification of the cytoplasm enhances the activity of pyruvic decarboxylase that then shunts pyruvate to form CO_2 and acetaldehyde. The acetaldehyde is converted by the enzyme alcohol dehydrogenase to ethanol with the regeneration of NAD^-. Two molecules of ATP and 21 kcal of heat energy are produced in anaerobic respiration (alcoholic fermentation) from each molecule of glucose. To maintain the supply of ATP at the aerobic rate. 19 times as many glucose molecules would be needed, and glycolysis would increase 19-fold but only by 6.3-fold (that is, 19 + 3). Concomitantly, there would be substantial accumulation of ethanol and smaller amounts of acetaldehyde. However, glycolysis usually increases only 3-to 6-fold.

The O_2 concentration at which a shift from predominantly aerobic to predominantly anaerobic respiration occurs varies among tissues and is known as the extinction point, the anaerobic compensation power and the fermentative threshold. Since O_2 concentration at any point in a fruit or vegetables varies with rates of gas diffusion and respiration, some parts of the commodity may because anaerobic while others remain aerobic.

Synthesis of New Compounds

Postharvest storage can be used either to prevent any reduction in quality or to promote changes that increase quality. The quality of most vegetables (for example, cucumbers and lettuce) and nonclimacteric fruit (for example. strawbethes) is maximal at harvest, and storage conditions are optimized to prevent quality loss. In contrast, many flowers (for example. carnations and roses). nonclimacteric fruit (for example. lemons and oranges). And climacteric fruit (for example, bananas and tomatoes) are harvested before they reach their best quality. and storage conditions are optimized to permit development of optimum quality. In the first case, the synthesis of new compounds is unnecessary because they lead to reduced quality (for example. enzymes that destroy chlorophyll in lettuce or promote lipification in asparagus), hi the second case, synthesis of pigments and volatiles (for example lycopene in tomatoes and amyl esters in banana). loss of chlorophyll (for example. chlorophyll-degrading enzymes in banana and lemons), and the conversion of starch to sugar (for example, sweetening, of apples and bananas) is necessary for development of maximum quality. These synthetic reactions require energy and organic molecules derived from respiration.

Oxidative Stress

Many exogenous factors such as pre-harvest production practices, held stresses, harvesting techniques, commodity maturity, storage conditions, processing protocols, and transport and marketing procedure can interact with endogenous

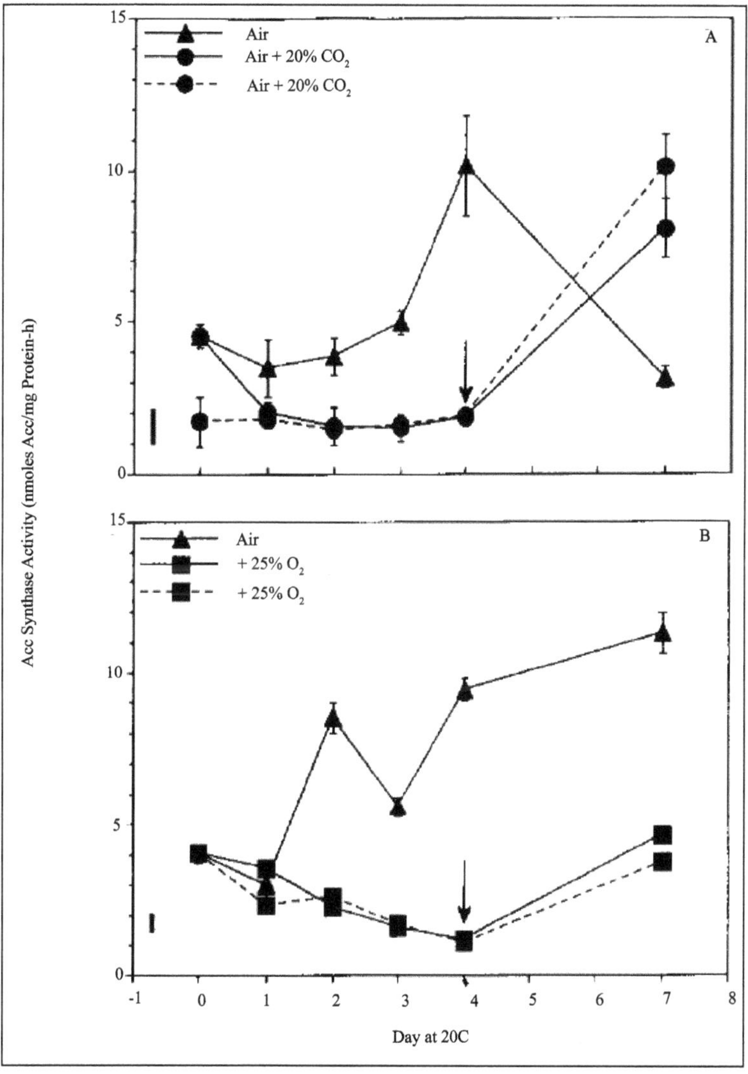

Figure 16.2

factors such as genotype, genetic up-or down-regulation, hormones, and/or general senescent/ripening processes to influence AOS levels, scavenging capacity, and ultimately postharvest oxidative stress of fruits and vegetables. Genotypes exhibiting faster rates of postharvest quality decline concomitant with increased oxidative stress also exhibit lower and/or rapidly declining antioxidant capacities. A survey of recent research does suggest that the H_2O and ascorbate metabolism are intrinsically associated with oxidative stress in postharvest storage situations. Ethylene synthesis is enhanced in association with environmental and biological stresses such as wounding or pathogen attack. Ethylene synthesis may be stimulated by AOS and lipoxygenase activity during membrane lipid peroxidation, further accelerating ripening and senescence process. Lipid peroxidation is also thought to play an important role in the biosynthesis of ethylene, a hormone involved in senescence regulation. The increase in ethylene due to lipid peroxidation may be attributed to production of hydroperoxy conjugated dienes from polyunsaturated fatty acids or perhaps through increased accessibility of its precursor 1-amino-cyclopropane – 1- carboxylic acid (ACC) to ACC oxidase.

Climacteric fruits when stored in atmosphere of air + 20 per cent CO_2 or 0.25 per cent 0_2 inhibited the ACC synthase activity significantly. However, western blot analysis revealed that ACC-O activity was not significantly reduced by atmosphere of air+20 per cent CO_2 or 0.25 per cent O_2.

Many chemical compounds which either control ethylene biosynthesis, ethylene perception and ethylene signaling may contribute to fruit ripening and quality

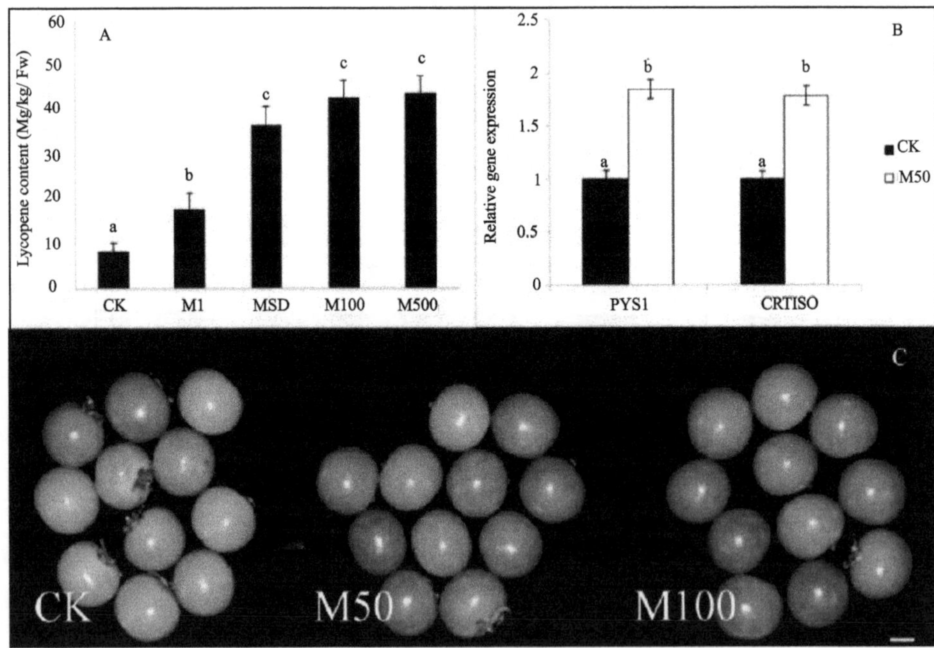

Figure 16.3

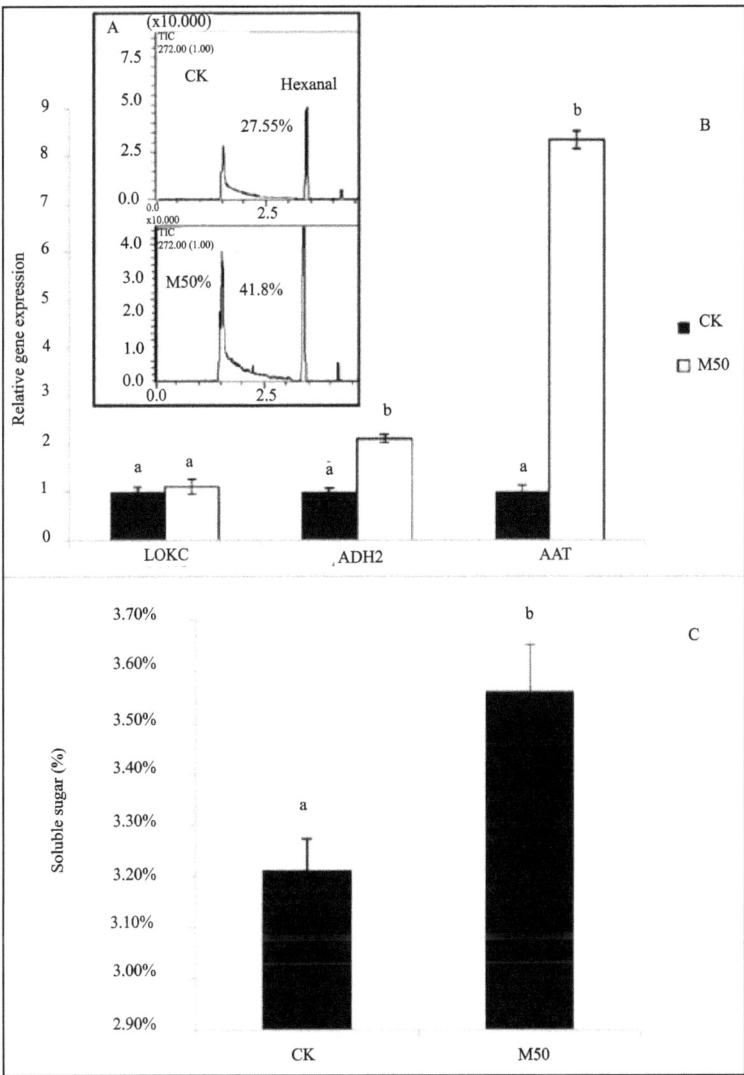

Figure 16.4

improvement. The treatment of fruits with 50uM melatonin has been reported to increase ethylene production by 27.1 per cent in tomatoes and expression of many genes related to cell wall modification *viz.* PG, PE and beta galactosidase with increased levels of lycopene by 5.8 fold with increased expression of genes involved in fruit colour development including phytoene synthase (*PSY1*) and carotenoid isomerase (*CRTISO*). These findings may promote the application of melatonin on postharvest ripening and quality improvement of other horticultural crops.

Understanding the metabolism of fruit crops during pre and postharvest management has resulted in targeting the major enzyme regulatory processes that either control respiration, anti oxidative stress as well as ethylene biosynthesis, perception and signaling.

Ethylene Synthesis and other Hormones Cross Talk and Storage Life

It has long been recognized that long storage life is associated with low ethylene production. Mutations of specific ACS genes appear to be responsible for the postharvest behavior of these genotypes. In apple fruit, the fruit-specific *Md*ACS1 gene is highly expressed during fruit ripening and is responsible for climacteric ethylene production. Low ethylene production is conferred by the presence in some cultivars of two alletic forms of the gene, *Md*ACS1-1 and *Md*ACS1-2 Cultivars homozygous for the *Md*ACS1-2 allele showing low transcription due to the insertion of a retro-transposon-like element have low ethylene production and longer shelf life.

Table 16.1: Inventory of References Related to the Genome Sequences of Fruit Species and TILLING Collections

Species	*References*
Sequence Data	
Grapevine (*Vitis vinifera*)	Jaillon *et al.* (2007)
Papaya (*Carica papaya*}	Ming *et al.* (2008)
Cucumber (*Cucumis sativus*)	Huang *et al.* (2009)
Apple (*Mains domestica*)	Velasco *et al.* (2010)
Strawberry (*Fragaria vesca*)	Shulaev *et al.* (2011)
Tomato (*Solanum lycopersicum*)	Tomato Genome Consortium (2012)
Banana (*Musa acuminata*)	D'Hont *et al.* (2012)
Melon (*Cucumis melo*)	Garcia-Mas *et al.* (2012)
Sweet orange (*Citrus sinensis*)	Xu *et al.* (2012)
Peach (*Prunus persica*)	Arus *et al.* (2012)
Pear (*Pyrus bretschneideri*)	Wu *et al.* (2013)
Water melon (*Citrullus lanatus*)	Guo *et al.* (2013)
TILLING Collection	
Tomato micro-torn (*Solanum lycopersicum*)	Saito *et al.* (2011)
Tomato (*Solanum lycopersicum*)	Minoia *et al.* (2010)
Melon (*Cucumis melo*)	Dahmani-Mardas *et al.* (2010)
Melon Piel de Sapo (*Cucumis melo*)	Gonzalez *et al.* (2011)

The expression of ethylene receptor(s) together with the expression of downstream signaling elements in non-climacteric fruit reinforces the concept that climacteric and non-climacteric fruit share a minimum of common regulatory pathways. The observations that some aspects of non-climacteric fruit ripening are inhibited by the ethylene perception inhibitor 1-MCP also supports a role for ethylene in non-climacteric fruit ripening.

The major emphasis of research on the regulatory mechanisms of fruit ripening has been on ethylene. However, functional characterization of several tomato (*Solanum lycopersicum*) mutants defectives in ethylene response, such as rin and Never Ripe, the transgenic fruits with reduced expression of ACS and ACO, and the use of 1-MCP, have strengthened the evidence for the existence of several ethylene-independent ripening events in climacteric fruits. Furthermore, the application of 1-MCP in non-climacteric fruit has demonstrated the existence of several processes in these fruits types for which ethylene performs regulatory roles, such as loss of firmness in strawberries.

TILLING (Targeting Induced Local Lesions IN Genomics) is a genetic tool that makes it possible to reap many of the benefits of genetic engineering without its disadvantages, either real or perceived (40,41). TILLING is a high-throughput reverse genetic method to obtain allelic series from a chemically mutagenised population making it possible to dissect the functn of the protein as well as to investigate the role of lethal genes. In investigating melon fruit maturation and fruit softening, a reference EMS mutant population under controlled conditions was developed; a TILLING platform established and key mutants isolated that enhanced melon fruit quality.

References

Bapat VA, Trivedi PK, Ghosh A, Sane VA, Ganapathi TR and Nath P. 2010 Ripening of fleshy fruit: Molecular insight and the role of ethylene. Biotechnology Advances (28): 94-107

Kendrick MD and Chang C, 2008 Ethylene signaling: new levels of complexity and regulation. Current Opinion in Plant Biology, (11):283-289

Owino W. O. and Ezura, H. 2011 Biotechnology in postharvest research of fruits: prospectus and achievements. Steward Postharvest Review, (1): 1-7

Pech JC, Purgatto E, Girardi CL, Rombaldi CV, and Latche A. 2013 Current challenges in postharvest biology of fruit ripening Current Agricultural Science and Technology (19): 1-18

The major emphasis of research on the regulatory mechanism of ripening has been on ethylene. However, functional characterization of several tomato ripening (transcription) factors in ethylene response, such as rin and Never Ripe, the transgenic lines with reduced expression of ACS and ACO, and the use of 1-MCP have strengthened the evidence for the existence of several ethylene-independent ripening events in climacteric fruits. Furthermore, the application of 1-MCP in non-climacteric fruit has demonstrated the existence of several processes in these fruits types for which ethylene performs regulatory roles, such as loss of firmness in strawberries.

TFT [illegible]

Chapter 17

Recent Advances in Postharvest Handling of Fruit Crops

J.K. Dhemre, V.P. Kad and D.G. Kadam

Postharvest Technology Centre,
Department of Horticulture,
Mahatma Phule Krishi Vidyapeeth, Rahuri – 413 722, Dist. Ahmednagar, M.S.

Importance of Postharvest Technology

Postharvest Research Since, fruits and vegetables are highly perishable, efficient Postharvest Management has become an absolute necessity. It is also important for effective exploitation of the export potential of fruits and vegetables. Joint effort of R&D institutions, farmers, government agencies and traders has resulted in India emerging as a major producer of fruits and vegetables in the world. But the magnitude of loss in food grains is to the tune of 10 per cent whereas for fruits and vegetables losses are estimated at 35- 40 per cent due to improper Postharvest Management (PHM). It amounts to a loss estimated at Rs 40,000 crores per year! This is not only loss of produce of crores of rupees but also wastage of labour, energy and inputs involved in production.

India wastes fruits and vegetables every year equivalent to the annual consumption of the United Kingdom. So there is need to have a strong postharvest infrastructure for postharvest management of these perishables. Postharvest technology (PHT) is inter-disciplinary "Science and Technique" applied to agricultural produce after harvest for its protection, conservation, processing, packaging, distribution, marketing, and utilization to meet the food and nutritional requirements of the people in relation to their needs. Use of appropriate PHT reduces the postharvest and storage losses; adds value to the product, generates employment in the village and re-establishes agro-industries in rural sector. For every one per cent reduction in loss will save 5 million tons of fruit and vegetable per year. Fruit and vegetable are living commodities as they respire. Hence, proper postharvest management handling and processing is required in horticultural crops. A variety of fresh fruit and vegetable in India can be made available in plenty due to favorable agro-climatic situations. Storage, Processing and preservation technology helps to save excess fruit and vegetable during the glut season.The technology has become a necessity to improve the food safety and strengthen nation's food security.

Postharvest handling includes all the steps such as grading, sorting, washing and packaging. It also covers storage and transportation and the intermediate processing of fruits into a form for increased storage. Correct postharvest handling and storage of fruits is therefore essential to ensure the raw material is of top quality and to reduce losses through wastage. All fruits have different requirements in terms of storage temperature and humidity. However, there are several basic steps that are common to all fruits. Immature fruits should be removed and set aside to ripen. Some fruit species will ripen after harvest while others will tend to rot. Diseased and damaged fruits should be removed as these will infect the good fruit. Fruits should be graded according to colour, size and maturity. Fruits should be handled gently to avoid bruising.

Postharvest management of some of the important dryland fruit crops are given below.

1. Pomegranate

A. Maturity and Maturity Standards

For getting good quality fruits they are harvested at proper stage of maturity. Distinctive fruit colour characteristic of the variety, metallic sound on tapping, the calyx at the distal end of the fruit get closed at the time of maturity. Properly matured fruit is easily scratched with fingernails, Ganesh cultivar – TSS reaches 14°B, acidity per cent 0.3 to 0.35 per cent, Mridula, Bhagwa cultivar – TSS – 16°B, acidity 0.45 per cent, Phule Bhagwa Super cultivar- TSS – 15.79°B, acidity 0.39 per cent.

B. Quality Requirements of Importing Countries

i. Shining dark rose pink colour of the fruit
ii. Fruit weight around 200 - 400 gms
iii. Round shape of fruit
iv. Smooth cutting at the stem end
v. Uniform size and shape of fruit in a pack
vi. Dark rose pink arils (kernels)
vii. Softness of seeds
viii. High total soluble solids near about 16°B to 17°B
ix. Freedom from scars, blemishes, scratches, rusting, rough surface, disease spots and fruit borer
x. Bracts of calyx having freshness and without any damage

xi. Pleasant flavour and aroma

xii. Without blackening of arils

C. Methods and Time of Harvesting

Harvest pomegranates early in the morning or in evening hours. Use clippers and bags for harvesting. Do not pull the fruits. Cut the stem end close to the base. Do not bruise the base while cutting the fruits. Handle carefully at harvest. Do not keep longer stem end, as it can cause damage to other fruits. Keep the fruits in plastic crates for emptying the bags. Do not keep fruits on the ground. Do not keep empty crates in the sun but in shade. Do not overfill the crates with fruits and keep them in shade and not in open sun. Do not damage calyx while harvesting or while keeping in crates or during field storage.

D. Transporting Fruits from Field to Pack-house

Transporting immediate to pack-house without any loss of time. For higher value realization – shiny appearance, bright colour of rind. Fruit should not be spoiled during loading, unloading and transportation. Vehicle should be clean to avoid infection. No jerks or bumps in the vehicle. No other product should be transported.

I. Sorting, washing and treatment with fungicide and wax on a packing line is given below.

A) Sorting: Inspection roller-cum-conveyer, Fruits showing greenish colour, fruits with sunspots, affected by fruit borer, disease spot, rusting, rough surface be eliminated.Having cuts, blemishes, damages must be rejected.

B) Washing: Use of disinfectant – sodium hypochlorite @ 100-200 ppm, rinsing with clean Water, Potable water should be used

C) Elimination of excess water on the fruit surface: Use of absorbent roller

D) Waxing and fungicide application: Thiabendazole or Benomyl @ 1000 ppm concentration, Only good quality edible wax must be used, Monitor edible wax quantity, Monitor concentration of the fungicide to avoid crossing of international residue limit, effective wax coating is very important.

E) Drying the wax coating: Temperature of drying tunnel must not to beyond 30-35°C, Bulbs, heaters, hot air circulation along with fans, air knives must work properly for proper drying.

II. Sizing of Fruits

Sl.No.	*Different Size Ranges (g)*	*No. of Fruits in*	
		4.0 kg box	*5.0 kg box*
1.	750 –Super size	4-5	5-6
2.	500-750- King size	6-7	7-8
3.	450 to 500 – Queen size 400 to 450	8-9	9-10

Sl.No.	Different Size Ranges (g)	No. of Fruits in	
		4.0 kg box	5.0 kg box
4.	350 to 400 –Prince size	10 – 12	12 – 15
	300 to 350	12 – 15	15 – 18
5.	250 to 300 -12 A grade	12 – 15	15 – 18
6.	200 to 250 -12 B grade	15 – 18	18 - 20

III. Nose Cleaning

Before packing a box, fruits at stylar end or calyx side are wiped with a cloth soaked in a fungicide solution (Benomyl or Thiabendazole 1000 ppm) to avoid cross contamination in a packed box.

E. Packing and Coding

All the graded fruits are wrapped in paper and packed in corrugated fiberboard (CFB) boxes. Brown coloured 3 fold CFB boxes are used for local market while, white coloured 5 fold CFB boxes are used for distant markets. Paper shreds are used as padding material. For 'Super' and 'King' grades, boxes of size 32.5 x 22.5 x 10 cm are used. 'Queen' grade fruits are packed in 37.5 x 27.5 x 10 cm size boxes while 'Prince' and '12-A' grades are packed in 35 x 25 x 10 cm size boxes.

F. Pre-cooling

Before storing, the fruits need to be pre-cooled by forced air. The ideal pre-cooling temperatures must be 5–7°C at least within 10 hours after harvesting. Chilling damage of fruit is possible if stored below 5°C. Humidity should be around 90 – 95 per cent during pre-cooling. Prompt pre-cooling is very important for avoiding postharvest losses. Filled cartons should be shifted to pre-cooling chamber without any delay. On transfer of product, outflow temperature and humidity should be properly recorded. Product should remain in the pre-cooling chamber till the temperature of the fruits reaches to required stage. Do not allow to fall the temperature otherwise chilling injury will make the fruits unfit for consumption. Temperature, humidity and air circulation of the room shall be properly maintained. After pre-cooling and palletization, strapping should be done. This needs to be done only in cold room to maintain the cool chain.

G. Storage

Fruits can be stored at 5°C with 90-95 per cent relative humidity for 2 months. In case of storage beyond two months, temperature should be maintained at 10°C to avoid chilling injury. Pomegranates are very susceptible to water loss resulting in shriveling of the skins. Storing fruit in plastic liners and waxing can reduce water loss, especially under conditions of lower relative humidity. Pomegranates subject to low temperature develop brown discolouration of skin, pale colour of arils and brown discolouration of white segments separating the arils.

The export specifications for Pomegranate are given below.

Sl.No.	Variety	Particulars	Countries		
			Middle East	Netherlands/ Germany	U.K.
1.	Ganesh, Bhagwa	Weight	Wt. 300-450 g	Wt. 250-300 g	Wt. 250-300 g
		Colour	Red	Red	Red
2.	Arakta, Mridula	Weight	Wt. 200-250 g	Wt. 200-250 g	Wt. 200-250 g
		Colour	Deep red	Deep red	Deep red
		Packing	5 kg	3 kg	3 kg
		Storage temperature	5°C	5°C	5°C
		Export	By sea	By sea	By sea

2. Mango

A. Maturity Standards

Mango takes about 3 months from the time of fruit set to mature and ripen. Some varieties may take 3-1/2 to 4 months. Fruits are plucked from the tree when they are still hard and green. When a few semi-ripe fruit fall from the tree, it is traditionally considered that the fruits are mature for harvest. If picked immature, fruits develop white patches or air pockets and show lower amount of Brix, acid ratio, taste, and flavor, whereas over- mature fruits lose their storage life. Such fruits present numerous problems during handling. It is therefore, desirable to pick the fruits at the correct stage of maturity to facilitate ripening, distant transportation and maximum storage life, and thus to increase their quality and market value. In Konkan region, mango harvesting season is from March-May, while rest of the Maharashtra it is from May-August. The harvest indices are important criteria for deciding when to harvest so as to provide some flexibility in marketing and to ensure the attainment of acceptable eating quality. The starch content and acid content and specific gravity of the hard fruit are associated with maturity. Picking of fruits when their specific gravity is 1.01- 1.02 is reported to give satisfactory results in respect of uniform ripening and keeping quality. The following main criteria can be used for judging the maturity of mango fruits.

i. The height of the shoulders of the fruit with respect to the point of stem attachment; when shoulders are above the stem attachment, the fruit is mature

ii. The fruit colour changes from dark green to light as the fruit matures; yellowing of the green fruit occurs as ripening begins; The pulp colour changes little yellowish.

iii. When the cheeks of the fruit are full, it is mature

iv. In mature fruit the sap flowing from the cut stem at harvest is thich and does not flow freely.

v. Oil glands of the fruit skin becomes distinctly clear(Kesar variety)

vi. The T.S.S. 7-8.5 per cent for distant market and 8-9 per cent for local market.

vii. Specific gravity: 1.01-1.02.

Harvesting Methods

The general method of harvesting is with a tool called mango picker (Khoodi and Zela). Injury caused by rough handling contributes to fast ripening, fungus infection and rot. Detachment of stem at the time of harvest causes fruit latex to trickle down the skin, which imparts shabby appearance to the fruit by blackening the skin. Hence the fruit should be harvested with few centimeter stalk attached, which helps in avoiding of latex, restricting the entries of pathogens from stalk and injuries responsible for stem rot diseases and reducing the physiological loss in weight after harvest. Dipping the fruits in hot water at 50-53°C for 15 min is the most effective postharvest treatment.

Grading

It consists of separating the fruits into different lots according to their size, colour and stage of ripeness. It is of importance in successful marketing. Systematic grading is desired to get better price and good preservation value. Grade standards are generally based on physical characteristics such as size or weight. The produce is generally sorted out into three to four grades according to the size and packed in same container. This practice of sorting the fruits in several grades is common in Gujarat and Maharashtra for Alphonso. There are certain growers and contractors who use sieves having circular holes of 8 cm, 7.5 cm, and 7.00 cm. on diameter to distinguish fruits by simple marks on the package. However, grading according to the sp. Gravity definitely helps to separate out fruits of same maturity stage. Following grades have been recommended as suitable for the Indian varieties.

Grade	*Weight (g)*
Grade I	301to 350
Grade II	251 to 300
Grade III	200 to 250

Postharvest Treatments

Fruits for urban supermarkets may need to be washed if they are heavily contaminated with latex or dust. If they are washed, they should be dried at once by spreading them in a single layer on a raised mesh or slatted rack, in the shade but with good air circulation. In no circumstances should the wet mangoes be piled up on the ground or left exposed to the sun to dry. Anthracnose is the principal cause of postharvest decay in mangoes. It is a latent infection, spread by raindrops that collect spores from plant branches and spread them on the fruit, where they germinate only after harvest as the fruit ripens.The disease does not respond to a postharvest fungicide dip alone. On a commercial scale, mangoes for export are sometimes dipped in hot water containing fungicide for the control of this disease. The postharvest treatment of the matured green Alphonso mangoes with 500 ppm carbendazim by dip for 5 minutes reduces the postharvest decay.

Methods of Ripening

The general method of ripening of mango is with the paddy straw or paper cuttings or held in wooden boxes. Recommended ripening temperature for mango is 29-31 °C using ethylene at 10 micron 1/lt for 24 hours in 85-90 per cent RH.

Ethylene Application

Ethylene is applied at a proper concentration (50 ppm) in a closed chamber or applied as an ethrel dip (500 ppm) method to enhance ripening and degreening the fruits.The externally applied ethylene help to hasten the faster ripening but may increase the susceptibility particularly of immature fruits to microbial infection. Hence, artificial ripening should be followed with caution.

Heat Treatment

Use of hot water treatment accelerates ripening and is safer to use.

Packing and Transport

The mangoes are packed in cylindrical or semispherical bamboo baskets (holding 50-100 fruits) boxes or crates for transport to distant markets. Nearly 40 per cent loss was observed due to defective picking and packing. Baskets made of bamboo or wooden crates with paddy straw as cushioning material is commonly used in the packing of fruits for transport over short distances. Now days fruit wrapped in tissue paper or without wrapping are being packed in corrugated card board boxes having 3 or 5 ply are being used on large scale they are safe and convenient to use. The fruits for export should be packed in single layer fibreboard carton, net weight 4-5 kg with 8 to 20 fruits per carton. The box must have at least 5 per cent holes for adequate ventilation and should be fully lidded to give better protection. The individual fruits can be wrapped with or without cushioning. The boxes are branded and all the information should be written on the boxes.

Methods for Extension of Shelf Life

1. Cold Storage

The shelf life of fruits gets extended when stored at low temperature as the low temperature suppresses the respiration rate of fruits. These ultimately slower down the metabolic changes which results in extending the shelf life of fruits. Low temperature storage is one of the most effective means of extending the shelf life of mangoes. It reduces the rate of respiration and lowers the rate of ethylene evolution. The low temperature must be accompanied with high relative humidity so as to reduce the physiological loss in weight. The best temperature for storage of maximum varieties of mangoes for export and domestic market is as low as 12.5°C with 85-90 per cent RH. If temperature goes below12.5°C there are chances of chilling injury. The storage period of 18-21 and 23-26 days, respectively are for ripe and mature green mangoes stored at 12 °C.

2. Pre-cooling

Pre-cooling means the removal of field heat from the produce. The pre-cooling of mangoes is done at 12.5 °C for 6-8 hours. It helps in reducing the load on cold storage machineries, in reducing the microbial load and in extension of shelf life and gives the shining to the produce after ripening.

3. Modified Atmospheric Packaging

Fruits are packed in poly bags of suitable thickness with 5 per cent vents. By this means the suitable increase in CO_2 and the decrease in O_2 is observed in the vicinity of fruits as a result, the respiration gets suppressed and fruits can be kept longer.

4. Waxing

In waxing, 3-6 per cent wax is used with suitable fungicidal formulations and smeared with dipping fruits in emulsions or by spraying on fruits. The application can be made during handling time. The waxing also creates modified atmosphere by closing the stomata's of fruits which help in suppression of respiration resulting in extension of shelf life. G.A.3 treatment and use of ethylene absorbent can also extend the shelf life of mangoes.

Handling of Mango Fruit for Export

1. Fruits should be harvested at the beginning of maturity *i.e.* 7-8.5 per cent T.S.S.
2. Fruits should be harvested very gently by keeping 3 to 4 cm stalk to fruits. The harvesting should be carried out in the morning between 6 to 9 am and in evening after 5 pm when the temperature of atmosphere will not be more than 30 °C.
3. Immediately after harvest the fruits should be kept in single layer in plastic crates with sufficient cushioning at the bottom.
4. Crates should be loaded in reefer trucks by keeping 25-30 °C temperatures inside the trucks and be transported to pack house.
5. The crates should be downloaded at pack house, weighed and kept.
6. The desapping should be done by cutting down the excess stalk.
7. The desapping of the fruits need to be kept in inverted position for 1.5 hrs on specially designed sketches.
8. For de-sapping fruits are washed with detergent.
9. After washing the second washing is also given with 250 ppm benomil.
10. After second washing the fruits are subjected to hot water treatment for 2.5-5 min with 500-1000 ppm benomil solution.
11. After hot water treatment the fruits are air dried and then graded on the basis of weight and packed in CFB boxes.
12. Then the boxes are loaded on pallets and subjected for pre-cooling at 12.5 °C in pre- cooling chamber.
13. The boxes are to be kept in cold storage to wait for reefer container. The cold storage temperature should be kept at 12.5 °C.
14. The pallets are loaded in reefer container and transported to port.

The export specifications for Mango are given below.

Variety	*Particulars*		*Countries*			
		Middle East	*Netherlands/ Germany*	*U.K.*	*Japan*	*USA*
Alphonso	Weight	200-250 g	250-300 g	250-300 g	250-300 g	250-300 g
Kesar	Weight	200-250 g	225-250 g	225-250 g	250-300 g	250-300 g
	Packing	1 Doz./2.5 kg	1 Doz./2.5 kg	1 Doz./2.5 kg	1Doz./3.5 kg	1 Doz.3.5 kg
Storage temperature		13°C	13°C	13°C	13°C	13°C
Export		By sea	By Air	By Air	By Air	By Air

3. Grapes

The following points should be kept in view to minimize postharvest losses and to get high quality berries:

- Grape berries which contain more pulp percentage have longer shelf life.
- Glucose sugar helps in better storage than fructose sugar. The sugar content of berries is related directly with postharvest life.
- If it is more, the shelf life is more and vice-versa.
- The berry obtained from low nitrogen applied vines show better shelf life.
- Grape berries which retain shining for more time have better storage life.
- Green stemmed bunches of berries last longer.
- Healthy grape berries last longer in storage.
- Grapes harvested at low temperature and kept at low metabolic/ physiological activities have more shelf life.
- By using grape guards, the freshness of berries can be retained which ultimately prolongs shelf life.

Maturity

As grape is a non-climacteric fruit, it should be harvested only when the berries are fully ripened on the vine itself. The optimum stage for harvesting of grape bunches is indicated by the characteristic colour of fruits, variety, change in colour of bunch stalk from green to yellow or brown, aroma/flavour, softening of berries, sweetness of pulp and thickening of juice. Stage of maturity at the time of harvesting determines the suitability of fruits for processing, domestic or export market.

Generally, all bunches in the vineyard do not ripen evenly at the same time and all the berries in the same bunch do not ripe at the same time. Therefore, it is desirable to thin them out by removing small sized berries at the developing stage. It will bring uniformity in maturity and size of berries.

Overall harvesting starts from December and lasts upto June. Peak period of harvesting is from February to March. However, harvesting of grapes can be advanced or postponed for about 10 to 15 days depending on the market situation.

In case of early pruning, grapes become ready for harvest in about 130 days, in case of late pruning in about 150 to 180 days.

Ripening period may be enhanced by spray of ethephon at the rate of 100 to 500 ppm, 10-20 days before harvesting. The spray of 1 per cent calcium nitrate 10 days before harvest decreases postharvest loss. To bring uniform colour development in berries, treatment of Ethrel (500 ppm) is advised 7 days before harvesting.

The following methods are recommended for increasing the shelf life of grapes:

Sl.No.	Type of treatment	Period of Application	Effect
1.	Spraying of fungicides- Captan (0.2 per cent), aureofungin (500 ppm.), DCNA (2,6-dicholoro-4-nitroaniline)-0.2 per cent	3-4 days before harvest	Reduce decay of berries in storage.
2.	Application of growth retardants- NAA (500 ppm.), kinetin (50 ppm.)	Before harvest	Reduce berry drop during storage.
3.	Spraying of growth retardants- B-9 (2000 ppm.), CCC (2000 ppm.), Alar (500 ppm.), maleic hydrazide (500 ppm.), phosphon- D (500 ppm.)	At harvest or berry softening stage	Retain the freshness of grapes for a longer period.
4.	Sraying with Calcium nitrate (1 per cent)	Ten days before harvest	Reduce the physiological loss in weight.
5.	Spray of ethephon at the rate of 100 to 500 ppm	10-20 days before harvesting	Ripening period may be enhanced
6.	Treatment of Ethrel (500 ppm)	7 days before harvesting.	Uniform colour development in berries

Harvesting

Harvesting is the detachment of bunches from the vine at the proper level of maturity. Great care should be taken at time of harvesting. It should be performed without mechanical damage and product loss, as quickly as possible. The proper stage for harvesting the grapes is adjudged by the development of characteristic colour of a variety and also by touching the grapes from the apical portion of the bunch. In grapes three major colours, *viz.* white, red and black are found. The criteria used to judge ripening differ according to the use of the grapes. For making raisins, harvesting at a late stage is preferred, to get more sugar in grapes for increasing the weight of the dried product. For all other purposes, ripening is judged on the basis of sugar: acidratio for getting proper blend required for table purpose or wine making. The correct blend of sugar acid ratio should be between 25-30. For export market, the following parameters are considered during harvesting of grapes:

For export market, the following parameters are considered during harvesting of grapes:

- ✰ Berry size- Berry size should be more than 16mm in diameter.
- ✰ TSS- TSS should be more than 17° Brix.
- ✰ Bunch weight- Bunch weight should be between 300-750 g.
- ✰ Bunch colour- Bunch colour should be milky green.

- ☆ The selected bunch should not be compact.
- ☆ All the berries should be of uniform colour and size in a bunch.
- ☆ Less than 2 per cent sunburnt or sulphur bleached, bruised or crushed berries.
- ☆ Peduncle should be fresh and green.

For local markets, the grapes are harvested early in the morning by cutting selected bunches with long nose scissors. Care must be exercised not to injure other berries by the scissors. While harvesting, care should be taken to avoid erasing the waxy coating of berries by holding the stem of the bunch by hand. The harvested bunches are kept in perforated plastic trays. Bunches should be trimmed by removing the decayed or otherwise defective berries before packing.

For export purpose, harvesting has to be done in early hours of the day. It should be stopped when the berry temperature rises above 20°C. It is advisable to close the harvesting by 10 a.m. otherwise the berry temperature cannot be brought down to 4°C within the stipulated time of six hours by precooling. The selected bunches are harvested by giving a cut above the knot present on the stalk of the bunch. Any type of mechanical injury to berries should be avoided while harvesting and handling. The injured portion of skin serves as the entry point for many fungi causing decay. Harvested bunches are placed gently in clean perforated plastic trays in not more than two layers and shifted to the packing shed without loss of time.

At present, hand harvest is the only harvesting system for table-grapes. Bunch is held in one hand and the stalk is cut with a pair of sharp scissors or sharp knife about two inches above the berries. The bunches should be nipped in such a way that no injury/cut or damage is caused to the bunches and individual berries. Harvesting of grapes is preferably done during early morning cool hours or cool late afternoons when the fruit is as cool as possible.

Trimming and Handling

Grapes used as table purpose must attract the consumer attention to fetch better price. The postharvest quality of the fruit depends fully on the proper care taken after harvesting of the crop. The removal of undesirable, diseased and injured berries at initial stage improves the appearance and reduces the infection at later stage. The clusters after thinning/trimming must be handled with utmost care. Rough handling promotes more dry stems, brown stems, more weight loss and soft berries. Careless handling reduces shine and attractiveness of the berries.

Physio-chemical Changes during Development of Grape Berry

The growth of grape berry is three-phase process

i) During the first phase, volume of berry development increases rapidly and cell division of the pericarp attains highest rate.

ii) In second phase, auxin level reaches at the maximum and ascorbic acid which initially remains at a low level, increases towards the end of this phase.

iii) During the third phase, change in colour and consistency of berries further increase in volume due to the presence of inherent sugar water and decrease in ascorbic acid content. The state of colour change in berries is technically known as 'Verasion'.

The stage of berry development can be further classified into green stage, ripening stage, ripe stage and over ripe stage. The green stage begins with setting up of berries and extends up to ripening stage which is characterized with rapid increase in size of berry, low quantity of sugar and consistently high acidity in hard structured berries. Ripening stage starts with initial level of ripening to the fully ripe stage of grape, during this stage berries become soft and exhibit changes in colour. The green coloured varieties turns into white or yellow with the increase in sugar content and decrease in acidity. The colour change in red and black coloured varieties become more intense. In over ripe stage, the sugar contents increase due to evaporation and decrease in acidity. The berries are subject to fungal infestation, pest attack, shriveling and drying in this stage. The strength of the skin provides safeguard for postharvest handling at various stages.

Qualities Preferred in Table Grapes

1. The berry should be seedless.
2. The taste of the fruit should be a blend of sweet and sour with 18 per cent - 20 per cent sugar and 0.5 to 0.6 per cent acidity. The sugar, acid ratio varies from one variety to another and on climatic conditions.
3. The colour of berries should be uniform and turning green amber. The skin of berry should be thin, crisp and should dissolve with pulp at the time of consumption.
4. The berry should be fully developed, uniform and sound.
5. The bloom on the skin of berries should be retained uniformly.
6. The stem and peduncle should look fresh and should be of medium thickness, greenish yellow in colour of moderate length.
7. Berry drop is not preferred during handling and packing.
8. The fruit should be free from residual toxicity of pesticides, stains, diseases and pests.
9. Cracked berries, dried berriesand off coloured berries are not preferred.

Postharvest Management of Physiological Disorders

To minimize postharvest losses, certain measures are adopted to check physiological disorders. The postharvest control of certain physiological disorders is as under:

1. Stem Browning

The stem of grape bunches is important part of quality. It is essentially required for handling and trimming. It is the first part of the cluster to deteriorate. Thestem turns brown early if temperature is high. The treatment of 10ppm N.A.A. 7-10 days before harvesting helps retention of stems green colour. The spray of calcium chloride (1 gm./litre of water) at the time of sugar conversion increases the strength of the stem, reduces berry drop and also keeps the stem green. Its spray given during last 20 days before harvesting also helps in retaining bloom and giving attractive appearance. Calcium orthophosphate 1.1 gram or Superphosphate 10 gram per 10 litre of water can also be used for this purpose.

2. Water Loss

Humidity and temperature are the important factors which decide the degree of water loss. Low temperature and high relative humidity lower the water loss in grape berries. Due to water loss, the berries become soft shriveled and unattractive and quality deteriorates considerably. This water loss can be reduced and the quality can be maintained by quick movement of grape from the vineyard to pre-cooling after harvest. To minimize postharvest loss, irrigation one day before harvesting should be avoided. It will reduce fungal infestation in packing also.

3. Berry Cracking

It results due to heavy rain fall or irrigation at the final stage of development of grape berries. Formation of high pressure in the berries on the un-stretchable skin results in berry cracking. The moisture should be controlled at the harvesting stage to reduce this problem.

Grading

Grading is mainly done based on the size and colour of the grapes to maintain uniformity of berries in a package. Size is determined by the weight of bunches (in gms). The following minimum (in gms) requirements per bunch are laid down for large and small berries grapes.

Grade	*Large Berries (g)*	*Small Berries (g)*
Extra class	200	150
Class I	150	100
Class II	100	75

The export specifications for Grapes is given below.

Variety	*Characters*	*Countries*		
		Middle East	*Holland/ Germany*	*U.K.*
Thompson Seedless	Berry size	15 mm	16mm	18mm
	Colour-	amber	white/amber	white
Sharad Seedless	Berry size	15 mm	16 mm	18 mm
	Colour-	black	black	black
Flame Seedless	Berry size	-	16 mm	18 mm
	Colour-	-	pink	pink
Packing		1 kg	4.5/9 kg	4.5/9 kg
Storage temperature		0-1 °C	0-1 °C	0-1 °C

Pre-cooling

As grape is a non-climateric fruit with a very short postharvest life, pre-cooling is essential to minimize water loss, avoid decay and reduce the physiological and metabolic activities of grape by lowering the temperature. In pre-cooling, the temperature is quickly brought down by fast and prompt cooling to check drying and browning of the stems and maintaining the firmness of berries. Grapes should be pre-cooled promptly after harvest in separate rooms with large refrigeration capacity, high air velocity and high relative humidity. They are normally pre-cooled at 1-2° C and relative humidity at 90 per cent and above within 6 hours of harvest.

Packaging

Packaging is very important as good packaging protect the produce from any damage during storage, transportation and other marketing operations. It provides convenience in handling during transportation and storage.

The grapes meant for the local market are packed in ventilated corrugated boxes (2kg and 10kg capacity), bamboo baskets (4kg-15kg) and plastic crates (6-12 kg). These boxes are lined with newsprint paper. Fine paper shred or fine hay is spread at the bottom and top of the box for cushioning. The open flaps of the box are secured firmly by an adhesive tape. The cartons are printed with information such as name of the variety, date of packing, weight, name of the consignee and consignor, destination for domestic markets.

Table grapes for overseas market are packed in 5 ply corrugated boxes of size 500x300mm accommodating 5 kg of grapes. The graded bunches are weighed into 5 kg lots in plastic trays. One or two bunches weighing between 350-650 g are placed in small and thin polythene pouches. Before placing the pouches in the carton, the bubbled polythene cushion is placed at the bottom of the carton. A white and soft polythene lineris spread over the top of the bubble sheet. These pouches are arranged in a single layer in a slanting fashion in the carton. The flaps at the open end of the carton are folded before precooling. After precooling, dual purpose SO_2 releasing pads are placed over the pouches. A few opening points are also provided in the cartons for free circulation of air. Attractive printing and marking of cartonsis also done for the export packing

The Corrugated Fiber Board boxes made of craft paper is sufficiently available and also lighter in weight and recyclable. In addition, desired strength may also be reinforced in to the C.F.B. cartons through the judicious selection of lining material. The boxes which could be designed in varying forms are amenable to printing of quality parameters and other pertinent related information on the package itself. These cartons can also be made waterproof and therefore are ideally suited for pre-cooling treatment, storage in cold storage and transportation in insulated containers, especially for exports from the ports through sea route or by air from airports.

Grapeguard

After pre-cooling, the dual releasing Sulphur dioxide pads (Grape guard) are placed with their coated surfaces downwards on the filled plastic pouches and covered with the polythene liner. It is a paper, chemically treated with Sodium metabisulphate or Potassium metabisulphate, which when it comes in contact with moisture releases sulphur dioxide which helps in retaining freshness of fruits and checks fungal infection in transit and storage. Fast release grapeguard is used for transport and slow release grapeguard is used for storage; while dual release paper is used to take care of both. To eliminate discolouration and to avoid direct contact with fruits it should be wrapped in suitable material before keeping in cartons meant for export or domestic trade.

Cold Storage

The cold storage of grape is done at the temperature varying between 0-2° C and relative humidity ranging from 90 to 95 per cent. Cold storage of grapes extends the shelf life of grapes by 3 to 4 months. Besides, it also preserves the freshness of fruit, prevents decay and discoloration. In addition, it also minimizes the discoloration on account of moisture loss and maintains the stem in good condition. Use of slow release grape guards further enhance the storage period in a cold storage. Since the cold storage do not have pre-cooling air movement facilities for quick cooling, pre-cooling prior to cold storage is a basic requirement for rapid removal of field heat. During storage, shattering of berry can be minimized by avoiding rough handling. Freezing is more in less sugar containing clusters and water loss is more in fruits of heavy bearing vineyards. The uniform maintenance of temperature and humidity prevents berry cracking.

4. Custard Apple

Harvesting

Custard apple is harvested in several instalments, but the best harvesting stage is when the firm fruit begins to develop colour. It is generally picked when it becomes creamy yellow between the segments and begins to crack slightly. The fruit has the tendency to burst open if kept on the tree for a long time. Custard apple is highly perishable and cannot be stored for long time.

Postharvest treatments

Sugar apple fruits can be treated as follows prior to storage or transport: Rinse them in chlorinated water (100 ppm) at 10-12°C for 20 minutes, followed by 10 minutes in less chlorinated water (20ppm). To remove chlorine residues, immerse them in water containing 2 ppm of chlorine. Pre-cooling treatment prior to shipment does NOT improve the postharvest life of sugar apple fruits.

Storage

Physiologically mature sugar apples stored at 13°C for 12 days ripen within:- 2-4 days when transferred to room temperature (27°C);- 6 days when transferred to 20°C. Ripe fruits can be kept for only 1 day at room temperature. When packed in straw, they can be kept for 2 days. Ripe fruits can be stored for 5 days at 5°C. Kept at 4.5°C, they can be stored for 5-6 weeks in good condition for consumption. However, the skin shows injuries due to chilling and becomes brown. Fruits could be stored for

6 days with 2,4-D 30ppm + Bavistin 500ppm or 6 per cent wax emulsion + Bavistin 500ppm,7 days with NAA 30ppm + Bavistin 500ppm, 8 days with Bavistin 500ppm in polythene bag and 9 days with Bavistin 500ppm in polythene bag containing $KMnO_4$. So fruits dipped in Bavistin 500ppm and kept in polythene bag containing $KMnO_4$ may be transported to distant markets with less postharvest losses. Immature fruits stored below 15 C develop injury due to chilling, resulting in an unpleasant appearance.

At CRIDA, a simple technique has been developed for manual extraction of custard apple pulp by rotatory motion of a round hair comb in the scooped fruit held in stainless steel sieve. This pulp can be supplied to the ice-cream industries.

Aonla

Harvesting

Change in seed colour from creamy white to brown is indicative of maturity. Fully developed fruits, which show sign of maturity, are harvested. This helps in size gain of remaining fruit. Delay in harvesting results in heavy dropping of fruits particularly in varieties like Banarasi and Francis. It also adversely affects the following years bearing. Individual fruits are plucked by climbing on tree with the help of pegged bamboo or ladder. Harvesting should be done in early or in the late hours.

Postharvest Operations

1. Remove immature, damaged and diseased fruits.
2. Grade the remaining fruits according to their size, weight, colour and maturity into three grades
3. Wash fruits using chlorinated water (100 ppm) and drain them.

Grade	*Standard*	*Use*
large	> 40 mm	preserves
medium	30-40 mm	other products
small	< 30 mm	medicinal uses

Packaging of Fresh Fruits

For short distance transport: pack fruits in gunny bags or baskets (40- 45 kg capacity) made of pigeon pea stems lined with newspaper as cushioning material. For long distance transport: Pack fruits in corrugated fiberboard boxes which give better protection.

Storage

Aonla fruits are highly perishable in nature as its storage in atmospheric conditions after harvesting is very limited. Stored in low energy cool chambers, the shelf-life can be extended up to 12-18 days. Mature fruits can be stored up to 2 months at 5-7°C.Singh and kumar (1997) stored fully matured aonla fruits in cold storage for 7-8 days at 0-2° C and 85-90 per cent relative humidity. Kept in 15 per cent brine solution (150 g salt/l) at room temperature, the shelf-life can be extended for up to 75 days. Storage facilities such as cold storage and controlled/ modified atmospheric storage are very expensive and not in the direct reach of poor farmers (Kumar and nath, 1993). Nath *et al.* (1992) studied the effect of postharvest treatments on shelf life of aonla fruits with calcium nitrate (1 per cent) and borax (4 per cent) and found that the physiological loss in weight and pathological loss increased with the length of storage period. Calcium nitrate (1 per cent) minimized the weight loss during the storage period and no pathological loss was observed with borax upto 9 days of storage.

Ber

Harvesting

Ber matures 150-175 days after flowering. A pre-harvest spray of 750ppm 2 chloroethyl phosphoric acid (Ethephon) at colour turning stage induces early maturity. Fully mature fruits are harvested by picking. Picking should be done in the forenoon. In south India, the fruits are harvested during October-November, in Gujarat during December-March, in Rajasthan during January-March and in North India during February-April. In dry areas, under rainfed conditions, fruits lose their eating quality and storage life. Therefore, fruits which are just mature and have shining yellow colour should be harvested. At this stage, the fruits contain the desired sugar: acid ratio and ascorbic acid content.

Grading

Remove under-ripe, over-ripe, damaged and misshapen fruits. Grade the remaining fruits manually or by passing them through sieves into 2 or 3 levels on the basis of size and colour according to the grading standard given below. Wash graded fruits using chlorinated water (100 ppm) and drain them for further processing or pack them for storing.

Grade	*Standard*
A	Shining yellow, large (>35mm) to medium size (25- 35mm) fruits of uniform shape with no blemishes.
B	Uneven yellow or yellow red, large (>35mm) to medium size (25-35mm) fruits, of uniform shape with some blemishes.
C	Red, large (>35mm) to small (<25mm) fruits. Uneven yellow, small (<25mm) fruits.

Postharvest Treatments

To extend the shelf-life and to reduce decay losses in storage ber fruits can be treated as follows:

1. They can be dipped in cold water for 2 hours or exposed to cold air for 4 hours immediately after harvest to remove field heat.
2. They can be dipped in calcium chloride or ascorbic acid solution.

3. They can be treated with growth regulators (*e.g.* cycocel), waxed or fumigated.
4. They can be sprayed with fungicides, *e.g.* thiobendazole (at 500 ppm) or 0.2 per cent $ZnSO_4$.

Packaging

Pack ber fruits either for proper storage or for safe transport to local or distant markets. In small packages of 1-2 kg in perforated 150 gauge polythene or nylon-net bags or cardboard cartons.In large packages of 10-20 kg in gunny bags, net bags, cloth packages or wooden or plywood boxes with holes or slits. For transportation, corrugated cartons of about 10 kg are the most suitable packaging material. For short distances, cheaper materials can be used if cushioning and ventilation is provided.

Storage

Freshly harvested Ber fruits can be stored in containers such as gunny, net- or polythene bags, cloth packs and boxes for 4-15 days at room temperature (25- 35°C) without loss of organoleptic quality. Fruits can be stored in a cool chamber for 6-10 days. However, the high humidity which develops in the cool chamber is conducive to spoilage. Fruits can retain acceptable organoleptic quality for 3 weeks kept in polythene bags and baskets at 13°C in an incubator. Kept in polythene bags in a cold storage at 10°C, fruits of some cultivars can be stored for 28-42 days. Fruits can be stored frozen at -18°C for 6 weeks. The shelf-life depends mainly on the ber cultivar, storage temperature, packaging method and stage of harvest. During storage, the fruits loose weight and shrivel, change colour and become red, loose acidity and ascorbic acid, but gain in sweetness.

Fig

Harvesting

Harvesting starts from middle of June and continue till June. For best flavour, fruits should be picked when they are soft and wilt at the neck, hanging down from their own weight. Milky latex exuding from the stem when the fruit is pulled off indicates that the fruit is still immature. For distant markets, harvest the fruits when slightly immature.

Keeping Quality

Fresh figs are very perishable. At 40° to 43°F (4.44°- 6.11°C) and 75 per cent relative humidity, figs remain in good condition for 8 days but have a shelf life of only 1 to 2 days when removed from storage. At 50°F (10°C) and relative humidity of 85 per cent, figs can be kept no longer than 21 days. They remain in good condition for 30 days when stored at 32° to 35° F (0°-1.67° C). If frozen whole, they can be maintained for several months.

Jamun

Harvesting and Postharvest Management

Fruits ripen during June- July or with the onset of rains and disappear with onset of the monsoon. Fruits change their colour from green to deep red or bluish black. This is a non climacteric fruit hence it des not ripen after harvesting. Fully ripe fruits are harvested daily by hand picking or by shaking the branches and collecting the fruits on a polythene sheet. The fruits cannot remain on the tree in the ripe stage and start dropping to the ground immediately after ripening. As a result, a considerable amount of fruit is damaged and become unfit for fresh consumption. Jamun trees need a number of pickings, since all fruits do not ripen at a time. The average yield of fully grown budded and seedling trees are 50-70 kg and 80-100 kg/plant/year. The big fruits (small seeded) are oblong in shape, have more pulp and are suitable for fresh marketing. They have high °Brix ratio, low content of acid, tannins and total anthocyanins. The small fruits (large seeded) are round in shape, grow wild and have very little pulp. The small sized fruits are not suitable for fresh marketing but are ideal for processing. They contain high amount of acids, tannins and anthocyanins.

Kokum

Harvesting

Harvesting of fruits commences from March and continues upto first fortnight of June. The peak period of harvesting is April-May. At fruit maturity, colour changes from green to red. The ripened fruits are harvested frequently by using harvester. Immediately after the harvest, the fruits are cut opened and the seeds alongwith the pulp are separated. Rind can be dried under sun or stroked for about 1-2 weeks. At this time the rind turns into a coal colour and imparts its characteristic flavour. The dried rinds used for culinary purposes can be preserved by rubbing them with salt and coconut oil, which further improves appearance and quality.

Packaging and Storage

Freshly harvested fruit are reddish green in colour and turn into full Red- purple colour in a day or two. The flesh of the fruit is juicy and has a sweetish acid taste. The normal shelf-life of fresh fruit is about five days. The common method practiced for preservation is sun drying. For this, the fresh fruits are cut into halves and the fleshy portion containing the seed is removed. The rind (skin) is then repeatedly soaked in the juice of the pulp during sun drying. The product obtained after sun drying is referred to as amsul or unsalted kokum in commerce. The seeds yield a valuable, edible fat known in commerce as kokum butter. It is extracted mostly as a cottage industry by crushing the kernels, boiling the pulp in water and skimming off the fat from the top or churning the crushed pulp with water. The dried kokum is packed in gunny bags.

Future Postharvest Research Priorities

1. Objective determination of suitable harvesting date.
2. Postharvest treatments (heat, UV irradiation, CO_2, chemicals, wax) for storage ability.
3. Monitoring refrigerating systems.
4. Optimum storage conditions for storage of underutilized fruits, planting material, fresh pack and lightly processed produce.
5. Possibilities of modified atmosphere packaging (MAP), absorption layers, insert sensors, PP films with microspores.
6. Prevention of possible pathogenic organisms in MAP products.
7. Adaptive control of storage conditions with biological sensors.
8. Storage, during transport and as quarantine measures.
9. Humid forced air pre-cooling.
10. Minimum impact and vibration norms, for bruising during sorting and packaging.
11. Objective non destructive measuring of quality and maturity.
12. Environmental friendly packaging using nanotechnology.
13. Consumer and traders quality preferences in each Country.
14. Fundamental research on senescence, ripening, respiration, ethylene effect, chilling, fermentation, superficial browning.
15. Postharvest Physiology of horticultural produce.
16. Impact of pre-harvest and postharvest factors on shelf life and postharvest quality.
17. Ethylene management during ripening and storage of horticultural crops.
18. Regulation of fruit colour and aroma.
19. Controlled atmosphere (CA) storage of fruits.

Chapter 18

Recent Advances in Postharvest Management of Fruit Crops

J.K. Dhemre, V.P. Kad and S.D.Masalkar,

Department of Horticulture,
Mahatma Phule Krishi Vidyapeeth, Rahuri – 413 722, Dist. Ahmednagar, M.S.

Fruits crops play an important role in human nutrition and are vital components of our daily diet. Fruits are considered as the major sources of dietary fibres, vitamins and minerals. Fruits are characterized by their water content, tenderness and short life. Hence, the fruits need to be disposed off as early as possible. The dryland fruit crops are cultivated on light to medium type soils and where the rainfall is scanty and dry weather. The horticultural produce exhibits considerable variation in their quality this is on account of inherent variety of soil and climatic conditions. The fruit cultivation provides more employment throughout the year and pay higher returns to the cultivation than other crops.

The estimated production of fruits in the country in the year 2013-14 is estimated at 89 million tones *i.e.* 13.6 per cent of world's production. Therefore, India ranks second after in the world in production of fruits. The total area under fruits is 7.3 million hectares with 12.3 metric tons per ha. The important temperate fruits grown in our country are apples, apricots, pears, peaches and cherries. Among the tropical and subtropical groups mentioned are mangoes, banana, citrus, oranges, melons, pineapple, grapes, and leachy are cultivated. The dryland fruits like mangoes, pomegranates, aonla, fig, wood apple, tamarind, jamun, ber, custard apple, guava and phalsa grown in India. Besides, there are many exotic fruits like starburst, avocado, pears, kinnows, artichoke, sapota, olives and kiwis which are also grown but not commercialized.

India's biggest markets are South Asian and Middle East countries. India's share in the global market is 13.6 per cent (Figure 18.1). India's export of processed

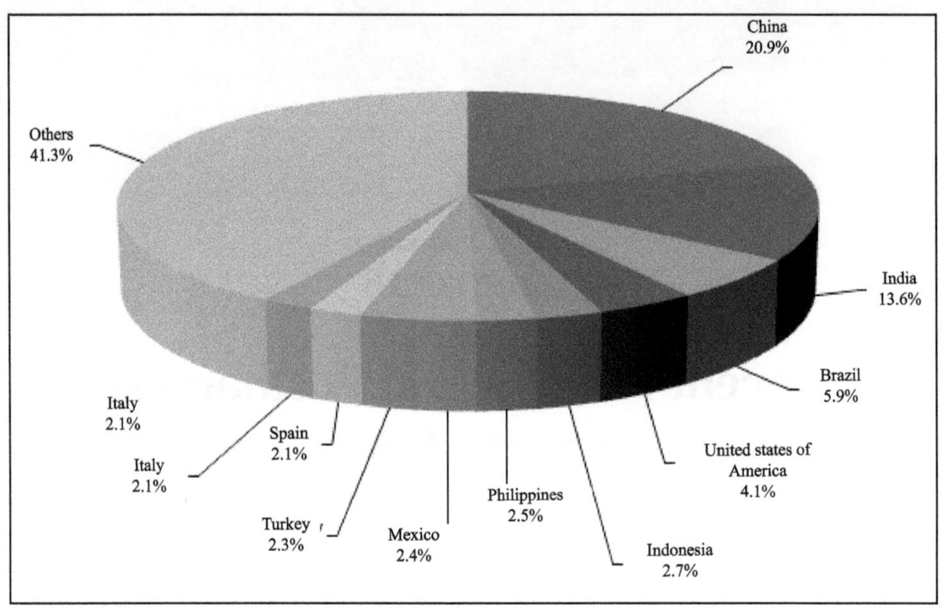

Figure 18.1: Leading Fruit Producing Countries.

food was Rs 38150.86 crores in 2013-14 which include the share of mango pulp-608.56 crores, dried and preserved vegetables- 637.96 crores other processed fruits and vegetables -1733.06 crores, pulses- 1285.00 crores, groundnut-4065.36 crores, guargum-21287.01 crores, jaggery and confectionery -917.60 crores, alcoholic and non- alcoholic beverage- 1909.37 crores. Despite being the world's second largest producer of fruits and vegetables yet the commercial processing of fruits and vegetables in our country is around 2.2 per cent as against 83 per cent in Malaysia, 80 per cent in South Africa, 78 per cent in Phillipines, 70 per cent in Brazil and 65 per cent in USA. The main reason being that domestic consumption of processed fruits is quite meager because of economic reasons and also a matter of habit.

The Indian consumers by and large very much prefer fresh fruits. The high cost of packaging pushes up the cost of processed items and thereby makes them out of reach of common man. Because of varied climatic conditions, fresh fruits are available throughout the year. The fruits like apple, oranges and grapes can be put in the cold storages for prolonging their shelf life and making them available during the off season. Some fruits like guava, oranges, pomegranates and aonla have two seasons so they are available in fresh form for six to seven months within a year. It is estimated that the annual loss of fruits and vegetables to the tune of 20-30 per cent and which results into loss of about one lakh crores annually. This has been attributed due to poor infrastructure and postharvest handling, lack of suitable storage systems, lack of adequate demand and immediate processing facilities. The lack of linkages between the producers, processors and marketers of fruits and uncertainty of prices, many times discourage the farmers for large scale cultivation of fruits.

On account of increasing urbanization and industrialization in India more people are employed in the urban agglomeration. They not only need quality food, but also new products that are convenient to use and easy to cook. Therefore, the demand for processed agricultural products is ever increasing. India's potential for export of processed agricultural product is very high (Figure 18.2).

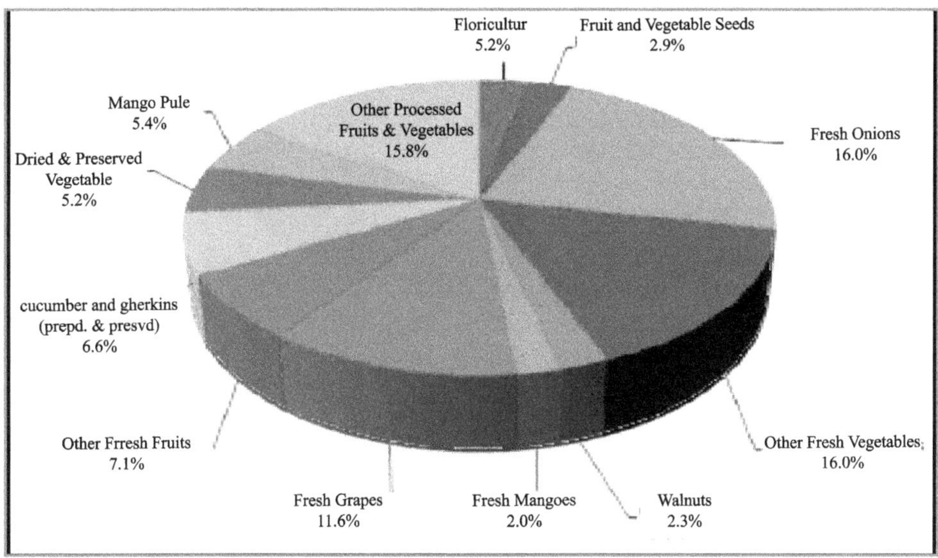

Figure 18.2: Commodity-wise Share of Exports of Horticultural Products from India 2013-14.

Present Status of Fruit Processing in India

In India, banana (33.4 per cent), mango (20.7 per cent) and citrus (12.5 per cent) are grown (Figure 18.3) and in Maharashtra the fruits are grown in 22 per cent area with 15 per cent production (Figures 18.4 and 18.5). The major fruits being processed are mangoes, pineapples, guava, papaya, lemon oranges, grapes, and also dryland fruits like pomegranate, aonla, fig wood apple, tamarind, jamun, ber, custard apple and Karonda. The breakup of processed fruit products is presented in Table 1. Out of the total production of processed fruit products, fruit pulp and juice constituted 27 per cent followed by the beverages (23 per cent), pickles (12 per cent), squashes and canned fruits (4 per cent) and wine (2 per cent).

The processed products prepared from various dryland fruits are presented in Table 19.2.

Prospects in Fruit Processing Industry

Demand for processed fruits comes from both domestic and foreign markets. Though Indians have a typical liking for fresh fruits over processed one the scenario is fast changing with the changes in life styles, urbanization increasing participation of women in work force breaking down of joint family system, growth in economy as well as aggressive market strategies by the producing firms. In

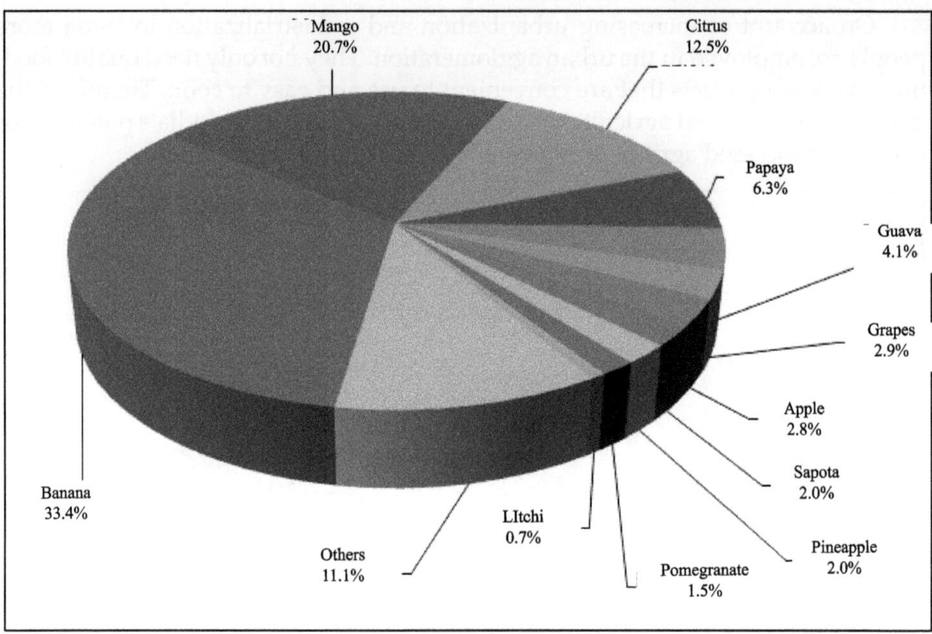

Figure 18.3: Production Share of Major Fruit Crops in India (2013-14).

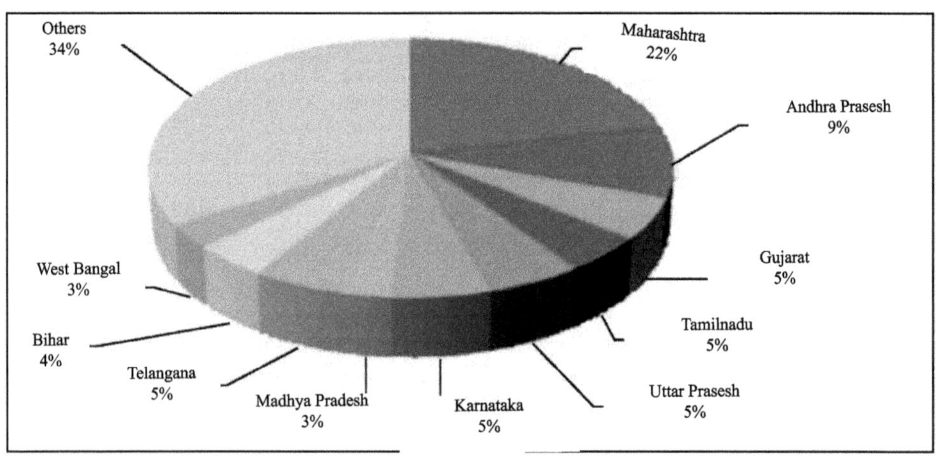

Figure 18.4: Leading Fruit Producing States in Terms of Area (2013-14).

India, per capita consumption of processed fruits have been rising fast and income elasticity of demand for processed food ranges between 1.0 to 3.0. It is expected that the domestic demand for processed fruits would increase mush faster with spurt income growth and changes in food habits. The fruit processing in India derives its significance from its high potential for employment at low capital cost, use local resources, possibilities, for forward and backward linkages and scope for earning foreign exchanges. It also stimulates cultivation of fruits to provide at least two to

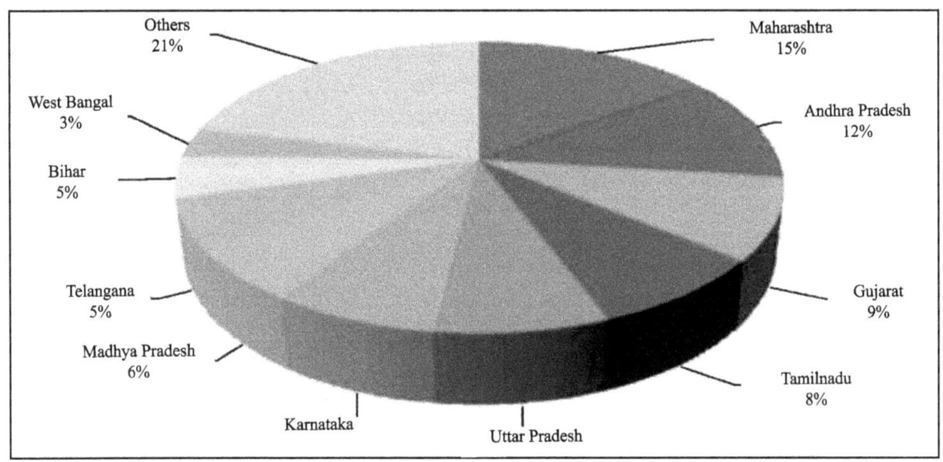

Figure 18.5: Leading Fruit Producing States in Terms of Production (2013-14).

three times higher employment opportunity and utilize poor quality land. Besides, India has vast tract of semi-arid and land that can be used for growing a variety of fruit crops. The production is being enlarged, modern methods of cultivation being adopted thus improving the productivity and cutting the per unit cost. To some extent cold chain being provided, this will help in retaining the quality, freshness and reduce postharvest losses. With invent of new hybrid varieties being added the production season is also being extended. These developments shall result in the greater availability of quality to the industry resulting in better capacity utilization and producing wider range of products and international standards. The rise in the per capita income particularly of the middle class a drastic change in food habits has been noticed. This will lead to an increased domestic consumption of processed food stuffs.

India has an outstanding competitive strength in fruit processing, being blessed with unsurpassed, natural advantages, plentiful sunshine, variable soil texture,

Table 19.1: Per cent Share of Processed Fruit Products

Sl.No.	*Products*	*Per cent Share*
1.	Fruit juice and pulp	27
2.	Beverages	23
3.	Pickles	12
4.	Jam and Jellies	10
5.	Syrups	8
6.	Squashes	4
7.	Canned fruits	4
8.	Wine	2
9.	Others	10
	Total	**100**

Table 19.2: Food Stuffs or Processed Products Prepared from Various Dryland Fruits

Mango	*Pomegranate*	*Ber*
Raw mango	Juices	Pulp
Raw mango slices	Concentrates	Juice
Mango sauce or chutney	Beverages	Beverage
Green mango or Panna	Wine	Dried Ber
Pickles	Jelly	Candy
Ripe mango	Syrup	Slices
Puree	Anardana(Arils)	Tutti-fruti
Concentrates	Anar Rub(jam)	Wine
Beverages	Nectar	Ber powder
Mango blends	Squash	Muramba
Mango slices	RTS	
Amba papad/poli/wadi		
Guava	*Tamarind*	*Aonla*
Canned guava	Pulp	Chawanpraysh
Dehydrated guava	Carbonated drink	Muramba
Powder	Juice	Anola supari
Pulp	Syrup	Pickles
Puree	Concentrate	Carbonated drinks
Jam	Seed powder	Aonla Candy
Toffee	RTS	RTS
Juice		Jam
Juice concentrate		Aonla tea
Jelly		Syrup
Wine		Squash
Pectin		
RTS		
Jamun	*Fig*	*Wood apple*
Juice	Dried fig	Jelly
Concentrate	Jam	Barfi
Seed powder	Candy	Juice
	Dried slices	
Custard Apple	*Phalsa*	*Karonda*
Powder	Juice	Chutney
Pulp	Squash	Pickles
Ice cream	RTS	RTS
Rabadi	Syrup	Syrup

varied agro climatic zones tropical, sub-temperate and rainfall ranging from less than 13 cm to 600 cm all these set the stage for the production of whole range of fruits. The strength on the demand side is India's huge domestic, fuelled by a consumers oriented growing middle class with younger age structure.

Constraints in Development of Fruit Processing Industry

Due to growing urbanization and changing lifestyle and spending pattern, the demand for processed fruit is increasing every year. There has been steep increase in foreign investment and technical coloration in last five years. But there are following constraints in development of processing fruit industry in India.

1. Poor domestic demand because of Indian people preferred home cooked food and also found processed foods are very costly.
2. Units have to pay heavy excise duty and taxes on plant and machinery, finished products and packing material.
3. The cost of packaging, advertising and brand building is very high due to smaller capacity of processing plants and seasonal nature of production and demand.
4. Non availability of right type material at right time and appropriate prices become a major constraint inspite of abundance production of fruits.
5. There exists a vast potential for fruit processing industry, there remain some serious impediments like technological backwardness, poor infrastructure, institutional problems and problems related to price and taxation, which prevents this industry from realizing its full potential.

Integrated Approach for Development of Fruit Processing Industry

1. To augment the supply of fruits here is need to promote the cultivation of off season fruits.
2. Processing units should be located in suitable production central place to reduce the postharvest losses.
3. Single window system for issuing the licenses
4. Fruit products need to be produced on commercial scale for international markets because delicately flavored products in the international markets as India has vast resources of indigenous fruits that have established their petite value apart from having high nutritive value.
5. Integration of production, storage and processing into a single system. Direct link between grower and processor will make it possible to reap the price benefits.
6. Fruit processors have great opportunity to develop their business in world competition. With removal of QRs they enjoy an option to import raw material of better quality and price.

SWOT Analysis

A. Strength

The strength of industry lies in:

1. The abundance of exotic tropical fruits available in India.
2. Matchless aroma, flavor and taste of Indian tropical fruits.
3. Diverse agro climatic condition in the country through which any raw material is available.
4. Availability of very large pool of qualified and experienced technocrats and of the research development laboratories dedicated to food processing industry.
5. Capability of designing, development and construction of process plant machinery matching international standards.
6. Nearly 40 per cent of India's export of processed fruit products is from mango above and India is uniquely placed with regard to mango producing over 40 per cent of the global output.

B) Weakness

The major weaknesses for fruit processing industry are:

1. Poor availability of suitable pavessmg variety of raw materials at present.
2. Lack of in-house quality control and testing facilities in conformity with international standards.
3. Lack of postharvest handling and long term storage facility.
4. Poor infrastructure facilities such as irregular power supply, high inland transportation cost and lack of cold chain facilities.
5. Existing technology obsolesce.
6. Lack of coherence and co-operation amongst the processor and exporters due to which the problems cannot be effectively addressed and tackled collectively.
7. Multiplicity of taxes which increases the tax burden consequently price of finished products.
8. Non-availability of suitable processing varieties of fruits for a longer period of time at reasonable price.
9. Inadequate linkage between fruit growers and processing units.

C) Opportunities

1. Based on exotic Indian raw materials the fruit salads, fruit bars, fruit based cereals and baby foods are gaining consumer's preference.
2. The pre prepared meals based on Indian fruits and vegetables for large Asian ethnic population settled in developed countries is very big opportunity for export.

3. For the region of emerging opportunity in the South East Asian countries which is presently catered too by USA and EU countries, India should take the location advantage that offers economy in cost of freight.
4. Dried tropical fruits going up in demand.
5. The present capacity of Indian fruit processing industry is to handle hardly about 2 per cent of its fruit production. To increase present level of 2 per cent to 10 per cent in next 10 years, the requirement of investment is to the tune of 1,50,000 crores.

D) Threats

The major threat to the industry is expected from the products offered by the neighboring countries such as China, Thailand. Indonesia and Philippines who have distinct price advantage accruing from the production cost which include the cost of power, liquid fuels, rate of interests on loans for fixed and working capital, import duty and export benefits.

Chapter 19

Recent Advances in Grading, Waxing, Packaging and Storage of Dryland Fruit Crops

V.P. Kad, J.K. Dhemre, S.D. Shelar and S.D. Magar

Postharvest Technology Centre,
Department of Horticulture
Mahatma Phule Krishi Vidyapeeth, Rahuri – 413 722, Dist. Ahmednagar, M.S.

Movement of produce from a field lug into a shipping container, or may include a variety of handling practices, from cleaning, waxing, sizing, quality grading to color sorting and packaging. The provision of different operations during the packing of all the agricultural commodities is extremely important. Pack house shade can be created as per the requirement of agricultural commodities and prepared different processed and value added products via a permanent roofed structure. When deciding upon where to locate a packinghouse, access to the field and market point, adequate space for vehicles to enter and leave the packinghouse and ease of access to labor will all be considerations. In the simplest packinghouse, produce is delivered in picking containers, immediately after harvest, directly to the packers. The packers then sort, grade, size and pack the produce directly into appropriate transport containers. In this case, each worker must be knowledgeable regarding produce defects, grade and size requirements, and packing methods. As the size and complexity of the packinghouse increases, more operations and workers trained in specific tasks might be added.

Pre-sorting

Pre-sorting produce is usually done to eliminate injured, decayed, or otherwise defective produce (culls) before cooling or additional handling. Pre-sorting will save energy in that culls will not be handled. Removing decaying produce items will limit the spread of infection to other units, especially if postharvest pesticides are not being used.

Cleaning

In some commodities, dry brushing may be sufficient to clean the produce and in some commodities require washing. The choice of brushing and/or washing will depend upon both the type of commodity and the type of contamination. Sanitation is essential, both to control the spread of disease from one item to another, and to limit spore buildup in wash water or in the packinghouse air. Chlorine treatments (100 to 150 ppm Cl) can be used in wash water to help control pathogen buildup during packing operations. Walls, floors and packing equipment can also be cleaned using quaternary ammonium compounds labeled as safe for food processing equipment during the handling the agricultural commodities.

Sizing/Grading

Sizing produce is optional but may be worthwhile if certain size grades receive a higher price than others. In most low-input packinghouses, manual sizing is still commonly practiced. Operators should be trained in selecting the size desired and to either directly pack the items into containers or place the selected produce gently into a bin for packing further down the line. Sizing/grading can be done subjectively (visually) with the use of standard size gauges. Examples of the smallest and largest acceptable sizes for each product can be placed within view of the operator for easy reference. Hand held sizers are used for a variety of products. Many products have established U.S. grades and standards that can assist packers in sorting and sizing produce. Several types of mechanical sizers are available for small scale operations. One type is composed of a long slanted tray with a series of openings which converge (largest at the top, smallest at the bottom). This type of sizer works best with round commodities. Other sizers are designed as conveyors fitted with chain or plastic belts with various sized openings, and are useful for sizing most commodities. Another simple method for mechanical sizing is to use a set of diverging bar rollers (see illustration below), where the smallest sized produce falls through the rollers first to a sorting belt or bin, and larger sized produce falls between successively more divergent rollers. In some of the market, grading was done on the basis of weight or colour. For this purpose weight sorter or colour sorter is also available to sort the commodity on the basis of weight or colour.

Waxing of Fruits and Vegetables

Food grade waxes are used to replace some of the natural waxes removed in washing and cleaning operations, and can help reduce water loss during handling and marketing. If produce is waxed, the wax coating must be allowed to dry thoroughly before further handling.

Waxes are esters of higher fatty acids with monohydric alcohols and hydrocarbons and some free fatty acids. Both vegetable and mineral waxes are used in varying compositions to make wax emulsions. Waxes are melted, emulsifiers are added and with addition of hot water, oil in water emulsions is prepared with continuous agitation. Wax emulsions have been developed to give a protective coating on the skin of fruits and vegetables to reduce transpiration losses, rate of respiration and prolong the storage life.

Waxing of fruits and vegetables was practiced as early as 12 and 13th Century in China. Dipping fruits in melton imparted thick coating to the fruits and lead to undesirable changes like fermentation. Hence, later improvements were made and a mixture of mineral oil and paraffin wax was applied to apples. This was further modified and experiments conducted with different vegetables waxes with water emulsions gave good results regarding reduction in physiological losses in weight, internal atmosphere, respiration rate, decay, shriveling, scald injury, palatability, taste and flavor in waxed apples during storage. Waxing of fresh fruits and vegetables has assumed a commercial importance in USA, Australia and South Africa.

The WAXOL-O-12 and WAXOL-W-12 (manufactured by the CFTRI, Mysore) each containing 12 per cent solids, for extending the storage life of fresh fruits and vegetables, both at ambient and low temperatures have been successfully tried. Edible Lac based wax SH-1, SH-2 and SH-03 are also available at Indian Institute of Natural resins and Gums, Ranchi (Jharkhand). The good results were found in case of lac based edible wax on different fruits stored at ambient, zero energy cool chamber and cold storage. The wax emulsion without fungicide in the required concentration did not protect fruits/vegetables against microbial spoilage. Hence, permissible concentration of fungicides should be incorporated in wax emulsions prior to its application or may be given as wash treatment prior to skin coating. All fruits and vegetables except a few do not respond to wax coating. They are cabbage, cauliflower, beetroots, spinach, radish, green peas and French beans.

The concentration of wax emulsion required to obtain maximum benefits vary from fruits to fruits depending upon the skin structure, from 4 to 12 per cent (Table 20.1). The approved fungicides incorporated in wax emulsion are SOPP 0.25-2.00 per cent, Flit 406 (CAPTAN 83) 0.2 – 1.0 per cent, sodium hypochlorite, NaGCL 0.1 – 1.0 per cent, Benlate 50: 0.1 – 2.0 per cent, Thiabendazol (Tecto 60) 0.2 – 0.5 per cent and Brassicol 75 (0.5 – 1.0 per cent).

Table 20.1

Name of Commodity	*Type of Wax Emulsion*	*Concentration of Wax Emulsion (Per cent)*				
		12	*9*	*8*	*6*	*4*
Carrots	W	Yes	–	–	–	–
Brinjals	W	Yes	–	–	–	–
Snake Gourds	W	Yes	–	–	–	–
Potatoes	W	Yes	–	–	–	–
Tomatoes (Green/firm)	O	–	Yes	–	–	–
Tomatoes (Pink or red/firm)	O	–	Yes	–	Yes	–
Mangoes	O/W	–	–	–	Yes	Yes
Musk melon	W	–	–	–	Yes	–

Name of Commodity	Type of Wax Emulsion	Concentration of Wax Emulsion (Per cent)				
		12	9	8	6	4
Cucumbers	W	Yes	–	–	–	–
Capsicum	W	Yes	–	–	–	–
Parwal	W	Yes	–	–	–	–
Sour limes	O	–	Yes	–	–	–
Pineapple	W	–	–	–	Yes	–
Mandarins	O	–	–	–	Yes	–
Oranges	O	–	Yes	–	–	–
Banana (75 per cent maturity)	W	–	–	–	Yes	–
Banana (80 per cent maturity)	W	–	–	Yes	–	–
Papaya	W	–	–	–	Yes	–
Apples	O/W	–	–	Yes	–	–
Guavas (Green/Firm)	W	–	–	–	Yes	–
Guavas (Ripe/firm)	O	–	–	–	Yes	–
Avocado	W/O	–	–	–	Yes	–

Fruits and vegetables should be dry prior to waxing and are waxed by dipping in wax emulsion for 30 to 60 seconds, removed, allowed draining and drying with the help of fan, as hot air blower, or under the shade of a tree by breeze. The fruits prior to dipping should be free of dust, dirt and any microbial infection or else it may contaminate the emulsion. Wax emulsion can be re-used and is stable up to one year. Proper application of the wax emulsion to fresh fruits and vegetables will not leave any residue or impart any undesirable odour or flavor, bad appearance or interfered with the natural appearance on the other hand, it imparts gloss to commodity and improves marketability very minute quantity goes on to the surface of the fruits and is inconspicuous and is found harmless.

Packaging

Packaging has been one of the most important tools at the disposal of the product manufacturer to make its product stand out in the shelf. Beyond shelf appear, packaging has also provided protection and containment for the contents. However, until recently there have been few revolutionary uses of technology to enhance these capabilities or to add to them. Packaging can be broadly categorized into the following types passive, active, intelligent, and smart.

Passive packaging refers to the traditional packaging that involves the use of a covering material, characterized by some inherent insulating, protective or ease-of-handling qualities. The most common example of this type of packaging is a simple plastic bag and it is the type of packaging that everyone is most familiar with.

Active packaging entails the concept of the package reacting to various stimulito keep the internal environment favorable for the products. A typical example would be a packaging with oxygen scavenger (an oxygen scavenger can absorb oxygen inside a package to increase the shelf-life of the item.) Both food and electronics use this type of packaging and, again, it is widely used. Well over 90 per cent of all packaging is either active or passive in the sense that we use the terms here.

Intelligent packaging refers to the concept of making innovations in the design of packaging that renders it more useful for the consumer. By way of example, intelligent packaging would include the packaging of automobile oil, where the very packaging structure makes it convenient for the user to pour oil into his automobile without getting his hands dirty. Thus, intelligent packaging refers to increasing the functionality of the package by simply changing the structure of the package, without the addition of any technology.

Smart packaging on the other hand refers to packaging that is made much more functional and useful; it involves the use of technology that adds features such that packaging becomes an irreplaceable part of the whole product. Smart packaging becomes an irreplaceable part of the whole product. Smart packaging performs additional functions, responds to stimuli generated by the environment or from the product being packaged, and reflects the change in a manner that makes the product more convenient and useful for the consumer or firms in the supply chain. Smart packaging relies on the use of chemical, electrical, electronic, or mechanical technology, or any combination of them.

Smart/Intelligent Packaging

Packaging is needed to keep foods, for convenience, protect foods from outside environment, to communicate information to consumers about the food inside the package, about the preparation and also information about the nutritional content. Moreover it provides protection of food from adulteration by water, gases, microorganisms, dust and punctures.

Twenty – first century food packaging may be characterised by innovation for the future, which resulted from the consumer's demand for packaging that is more advanced and creative than what is currently offered. Active and Intelligent packaging are the result of innovative thinking in packaging.

How Intelligent Packaging differs from Active Packaging

Intelligent packaging is defined as "Packaging that contains as external or internal indicator to provide information about aspects of the history of the package and/or the quality of the food" (Robertson, 2006). Intelligent packaging is an extension of the communication function of traditional packaging and communicates information to the consumer based on its ability to sense, detect or record external or internal changes in the product's environment where as Active packaging is defined as packaging in which subsidiary constituents have been deliberately included in or on either the packaging material or the package headspace to enhance the performance of package system (Robertson, 2006). Active packaging is an extension of the protection function of a package commonly used to protect against oxygen and moisture.

In short, active packaging senses change and modifies package properties in response and intelligent packaging senses and informs the change.

Working of Intelligent Packaging

An intelligent packaging senses change in the internal or external package environment to enable a user to offer benefits such as more convenience increased safety or better retention of quality. Intelligent packaging can play an important role in facilitating the flow of both materials and information in the food supply chain cycle. The outer circles represent the supply chain. cycle from raw material through manufacturing, packaging, distribution, product use, and disposal. The package in one form or another (such as pouch, container, drum, pallet), is traditionally used to facilitate the flow of materials from one location to another by performing the basic functions of containment and protection of the product. Furthermore, the package can also facilitate the flow of information (represented by the arrows in the figure) from one location to another by performing the basic functions of containment and protection of the product. Furthermore, the package can also facilitate the flow in information (represented by the communication links between the inner circle and outer circles), although this communication function has been largely over – or it can carry actual information in the direction of material flow (for example, via an indicator) or electronically (for example, via a barcode or the Internet) throughout every phase of supply chain cycle (Yam *et al.*, 2005)

Types of Intelligent Packaging

Gas Indicator

The gas composition within a packaging can easily change due to the interaction of food with its environment. For example, respiration of fresh produce, gas generation by spoilage microorganisms, or gas transmission through the packaging material or package leaks, may cause the gas composition inside the package to change. Gas indicators are helpful in monitoring the composition of gases inside a package by producing a change in the colour of the indicator through enzymatic or chemical reaction (De Jong *et al.*, 2005) The indicator must be in direct contact with the gaseous environment directly surrounding the food in a package or they may be used to verify the efficiency of an oxygen scavenger. Gas indicators mainly signal the presence or absence of oxygen or carbon dioxide. These indicators in the form of package label or printed on packaging film result in a colour change when oxygen is present and the presence of oxygen can indicate that the package has a leak or has been tampered with. They can also indicate improper sealing of the package. A number of oxygen indicators are designed to show color changes due to leaking or tampered packages. Krumhar and Karel (1992) used oxygen indicators to detect improper sealing and quality deterioration of modified atmisphere packages containing pizza or cooked beef. Gas indicators are also being developed to detect water vapour, ethanol and hydrogen sulphide (Yam et.al.2005)

Time Temperature Indicator (TTIs)

Time Temperature Indicators is useful way to tell the consumer when foods have been temperature abused. These are mechanical recorders. electronic recorders and physicochemical devices that sense temperature change as a function of time. Such devices are often directly correlated to product content, shelf life, quality and microbiological experience. For example, a TTI can be placed as a small adhesive label, and when TTIs experience abusive conditions, an irreversible change, like a colour change, will result. These are particularly useful with chilled or frozen foods, where the cold storage during transportation and distribution are important for food quality and safety. They are also used as freshness indicators for estimating the shelf life of perishable products.

Commercially available TTI Products

i) **Lifelines Fresh check indicators:** These are color changing labels that respond to cumulative exposure to temperature. They are attached to packages to help consumers determine that the products are still fresh at the point of purchase and at home. The indicator center gradually and irreversibly darkens rapidly at higher temperatures and slowly at lower temperatures. The TTI can be tailored to match a specific product's shelf life characteristics (Morris et.al., 2007)

ii) **3M Monitor Mark:** TTIs are activated by pulling an activation strip. Upon exposure to temperatures above the threshold the indicator's window irreversibly turns color.

iii) **Time strip:** This TTI is currently being employed by Nestle in their food service products in the UK. Time strip uses a steady diffusion of liquid through a membrane to measure the time that has elapsed at a particular temperature. This action can provide information about how long a product has been opened or in use. The Time strip is very useful for products like sauces that have to be refrigerated and used within a specific time period.

iv) **Thermo chromic inks:** Thermochromic inks are those which are temperature sensitive and can change colours based on temperature. These inks can be printed onto packages or labels such that a message can be conveyed to the consumer based on the color of the ink they are seeing. Thermochromic inks can let a consumer know whether a packages is too hot to touch or cold enough to drink. They are becoming a popular technology for beverages (Robertson, 2006)

Microbiological Growth/Spoilage Indicators

Food borne pathogens are of great concern to the food industry. There is need to rapidly and accurately detect even small amounts of pathogen or toxins in food which can be done by biosensor.

Biosensor

A biosensor is a compact, analytical device that detects, records, and transmits information pertaining to biochemical reactions. This device consists of 2 primary components a bio receptor that recognizes a target analyte (which is an organic or biological material such as antigen, microbe, hormone, nucleic acid and enzyme) and a transducer (which may be of electro chemical, optical and acoustic form depending on the parameter to be measured) that converts biochemical signals into a quantifiable electrical responses.

Toxin Alert (Ontario, California, USA) is developing a diagnostic system called Toxin Guard that incorporates antibodies into plastic packaging film to detect pathogens. Here when the antibody encounter a target pathogen the packaging material displays a clear visual signal to alert the consumer, retailer or inspector. This can be used to detect gross contamination but not sensitive enough for detecting low levels of disease causing pathogens (Bodenhamer *et al.*, 2004;Bodenhamer, 2002.)

Colour Sensor

COX Technologies has developed a colour indicating tag that is attached as a small adhesive label to the outside of packaging film, which monitors the freshness of seafood products. A barb on the backside of the tag penetrates the packaging film and allows the passage of volatile amines, generated by spoilage of the seafood. These are chemical sensor that turns fresh Tag progressively bright pink as the seafood ages (Butler, 2001)

Ripeness Sensor

Ripeness sensors marketed as Ripe Sense label senses aromatics or other valuable emissions such as ethylene, emitted from ripening fruit. The device signals ripeness by label visual cue/colour change for fruit that does not change external colour during ripening such as pears, melons and avocados. (White, 2005)

Bar Coding

Barcodes are the least expensive and most popular form of data carriers. To enable barcodes to communicate with scanners and printers, many standards have been developed over the years into commonly accepted languages known as "symbologies", The UPC barcode is a linear symbology consisting of a pattern of bars and spaces to represent 12 digits of data. Its meager storage capacity allows the containment of only very limited information such as manufacturer identification number and item number For encoding more data in a smaller space, a new family of barcode symbologies called the Reduced Space Symbology (RSS) is introduced (Uniform Code Council 2004) which are particularly well suited for product identification at point-of-sell and for product traceability in the grocery industry. The RSS-14 Stacked Omni-directional barcode encodes the full 14-digit Global Trade Item Number (GTIN), and it may be used for loose produce items such as apples or oranges where space limitation requires a narrow symbol. The RSS Expanded Barcode (also available in stacked format) encodes up to 74 alphanumeric characters, and it may be used for variable measure products (for example, meat and seafood that are sold by weight) where larger data capacity is required to encode additional information such as packed data, batch/lot number and package weight. As scanners are becoming more powerful and affordable, two dimensional barcodes are also gaining popularity. The PDF 417 is a 2-dimensional symbol that carries up to 1.1 kilobytes data in a space of a UPC barcode It allows the encoding of additional information not possible with linear barcodes, such a nutritional information, cooking instructions, Web site address of food manufacturer and even graphics. The advantage of portable data is that they are available immediately, without having to access an external database (Yam *et al.*, 2005).

Radio Frequency Identification (RFID)

RFID tags are an advanced form of data information carrier that can identify and trace a product. They are at present used for tracking expensive items and livestock. This RFID is widespread for distribution inventory control. In this method, a reader emits a radio signal to capture data from a RFID tag. The data is then passed to a computer for analysis. RFID tags contain a microchip connected to a tiny antenna. This allows for the tags to be read for a range of 100ft or more in more expensive tags to 15 feet in less expensive tags. The RFID does not need to be in a direct line – of – sight to be recognised by a scanner. This could revolutionise the way check – out works at a grocers store. Many RFID tags can be read simultaneously at a rapid rate. RFID tags could also store information such as temperature and relative humidity data, nutritional information and cooking instructions. They could be integrated with a time temperature indicator or a biosensor to carry time temperature information or microbiological data (Yam *et al.*, 2005)

Application of Intelligent Packaging

Food Quality and Convenience Enhancement

In addition to using Time Temperature Indicator labels to monitor temperature many other applications are also possible. For example it may be applied to cooking appliances such as intelligent microwave oven system. The uniqueness of this system is in its use of information sharing to enhance food quality and convenience (Yam, 2000). The PDF 417 barcode in the package carries data about the food product and the data processing system generates proper heating instructions for the microwave oven. The bar code in the food package could be scanned to enable the decision support system to match the power output and heating in the microwave over with food package. To achieve a higher level of quality, temperature and moisture sensors may be placed inside the microwave oven to provide the feedback for data processing systems. Scanning of barcode eliminates the need for manually entering heating instructions. this is especially useful for microwave oven that use complicated instructions. Apart from this, scanning of bar code is also helpful for people who are visually impaired or they cannot understand the language given in instruction provides convenient access to information relating to the packaged food such as the manufacturer's Web site, recipes, food allergen information, and product recall.

Food Safety and Bio-security Enhancement

Traceability, tracking and record keeping of product flow through the production process and supply chain is considered as a key in enhancing food safety and bio security (Golan *et al.*, 2004). Intelligent packaging especially when integrating with science based principles is a useful tool for tracking products and monitoring their conditions, facilitating real-time data access and exchange and enabling rapid response and timely decision making. Intelligent packaging concept could be effectively applied for HACCP systems (Yam *et al.*, 2005). When applying the HACCP principles, special attention should be given to design the decision support system. The hazard analysis should determine both flow of material and flow of information. The critical control points and the sensing devices required at specific locations should be identified. The limits of critical control points (such as time and temperature) may be encoded in bar codes or RFID tags to enable electronic data retrieval. Monitoring of the critical control points may be achieved at strategic locations. Read/write RFID tags and computer networks may facilitate record keeping and documentation. Decision for corrective actions may be facilitated using data processing software. Which should incorporate science-based knowledge such as risk assessment and predictive microbiology models (McMeekin et.al, 2002).

Storage

Most of the fruits and vegetables are seasonal in nature and there is glut of particular fruit or vegetable in the local market. The fruits and vegetables being highly perishable, the farmers have to sell their produce at the prevailing market price. It is, therefore, necessary to increase the shelf life and minimize the losses of fruits and vegetables by adopting advanced storage methods. Treatment of fruits and vegetable such as cleaning, grading, washing, waxing, pre-cooling *etc.* prior to their actual storage has been found to improve their shelf life.

Pre-cooling

Per-cooling is a process by which the field heat in the harvested fruit or vegetable is removed quickly to a temperature desired for transport and storage. Pre-cooling is done by using cold air, cold water, ice or vacuum.

1. **Air pre-cooling:** The most common practice followed for fruits is air pre-cooling. Cold air is forced through the product in the room, container or tunnel.
2. **Hydro cooling:** Hydro cooling is most commonly followed for vegetable. The vegetables are conveyed under a flood of cold water in the specially designed hydro-coolers. Hydro-cooling is 2 times faster than air cooling.
3. **Pre-cooling with ice:** Ice is kept within or on the top of the packages loaded in a refrigerated car or truck. It is commonly used for vegetables.
4. **Vacuum cooling:** Under low pressure, water boils at lower temperatures (0°C) at 4.6 mm Hg) one gram of water absorbs 540 calories of heat when it changes to vapor. Water in the fruit or vegetable, thus, acts as a refrigerant and the product is cooled. Vacuum cooling is 23 times faster than air cooling and 11 times faster than hydro-cooling vacuum cooling is particularly suitable for leafy vegetables as they are difficult to cool by air or water.

Methods for Storage

1. **Cold or refrigerated storage:** The optimum storage temperature, humidity and the shelf life very with the nature of fruits. Cold storage is the most commonly used method for storage of fruits because of its simple technology and economy. Some fruits suffer from chilling injury, if stored below the recommended temperature.
2. **Controlled atmosphere storage:** CA is the storage of fruits under conditions wherein the concentrations of O_2 and CO_2 (sometimes N) are controlled within 1 per cent. The proportion of O_2 is lowered while that of CO_2 is increased. The method is superior to cold storage. However, it is costlier and some fruits suffer from CO_2 injury.
3. **Modified atmospheric storage:** MA is similar to controlled atmosphere in principle. The control of O_2 and CO_2 concentrations are less precise or lacking in MA in CA.
4. **Hypobaric storage:** Application of vacuum reduces the temperature and O_2 supply. The increase in storage of fruits is more in hypobaric storage than in cold storage. However, the costs of equipment and operation are higher in case of hypobaric storage.
5. **Ventilated storage:** VA is cheaper and more convenient method of storage. There is, however, some loss in weight of the product and only limited improvement in the shelf life.
6. **Storage in cool chamber:** Storage in cool chamber is low cost technology for storage of fruits and vegetables. The temperature inside the chamber is lower than that of the outside atmosphere. The shelf life of fruits is increased by 2 to 3 days. The method is, however, may useful in summer than in other seasons.

References

Bodenhamer WT, inventor.2002. Toxin Alert, Inc. April 18, 2004. Method and apparatus for selective biological material detection. United States Patent 6,376,204.

Bodenhamer WT, Jackowski G, Davies E, inventors, 2004. Toxin Alert, Inc. Feb 17, 2004. Surface binding of an immunoglobulin to a flexible polymer using a water soluble varnish matrix. United States Patent 6,692,973.

Butler, P 2001. Smart Packaging – Intelligent Packaging for Food, Beverages, Pharmaceuticals and Household Products. Materials World, Vol.9, No.3pp. 11-13

De Jong, A. R. Boumans, H., Slaghek, T., Van Veen, J., Rijk, R. And Van Zandvoort, M. 2005.

Active and intelligent packaging for food: is it the future? Food Additives and Contaminants: Part A.22:97 5 – 979

Golan E, Krissoff B, Kuchler F, Calvin L, Nelson k, Gregory P. 2004, Traceability in the U.S. food supply: Economic theory and industry studies, Report no. 830. Washington, D.C.: Economic Research Service, U.S. Dept. of Agriculture.

Kader, AA (ed). 2002. Postharvest Technology of Horticultural Crops (3rd Edition). UC Publication 3311. University of California, Division of Agriculture and Natural Resources, Oakland, California 94608: pp 535.

Krumhar KC, Karel M. 1992. Visual indicator system. Patent 5,096,813. Massachusetts Institute of Technology. 17 March 1992

Marris,C., Brody, A. L. and Wicker, L. 2007. Nonthermal food processing/ preservation technologies: a review with packaging implications. Packag. Technol. Sci., 20: 275-286. doi: 10. 1002/pts. 789

Proctor, F.J. 1985. Postharvest handling of tropical fruit for export. The Courier 92: 83-86.

Robertson G. L. 2006. Active and intelligent Packaging. In Food packaging: principles and practice – 2nd ed. CRC Press, Boca Raton, Fl. Chap. 14.

Sarkar PC 2003. Applications of lac- past, present and emerging trends. In: Recent Advances in Lac Culture. (ed. Kumar KK, Ramani R and Sharma KK), ILRI, Ranchi, India:224 –230.

White, C., 2005. Shoppers wise up on ripeness with new label. America Fruit February/March,18.

Yam KL.2000. Intelligent packaging and the future smart kitchen. Package Technol Sci 13:83-5

Yam, K. L. Takhistov, P. T. and Miltz, J. 2005. Intelligent packaging: concepts and applications. Journal of Food Science. 70: R1-R10.

Zagory, D. 1999. Effects of post-processing handling and packaging on microbial populations. Postharv. Biol. Technol. 15:313-321.

Chapter 20

Cold Chain Management of Dryland Crops

V.P. Kad, V.S. Sonone, V.L. Kanawade and J.K. Dhemre

Postharvest Technology Centre,
Mahatma Phule Krishi Vidyapeeth, Rahuri – 413 722, Dist. Ahmednagar, M.S.

Fruits and vegetables production and consumption in Asia and the Pacific region have shown a marked upward trend over the past several years. Rising consumer demand in the region has come with greater awareness of food safety issues and increased need for convenience and quality. The fresh-cut produce sector has responded to these demands, and is currently at different stages of development across the region. It may be noted that India is a predominantly an agricultural economy and 60-70 per cent of its population live in villages/rural sector and earn their livelihood through agriculture. Rural population of India was 91 per cent in 1901 and now it is about 65 per cent and may reach to 50 per cent by 2020. Rural people migrate to urban areas for employment and better amenities of life. Such opportunities are presently not available in rural areas but could be created through selective mechanization of agriculture and appropriate postharvest management and value addition to the harvested biomass in the production catchments.

India products about 600 million tons of raw food materials of plant and animal origin which are refined, stored and transformed into various usable products using conventional and modern postharvest and food processing technology. It enhances and augments per capita food availability form a unit arable land and other resources by preventing avoidable postharvest losses and adding value to the fresh agro-produces. It also creates opportunities for employment and income generation. The present levels of post-production losses are 5-15 per cent in durables, 20-30 per cent in semi-perishables and 30-50 per cent in perishables. It should that various food items, to the tune of 68 million tons, approximately worth Rs.1,00,000 crores per annum, are lost during postharvest phase. A substantial amount of these losses could be prevented if appropriate agro-processing centers having backward linkage with farmers to ensure constant supply of quality raw food materials are established and operated. Deterioration of fruits and vegetables during storage depends largely on temperature. One way to slow down this change and so increase the length of time fruits and vegetables can be stored, is by lowering the temperature to an appropriate level. It must be remembered that if the temperature is too low, the produce will be damaged and also that as soon as the produce leaves the cold store, deterioration starts again and often at a faster rate.

Cold Chain

A cold chain is a temperature-controlled supply chain. Segment of supply chain where temperature sensitive goods are stored and supplied within 2° to 8°C. The equipment and people that keep the vaccines cold during their journey are together called the ***Cold Chain.***

The term "cold chain" and the components thereof, refer to steps from harvest to consumption that extends the natural shelf life of a product by controlling temperature. Typical components of a cold chain (Figure 20.1) may include postharvest handling, refrigerated transport, refrigerated storage, controlled atmosphere storage (CA), chilled or frozen processing, cold storage holding and/ or distribution, retail refrigeration, institutional refrigeration, zero energy chamber and home refrigeration.

Any food begins to deteriorate or lose quality upon harvest whether it is meat, poultry, seafood, dairy, fruit or vegetable. Most also continue to produce heat and in some cases ripening gases, even after harvest. Removing the heat from these products and maintaining product temperature and/or storage atmospheric composition, by chilling, refrigerated storage, CA storage or freezing reduces the rate of deterioration and extends the shelf-life of the product. In addition to protecting quality, application of the appropriate cold chain components provides flexibility by making it possible to market products at the optimum time.

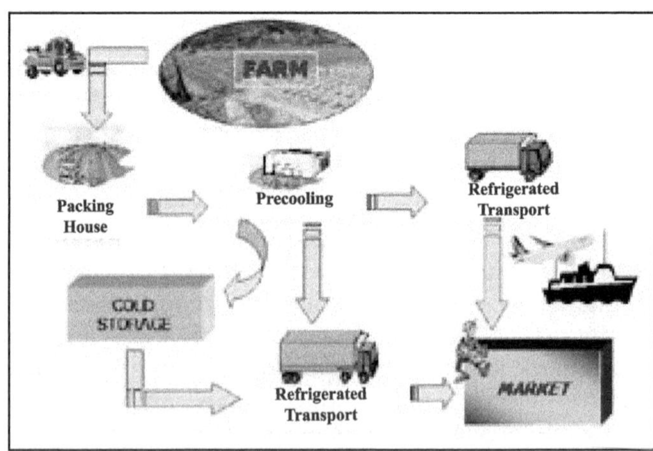

Figure 20.1: Commodity Flow in a Cold Chain Process.

Temperatures maintained in cold chain storage facilities may be divided into "refrigerated" and "frozen" categories. Refrigerated temperatures are typically those above 0°C (32°F) and frozen temperatures those lower than 0°C. Typically fresh meat, poultry, seafood, milk, flowers, fruits and vegetables are held at 38°F or 4°C while some products such as strawberries, cucumbers and tomatoes are held at higher temperatures due to sensitivity issues. Frozen storage temperatures normally include two common categories, -18°C (0°F) and – 29°C (-20°F); the latter often referred to as "ice cream temperature." Lower temperatures may be specified for specific products such as sashimi grade tuna -62°C (-80°F) but are not common. Temperatures used to freeze products are normally lower than storage temperatures.

Proper storage and warehousing is not only integral to maintaining quality, but to increasing prices for producers and/or distributors and providing consumers the benefit of longer consumption seasons.

The cold chain is the management of produce temperature, from harvesting through to consumer, to maintain the quality of the product. Maintenance of the cold chain is the best way to minimise all forms of deterioration after harvesting, including – Weight loss resulting in wilting and limpness, softening, unwanted ripening, bruising, colour changes, texture degradation, development of rots and moulds. Good cold chain management results in the consumer receiving a product of "fresh" quality, leading to greater satisfaction and increased demand.

Supply Chain Management/Cold Chain Maintenance

The **supply chain** (SC) encompasses all activities associated with the flow and transformation of goods from the raw materials stage (extraction), through to the end user, as well as the associated information flows. Materials and information flow both up and down the supply chain.

Supply chain management (SCM) is the management of a network of interconnected businesses involved in the ultimate provision of product and service packages required by end customers or "design, planning, execution, control and monitoring of supply chain activities.

Components of Cold Chain and Integration

Cold Chain Maintenance

Cold chain is a logistic system that provides a series of facilities for maintaining ideal storage conditions for perishables from the point of origin to the point of consumption in the food supply chain. The chain needs to start at the farm level (*e.g.* harvest methods, Pre-cooling) and cover up to the consumer level or at least to the retail level. The Cold chain logistics infrastructure generally consists of: Pre-cooling facilities, Cold Storages, Refrigerated Carriers, Packaging, Warehousing, Information Management systems (Traceability and Tracking *etc.*)

Start the Cold Chain at Harvest

It is essential that fruits and vegetables are not damaged during harvest and that they are kept clean. Damaged and bruised produce have much shorter storage lives and very poor appearance after storage. Dirty produce can introduce pests and moulds into the store. The produce should be harvested carefully using a sharp stainless steel knife or proper harvesting devices.

It is essential that the fruits and vegetables are harvested at the correct harvesting time.

- ☆ Harvest produce early, in the cooler hours of the day.
- ☆ Irrigation can be used to minimise field heat in some products, but only if it does not hinder harvest access or reduce the quality of the produce.
- ☆ Protect the harvested produce from the sun by placing in the shade and covering with reflective foil blankets.
- ☆ Construct a shade cover on picking platforms to keep produce in the shade.
- ☆ Transport produce to the packing shed regularly so that it can be cooled within the times given in the table below. Cover produce during transport to reduce air flow over the produce.

How Soon Need to Start Cooling?

This table provides a guide to the maximum hours from harvest to start of cooling. It is based on the perishability rating of the produce and the difference between produce harvest temperature and recommended storage temperature.

Temperature Difference	*Product Perishability Rating*			
	Very High	*High*	*Moderate*	*Low*
0 - 5°C	4hrs	8hrs	18hrs	36hrs
5 - 12°C	3hrs	5hrs	12hrs	24hrs
12 - 25°C	2hrs	3hrs	8hrs	18hrs
> 25°C	1hrs	2hrs	4hrs	12hrs

Handling

It is important that the produce does not get dirty or damaged during handling. Careful handling should be the rule. The best option is for the produce to be prepared for storage in the field and placed carefully in the storage containers used in the cold store. This considerably reduces the amount of handling and will keep damage to a minimum.

Preparation

If the produce is dirty it should be cleaned before storage. The water used has to be kept clean or fungus spores will be spread throughout the produce. Some fruit and vegetables need their outer leaves removed before storage. However, usually it is better to leave the leaves on during storage to reduce moisture loss, and then remove them before sale.

Preliminary Cooling (Precooling)

Pre-cooling is the rapid removal of heat from freshly harvested produce. This process is typically done before the produce is shipped to market or put into cold storage. Benefits resulting from pre-cooling fruits and vegetables include minimized production losses, improved economics of harvest operations, minimized losses during marketing, improved utilization by consumer, expanded market opportunities.

Rapid pre-cooling and storage at a stable temperature will extend this shelf life for most produce subject to rapid spoilage. For higher storage temperatures, it is a generally accepted rule of thumb that "deterioration of fresh produce doubles for every 10 degrees centigrade above the optimum storage temperature." Types of produce with high spoilage rates are often of higher value than those with low spoilage rates. This aspect makes control of spoilage an especially important factor for them.

Significant moisture loss affects the appearance and firmness of produce. Table 20.1 shows the maximum amount of moisture the produce may lose before losing its appearance and firmness.

Table 20.1: Percentage of Weight (Moisture) Loss from Fruits and Vegetable that Affects Produce Quality after Harvest

Sl.No.	*Product*	*Weight Loss (per cent)*
1	Asparagus	8.0
2	Beans (broad, runner, snap)	6.0, 5.0, 41.0
3	Beets	7.0
4	Broccoli	4.0
5	Brussel sprouts	8.0
6	Cabbage	8.0
7	Carrots (mature)	8.0
8	Carrots (young)	4.0
9	Cauliflower	7.0
10	Celery	10.0
11	Cucumbers	5
12	Leeks	7.0
13	Lettuce	3.7
14	Onions	10.0
15	Parsnips	7.0
16	Peas	5
17	Peppers	7.0
18	Potatoes (mature)	7.0
19	Potatoes (new)	7.0
20	Raspberries	6.0
21	Rhubarb	5

Sl.No.	Product	Weight Loss (per cent)
22	Spinach	3.0
23	Squash (summer)	23.9
24	Sweet Corn	7.0
25	Tomatoes	7.0

Pre-cooling rapidly lowers the temperature of freshly harvested produce and is done immediately following harvest to minimize spoilage. There are five principal methods of pre-cooling fresh produce: Room cooling, Forced-air cooling, Hydro-cooling, Ice cooling and Vacuum cooling.

Room Cooling

Room cooling is one method used for produce sensitive to free moisture or surface moisture. Because this type of cooling is slow, room cooling is only appropriate for very small amounts of produce or produce that does not deteriorate rapidly.Produce is loaded into a refrigerated room or transportation truck. Cold air is circulated within the room and around the produce by the refrigeration fan/s. This is a slow method of pre-cooling because the cold air does not circulate readily through the produce (Figure 20.2). This situation is particulary true for packaged produce. Room cooling systems are also used for the curing of produce such as winter squash.

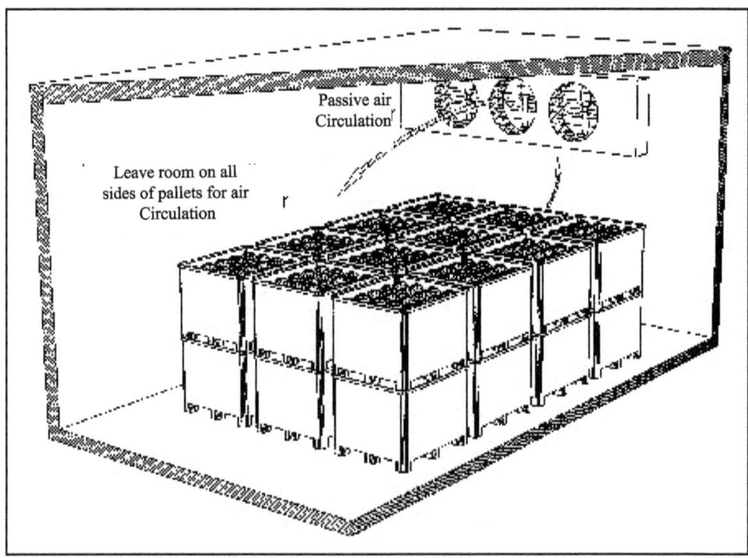

Figure 20.2: Room Cooling is a Slow Method.

Forced-Air Cooling

Forced air cooling is used mainly for bulk produce and palletized produce. It is the most versatile and widely used of all cooling methods. In forced air cooling, chilled air is forced to flow around each piece of produce. This close contact of chilled air with the produce results in rapid, even cooling throughout the mass of produce.

For piled bulk produce, such as carrots and potatoes, air ducts are used to distribute the chilled air through the produce. For palletized produce, pallets loaded with bulk or containerized produce are aligned with air ducts (plenums) that direct chilled air through them.

The air can be channeled to flow either horizontally or vertically. In a horizontal flow system, the air is forced to flow horizontally from one side of the pallet load to the other through holes in the sides of the pallet bin or containers (Figure 20.3). Only two sides that are opposite can be open in the pallet bin or containers. In stacking containers, the side holes must line up for the air to pass from one side of the stack to the other. In this system, the top and bottom of the pallet or containers must be sealed to prevent air from bypassing the produce.

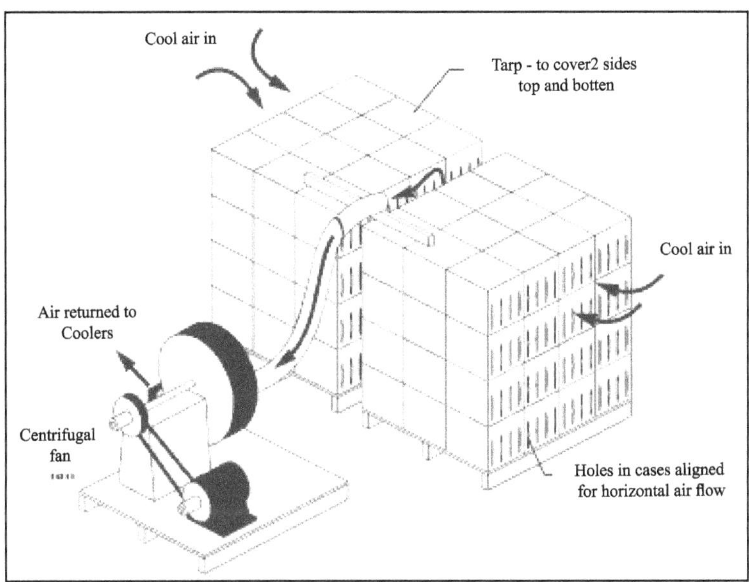

Figure 20.3: Forced Horizontal Air Flow.

In a vertical flow system, the air is forced to flow vertically from the bottom to the top of the pallet through holes in the bottom of the pallet, and containers if used, then out the top (Figure 20.4). In this system, the sides must be sealed to prevent the air from bypassing the produce. Also, if containers are used, the holes in the tops and bottoms of the containers must line up, so the air can travel vertically from one container to the next. This method is faster than room cooling because a flow of chilled air is in direct contact with the produce.

In these systems, condensation on the produce can be minimized by a simple cover placed on top of the stack of containers, which prevents the entry of ambient air during handling. To minimize moisture loss in room and forced air cooling,

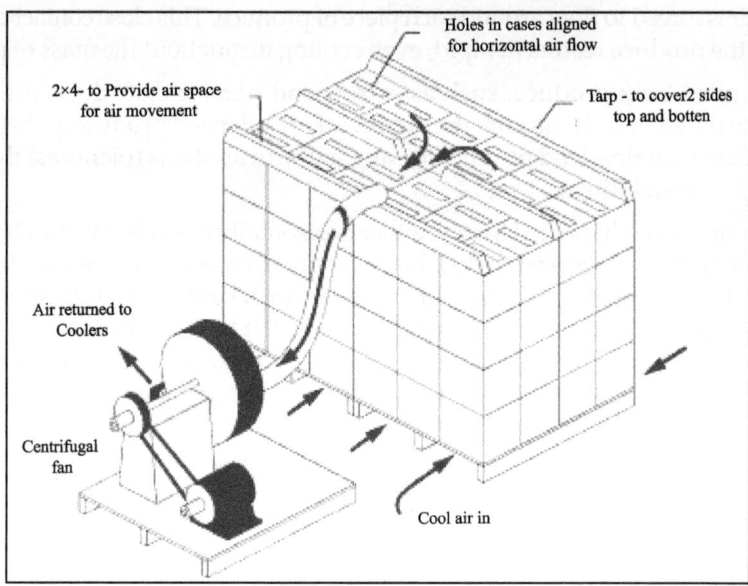

Figure 20.4: Forced Vertical Air Flow.

the evaporator of the refrigeration system should have as large a surface area as practical. Also, the evaporator should not be run at a temperature much below freezing as this practice causes undue drying of the air. Excessive air flow should also be avoided to minimize moisture loss.A bed of chunk or cube ice can be used to provide cooling in place of a refrigeration system. The temperature of the air coming through a bed of melting ice will be in the range of 1 to 4 degrees centigrade and near 100 per cent humidity.

Room and forced air cooling systems are also used for the higher temperature curing of products such as onions, potatoes and squash, before cooling for extended storage. Some market gardeners own a cold storage facility. Typically, these cold rooms are designed to remove heat from produce and other sources in 24 hours or maybe even longer. With increased refrigeration capacity and a system to force air through the produce, these facilities can also be used to rapidly pre-cool produce.

Mobile versions of forced air systems can be used in the field as an integral part of the harvest operations. Because produce can be loaded into the cooler as soon as it is harvested, cooling can be done within minutes of harvest. By minimizing the holding time before cooling, these systems can be very effective and competitive with other cooling systems that are currently considered as being fast. Mobile systems can also serve as refrigerated shipping systems for local needs.

Hydro-cooling

In hydro-cooling, cold water is used to pre-cool produce. Produce may be immersed in a tank of circulating cold water or as in other techniques, cold water is sprinkled or sprayed over the produce. The equipment used for hydrocooling is often equipped with a water chiller.

This method of pre-cooling is faster than forced air cooling because water in direct contact with the produce has a higher heat removal capacity than air. If ice is available, it can be used to chill the water. "However, not all types of produce can be hydrocooled. Strawberries and cauliflower, for example, should not be cooled by this method because it results in rapid spoilage of the produce. Hydro-cooling requires particular attention to water quality and sanitation. Chemicals may be used to kill molds and bacteria in the water and on the produce.

Ice-cooling

In ice cooling, crushed or fine granular ice is used to cool the produce. The ice is either packed around produce in cartons or sacks, or it is made into a slurry with water and injected into waxed cartons packed with produce. The ice then fills the voids around the produce.

Ice cooling is faster than hydrocooling because contact with the produce is good, and ice has a higher heat removal capacity than water. It takes 144 BTUs of energy to melt 1 pound of ice. This amount will theoretically lower the temperature of 4 pounds of produce by at least 36 degrees Fahrenheit.

Ice can also provide cooling for extended periods. It is often a marketing requirement that there be ice remaining in the sacks or cartons of produce at the time it is received by the retailer. As in hydro-cooling, ice cooling requires particular attention to water quality and sanitation.

Ice has characteristics that make it very effective in pre-cooling fruits and vegetables, especially for small-scale market gardeners. These characteristics include its versatility as a cold source for several pre-cooling methods, its thermal storage capacity and its portability.

.Ice may be considered stored refrigeration. A small ice-making machine can produce ice continuously for use at a high rate during a short period. The capital cost of the refrigeration system can therefore be a fraction of the cost of a conventional refrigeration system of the same capacity.

Cooling provided by melting ice is naturally regulated at just above freezing with the humidity of air around the ice near 100 per cent. The ice can be used several ways: directly on the produce, or for chilling the water for hydro-cooling, or in an ice bed to chill the air for forced air cooling.

Ice is also very adaptable to pre-cooling in the field as an integral part of the harvest operation. Residual ice in a mobile ice based cooler can also help maintain the temperature of produce during transport. These characteristics make ice ideal for use in cooling most fresh produce.

Many market gardeners do not pre-cool their produce because of the economics involved. By using ice as a cold source, the fixed cost of refrigeration can be substantially reduced, and the benefits of quality produce realized. The adaptability of ice to provide refrigeration for both wet and dry types of pre-cooling can be the basis for low cost pre-cooling of a wide range of fresh produce grown by market gardeners.

If cube or chunk ice is to be used for direct icing of produce, it should be crushed first. Crushed ice can also be used in a slurry with water for injection into cases of produce.

If ice is to provide refrigeration capacity for a forced air system, the air must be able to flow uniformly and easily around each piece of ice in a bin for the air to be chilled efficiently. Cube or chunk ice is best for this purpose. Flake and crushed ice tend to nest in a way that restricts the flow of air.

Pressure drop through an ice bin can be reduced by increasing the area of the bed. The reduced pressure requires less fan energy and, thus, less heat is generated by the fan.

Data and Rules of Thumb for Developing an Ice-based System

- ☆ Air required for pre-cooling is about 2 cubic feet per minute per pound of produce. Allow for at least 2 inches water pressure to move the air through the system.
- ☆ Air velocity through the floor should be kept below 200 feet per minute to avoid undue pressure drop.
- ☆ Specific heat of air is 0.23 BTUs per pound per degree Fahrenheit.
- ☆ 1 pound of air is about 14 cubic feet.
- ☆ The bulk density of cube ice is about 32.8 pounds per cubic foot.
- ☆ Latent heat of ice is 144 BTUs per pound.
- ☆ Latent heat of vaporization of water is about 1000 BTUs per pound.
- ☆ Specific heat of water is 1 BTU per pound per degree Fahrenheit.
- ☆ Maximum depth of ice should not exceed about 18 inches to avoid undue pressure drop.
- ☆ Minimum depth of ice at the start of a cooling cycle should not be less than 8 inches to ensure adequate cooling during the time of high temperature difference across the bed.
- ☆ Ice bin floors should be 60 per cent open to contain the ice while minimizing pressure loss.
- ☆ The shape and size of ice bin floor openings should be such as to contain the shape and size of ice used.
- ☆ Cube or chunk ice should be about 1 inch across to minimize pressure loss and provide sufficient surface area for good cooling.
- ☆ It takes about 1 pound of ice to cool 2 to 3 pounds of produce. The actual amount depends mainly on temperature of the produce, how well the cooler is insulated and sealed and the ambient temperature.
- ☆ Specific heat of high moisture produce is about 0.85 BTUs per pound per degree Fahrenheit

Vacuum Cooling

In vacuum cooling, containers of produce are put in a vacuum chamber. Air is drawn out of the chamber creating a high vacuum. Under the high vacuum, the boiling temperature of water is substantially lowered.

For example, water at the freezing point boils when the pressure is reduced to 0.08859 psi absolute. Normal atmospheric pressure is about 14 psi absolute. As a result, a small amount of water in the warm produce evaporates causing a corresponding lowering of the produce temperature. In some systems, the produce is misted with water to minimize weight loss.

Vacuum cooling is the fastest and most uniform method of pre-cooling because of the high amount of heat required for evaporating the water (about 1000 BTUs per pound of water evaporated). It is, however, one of the most expensive systems to set up. Because vacuum cooling is a batch process, sufficient produce must be harvested to fill the vacuum cooler before cooling can begin. For the system to be effective, it is imperative that the produce not be held any longer than absolutely necessary after harvest and before cooling. As a result, harvesting and vacuum cooling systems must be very carefully matched.

Keep Produce Cool during Handling

- ☆ Keep the packing shed as cool as possible. Depending on conditions, consider installing: A refrigerated packing area, Air conditioning, Evaporative cooling, Shed insulation, Sprinklers on the roof, Shade trees and grass
- ☆ Minimise the time the product spends out of the cool room during grading and packing.
- ☆ Measure produce pulp temperature, after handling procedures to determine whether rapid re-cooling is necessary. If not, produce should be placed directly in the cool room.
- ☆ Set a maximum "on packing shed floor" time for the produce being handled, taking into account product perishability and packing shed conditions.
- ☆ For some products (especially slow accumulating sizes and smaller volume packing operations) consider moving packed produce from the packing line directly to the cool room and palletising to common sizes inside the cool room.

Refrigerated Transport

This may include long haul refrigerated equipment including truck/trailer combinations, containers, specially designed rail cars, and equipment appropriate for air transport. Local delivery equipment may include smaller truck/trailer units, truck units, or even units that are well insulated but lack integral refrigeration equipment. Shipments of frozen or refrigerated meats by sea occurred as early as the nineteenth century using a combination of salt, ice and sawdust.

However, railroads using ice and specially insulated railcars really pioneered the transportation of significant quantities of meat and produce beginning in the mid to late nineteenth century. By the mid-twentieth century, the rail cars were converted to mechanically refrigeration using diesel or electric powered units and currently many of these rail cars, as well as newly constructed ones have had those units removed and now employ end mounted truck/trailer units. The advent of intermodal units (refrigerated trailers or containers) markedly increased the capacity of rail systems to economically transport large volumes of refrigerated and frozen products throughout much of the world. While a few "refrigerated break-bulk" freighters still exist, the vast majority of sea borne refrigerated transport is now done in containers. Once the initial heat is removed from the produce it should be immediately transported to the customer, packing house or cold storage facility, preferably in refrigerated trucks and/or trailers. Every effort should be made to insure the product does not regain heat during transportation.

Refrigerated Storage

Often associated with packing houses, these facilities hold the harvested products until distributed. Distribution may occur concurrent with the harvest, or may be at a later date, when the commodity is no longer in season and its price has increased thus permitting the grower, or the trader, a greater return on investment. This latter period not exceed six months, but in more developed economies product, particularly frozen product, may be held for a year or more. Environmental humidity is also a critical factor in storing fruits and vegetables for an extended time. Fruits and vegetables should be stored in storerooms with humidity in excess of 85 per cent; the exception being onions and some other products that require lower humidity in the range of 75 per cent.

During the distribution process one should monitor that process until one builds a sufficient data set that clearly demonstrates the process is in compliance and in a state of control. Each time the process does not conform to the process, the event should be properly documented, investigated and corrected so that the temperature excursion do not occur on future shipments. Thus the process is continually evolving and correcting for anomalies that occur in the process. Eventually the process can evolve into periodic monitoring once sufficient data demonstrates that the process is in a state of control. Any anomaly that occurs once a process is in a state of control will result in the process being invalidated and not in control and result in product withdraw from the market to ensure patient safety.

It is necessary to develop an internal documentation system as well as multi-party communication standards and protocols to transfer or create a central repository or hub to track information across the supply chain. These systems would monitor equipment status, product temperature history, and custody chain, *etc.* These help ensure that a food is safe and effective when reaching its intended consumer.

References

Evans, J. no date. Cold storage of food: Review of available information on energy consumption and energy savings options. Food Refrigeration and Process Engineering Research Centre (FRPERC), UK. http://www.grimsby.ac.uk/documents/defra/stor-coldstorescasestudy.pdf

Gustavsson, J., 2011. Global Food Losses and Food Waste: Extent, Causes and Prevention. UN FAO: Rome http://www.fao.org/fileadmin/user_upload/ags/publications/GFL_web.pdf

Kader, A.A. 2006. The return on investment in postharvest technology for assuring quality and food safety in horticultural crops. Journal of Agricultural Investment 4: 45-54.

Kitinoja, L., S. Saran, S.K. Roy, and A.A. Kader. 2011. Postharvest technology for developing countries: challenges and opportunities in research, outreach and advocacy. J. Sci. Food Agric. 91:597-603. http://ucce.ucdavis.edu/files/datastore/234-1922.pdf 15

Kitinoja, L. and J.F. Thompson. 2010. Pre-cooling systems for small-scale producers. Stewart Postharvest Review, 2010, 2:2, 14p. http://ucce.ucdavis.edu/files/datastore/234-1594.pdf

Lipinski, B., C Hanson, R Waite, L Kitinoja, T Searchinger, J Lomax,. 2013. Creating a Sustainable Food Future, Installment Two: Reducing Food Loss and Waste. World Resources Institute Working Paper: June, 2013. http://www.wri.org/publication/reducing-food-loss-and-waste

Mrema, G.C. and R.S. Rolle. 2002. Status of the postharvest sector and its contribution to agricultural development and economic growth. 9th JIRCAS International Symposium on 'Value-Addition to Agricultural Products', Ibaraki, Japan, pp.13-20.

Thompson, J.F., Brecht, P.E., and Hinsch, T. 2002a, Refrigerated trailer transport of perishable products. University of California, Division of Agriculture and Natural Resources Publication no. 21614.

Thompson JF, Mitchell FG and Kasmire RF. 2002b. Cooling horticultural commodities. In Postharvest technology of horticultural crops, Third edition, Kader AA (Tech. Ed). University of California, Division of Agriculture and Natural Resources, Publication 3311. 2002: 97– 112.

USDA Handbook 66. 2004. Commercial storage of fruits, vegetables and nursery stocks. http://www.ba.ars.usda.gov/hb66/

Winrock International. 2009. Empowering agriculture: energy solutions for horticulture. USAID Office of Infrastructure and Engineering and the Office of Agriculture; 79p. Available online: http://ucce.ucdavis.edu/files/datastore/234-1386.pdf

Yahia, E.M. 2010. Cold Chain Development and Challenges in the Developing World. Acta Hort. 877:127-132.

References

Evans, J. Rundale. Cold storage of food. Review of available information on energy consumption and energy saving options. Food Refrigeration and Process Engineering Research Centre (FRPERC). UK. http://www.grimsby.ac.uk/documents/defra/stor-coldstore-stats.pdf.

Gustavsson, J. 2011. Global Food Losses and Food Waste: Extent, Causes and Prevention. UN FAO: Rome. http://www.fao.org/fileadmin/user_upload/ags/publications/GFL_web.pdf.

Kader, A.A. 2005. The return on investment in postharvest technology for assuring quality and safety of horticultural crops. [illegible]

[illegible]

Chapter 21

Recent Advances in Storage of Fruits Crops

V.P. Kad, V.S. Sonone, J.K. Dhemre and A.P. Patil

Postharvest Technology Centre,
Mahatma Phule Krishi Vidyapeeth, Rahuri – 413 722, Dist. Ahmednagar, M.S.

Postharvest technologies have allowed horticultural industries to meet the global demands of local and large-scale production and intercontinental distribution of fresh produce that have high nutritional and sensory quality. Harvested produce are metabolically active, undergoing ripening and senescence processes that must be controlled to prolong postharvest quality. Inadequate management of these processes can result in major losses in nutritional and quality attributes, outbreaks of foodborne pathogens and financial loss for all players along the supply chain, from growers to consumers. Optimal postharvest treatments for fresh produce seek to slow down physiological processes of senescence and maturation, reduce/inhibit development of physiological disorders and minimize the risk of microbial growth and contamination. In addition to basic postharvest technologies of temperature management, an array of others have been developed including various physical (heat, irradiation and edible coatings), chemical (antimicrobials, antioxidants and anti-browning) and gaseous treatments.

The maintenance and improvement in the postharvest quality and postharvest life of fresh fruits and vegetables is become increasingly important. This has been partly the respose to the free market situation where supply of good quality fruits and vegetables constantly exceed demand. Therefore to maintain or increase the market share there increasing emphasis on quality. Also consumer expection in supply of all types of fruits and vegetables throughout the year often taken for granted. This latter expection is partly supplied by long term storage of many crops and also by long-distance transport. With growing awareness and concern for climate change, long distance transport of fruits and vegetable being questioned (Thomsons, 2010).

Proper marketing of perishable commodities such as fruits and vegetables often required some storage to balance day-to-day fluctuations between harvest and sale or long term storage to extend marketing beyond the end of the harvesting season. Storage of fresh fruits and vegetables prolong their usefulness and in some cases improves their quality. It also controls the market glut. The principle role of storage is to control the respiration, transpiration, deseases and the insect infestation and to preserve the commodity in its most usable form for the consumer. Storage life can be prolonged by harvesting at proper maturity, control of postharvest diseases, chemical treatment, irradiation, refrigeration, controlled and modified atmosphere and several other tratments. The main goal of storage include:

1. Slow the biological activity of fruits and vegetable without chilling injury
2. Slow the growth of micro-organisms
3. Reduce transpirational losses to avoid the undesiration processes that may occur in certain fruits and vegetables.

Different Storage Techniques

1. Control Atmosphere Storage

Control atmosphere storage is the most important innovative stoarge in the fruits and vegetable. CA storage is now used worldwide on a variety of fresh fruits and vegetables. The stimulation for its development was argubly the requirement for extended availability of fruits and vegetables, especially certain cultivar of apple that were subjected to chilling injury, which reduced the maximum storage period. CA storage has been subject the subject of an enormous number of biochemical, physiological and technological studies, inspite of which it is still not known precisely why it works.

First study regarding fruit presevation in gaseous were carried out in an atmosphere consists of more CO_2 and less O_2 than normally present in air. CO_2 plays the dominant role, therefore called the "Carbon dioxide preservetion method". It was found that CO_2 concentration in the regime not only reduced the vital activities of fruits but also provide strong antiseptic nature.

CA storage refers to the monitoring and adjustment of the CO_2 and O_2 levels within gas tight stores at optimum storage temperature. Thus, the atmosphere is controlled rather than established passively as in the case of MAP, though the effects of altered atmospheres on metabolism of fruits and vegetables are essentially the same. In most cases, the concentrations of CO_2 are higher and those of O_2 are lower, optimum concentrations depending on the specific product and the purpose of the CA storage conditions. Reduced O_2 and elevated CO_2 levels affect both primary (glycolysis, fermentation and aerobic respiration) and secondary (*e.g.* processes involved in ethylene production and action, pigments, phenolics and volatiles) metabolism (Beaudry, 1999).

The each fresh fruits and vegetable has an optimal range of O_2 and CO_2 for maintaining quality and extending shelf life (Saltveit, 2003), and these can differ for whole and fresh-cut products of the same fruit or vegetable. Benefits of CA as well as detrimental effects resulting from exposure of fruits and vegetables to atmospheres outside of the safe range were summarized by Kader (2004). The beneficial effects of CA include:

1. Retardation of senescence and associated biochemical and physiological changes, *e.g.* slowing down rates of respiration, ethylene production and softening
2. Reduction in sensitivity to ethylene action at O_2 levels less than 8 per cent and/or CO_2 levels more than 1 per cent
3. Alleviation of certain physiological disorders, *e.g.* chilling injury of avocado and superficial scald of apples and pears
4. Direct or indirect effect on postharvest pathogens (bacteria and fungi) and consequently decay incidence and severity *e.g.* CO_2 at 10–15 per cent inhibits development of *Botrytis* rot on strawberries and cherries and
5. Low O_2 (less than 1 per cent) and/or elevated CO_2 (40–60 per cent) can be a useful tool for insect control in some fresh and dried fruits, flowers and vegetables; and dried nuts and grains.

The Detrimental Effects of CA include

1. Initiation and/or aggravation of certain physiological disorders, such as internal browning in apples and pears, brown stain of lettuce and chilling injury of some commodities
2. Irregular ripening of fruits, such as banana, mango, pear and tomato, can result from exposure to O_2 levels below 2 per cent and/or CO_2 levels above 5 per cent for more than one month
3. Development of off-flavours and off-odours at very low O_2 concentrations (as a result of anaerobic respiration) and very high CO_2 levels (as a result of fermentative metabolism) and
4. Increased susceptibility to decay when the fruit is physiologically injured by insufficient O_2 or too-high CO_2 concentrations.

Use of CA technology is limited to relatively few fresh fruits and vegaetables, the major crop being apples. Atmospheric modification during long-distance transport is used on apples, asparagus, avocados, bananas, broccoli, cranberries, cherries, figs, kiwifruits, mangos, melons, nectarines, peaches, pears, plums and strawberries.

2. Dynamic Controlled Atmosphere (DCA)

A major problem with long term storage of apples and pear is the development of superficial scald on the fruit. This has been controlled by Diphenylamine (DPA) and more recently by Smart Fresh. Smart Fresh has radically changed fruit storage, but it is not currently certified by the organic movement and consumers are becoming increasingly concerned about chemicals, such as DPA, on fruit, and are seeking a 'no chemicals' approach to production and distribution. In many developed markets, supermarkets are responding to this by dictating residue levels well below those officially allowable. As a result, apples in some of these markets are not allowed to be treated with DPA or to be waxed and there is a preference for fruit certified as organic. Even before the commercialisation of Smart Fresh in Europe and North America in the 1990s, there was strong consumer demand for DPA-free fruit.

Dynamic Controlled Atmosphere (DCA) storage technology emerged as an appealing option, because it is non-chemical and uses existing controlled atmosphere (CA) technology. The main thing we are controlling is the oxygen levels in existing CA rooms. DCA and CA storage are similar, in that both maintain quality by storing products in a gas tight container that controls temperature, oxygen and carbon dioxide concentration and relative humidity. DCA technology uses fluorescence detection sensors to dynamically control the atmosphere. It makes the storage of fruit at its lowest respiration rate possible. The sensors measure the amount of fluorescence light emitted by the fruit's skin, allowing operators to pin-point the anaerobic point. This is an essential parameter in determining the minimum level of oxygen to be maintained in the room during storage. With DCA, it is possible to create ideal conditions for the medium to long-term storage of fruit, as it allows the operator to lower the oxygen levels in storage rooms without the risk of physiological disorders. It is thus possible to set and adapt the values of gases in a dynamic manner, based on the maturity, climatic variations, different origins and varying respiration rhythms of the monitored fruit.

Dynamic Controlled Atmosphere (DCA) technology is a healthy, environmentally-friendly way to keep fruit fresh using low oxygen levels to prevent the development of scald. DCA works by monitoring the stress level of the fruit in storage and adjusting the oxygen (O_2) and carbon dioxide (CO_2) levels by using ethanol to maximise fruit storage potential while maintaining sound, marketable fruit.

3. Modified Atmosphere Storage (MAP)

MAP can be broadly defined as any process which significantly changes the environment around the product from the normal composition of air in a package. The generic term also includes applications such as vacuum packaging. Map can be used in combination with a number of other food processing, preservation systems and packaging techniques to maintain product quality longer and extend shelf life. MAP benefits to the food product are extended shelf life and maintenance of flavour, colour, taste, texture and nutritive value. Shelf life extension is not a fad or marketing but a means to maximize the use of food chain. Extension of shelf life provides greater opportunities to sell the product may be able to remain in the distribution chain longer allowing it to reach new geographical markets which in turns open up new consumer markets.

Storage of fruits in packaging of polymer materials with selected permeability for gases is what it is called modified atmosphere storage. This technique was found to enhance the shelf life of produce stored in polymeric pouches, bags and containers. Such storage is done in conventional cold storage chambers and therefore does not involves large capital outlay for construction and equipping of specially hermetically sealed chambers with controlled atmosphere.

The growth in MAP fresh-cut (minimally processed) produce is one of the most dynamic in the last decade. Fresh produce are living, respiring organisms. Produce packaging involves a delicate balance of the products respiration rate, product temperature and the permeability of packaging materials. Selection of correct packaging material is crucial to maintain the desire balance of O_2 (2 per cent) and CO_2 (<15 per cent) which in turn, slow the product metabolisms and thus product senescence

Although gas atmosphere is important. Other criteria must be considered when contemplating MAP technology, which include temperature and relative humidity control, packaging equipment, packaging materials and food safety issues. High performanace packaging and packaging equipment are continually evolving. Technological advancements in packaging machinery have focused on higher speeds and lower residual oxygen levels.

Emerging packaging technologies are continually evolving to prolong the quality of fresh refrigerated and minimally processed foods. These emerging technologies involve active packaging including oxygen scavenging films; high oxygen barrier polymers, high oxygen permeable films and polymers with refined properties. Active packaging involves technologies in which the package interacts with the internal package atmosphere and/or the food in the package. The active system can be an integral part of the package or be a separate component placed inside the package.

Methods of Creating Modified Atmosphere Conditions

Modified atmospheres can be created either passively by the commodity or intentionally via active packaging (Kader *et al.*, 1989; Zagory and Kader, 1988)

Passive Modified Atmosphere

Modified atmospheres can passively evolve within a hermetically sealed package as a consequence of a commodity's respiration, *i.e.* O_2 consumption and CO_2 evolution. If a commodity's respiration characteristics are properly matched to film permeability values, then a beneficial modified atmosphere can be passively created within a package. If a film of correct intermediary permeability is chosen, then a desirable equilibrium modified atmosphere is established when the rates of O_2 and CO_2 transmission through the package equal a product's respiration rate.

Active Packaging

By pulling a slight vacuum and replacing the package atmosphere with a desired mixture of CO_2, O_2 and N_2, a beneficial equilibrium atmosphere may be established more quickly than a passively generated equilibrium atmosphere. Another active packaging technique is the use of O_2, CO_2 or ethylene scavengers/emitters. Such scavengers/emitters are capable of establishing a rapid equilibrium atmosphere within hermetically sealed produce packages.

Advantages and Disadvantages of MAP

Advantages of MAP

a. Increased shelf-life allowing less frequent loading of retail display shelves.
b. Reduction in retail waste.
c. Improved presentation-clear view of product and all round visibility.
d. Hygienic stackable pack, sealed and free from product drip and odour.
e. Easy separation of sliced products.
f. Little or no need for chemical preservatives.
g. Increased distribution area and reduced transport costs due to less frequent deliveries.
h. Centralized packaging and portion control.
i. Reduction in production and storage costs due to better utilisation of labour, space and equipment.

Disadvantages of MAP

a. Capital cost of gas packaging machinery.
b. Cost of gases and packaging materials.
c. Cost of analytical equipment to ensure that correct gas mixtures are being used.
d. Cost of quality assurance systems to prevent the distribution of leakers, *etc.*
e. Increased pack volume which will adversely affect transport costs and retail display space.
f. Potential growth of food-borne pathogens due to temperature abuse by retailers and consumers.
g. Benefits of MAP are lost once the pack is opened or leaks.

4. Ultra Low Oxygen Storage

It was started back in 1819 when Frenchman Jacques Etienne Berard discovered that fruit absorbs oxygen and produces carbon dioxide. He also proved that the fruit would not ripen without oxygen but continued to do so as soon as oxygen conditions were restored. In 1962, again in France, commercial scrubbed, low oxygen storage began in earnest, offering a natural way to slow the ripening of the fruit. Today's highly sophisticated ultra low oxygen stores ensure that Bramley apples are still available and in great condition right the way through to next season's crop. Another major advantage of ultra low oxygen storage is that it is a natural process that totally eliminates the need for any other chemicals.

This concept is based on the maintenance of controlled atmosphere chambers. The abbreviation comes from ULO "Ultra Low Oxygen," which means "extremely low levels of oxygen." Controlled atmosphere allows extended storage of fruits and vegetables. It is understood very well the performance in terms of chamber gas leak, sealing and the use of specially designed elements (of cooling doors, window control and service editorials for pipes and cables). The system is designed to control O_2 and CO_2, lowering content of oxygen (O_2) slows fruit respiration and allows a longer period of storage. However, too low content of oxygen leads to the "suppression" of the fetus, therefore it is necessary to provide adequate ventilation to ensure that oxygen levels in cold storage does not fall too low. During breathing, fruits and vegetables convert oxygen into carbon dioxide (CO_2). If the CO_2 level is too high, it can damage your fruit and therefore must be eliminated. It is therefore recommended that the oxygen in cold storage decreases more rapidly and controlled manner - by injecting nitrogen (N_2). The quality of the equipment to control the atmosphere is an extremely important aspect. In the developed world is the principle of ULO already become a norm for the storage of pears and apples.

5. Edible Coating

Edible coating technology is a promising method to preserve the quality of fresh and minimally processed fruits. Research and development efforts are leading to an improvement of the functional characteristics of the coatings, which depends on the properties of the fruit to be preserved or enhanced. These can be achieved by a precise optimum control of gas permeability, texture, and color changes by means of quantitative or qualitative changes in coating formulation.The application of edible coatings on fresh produce provides a partial barrier to the movement of moisture on the surface of fresh produce, thereby minimizing moisture loss during postharvest storage; a gas barrier, thereby establishing a modified atmosphere around the product, which slows down respiration, senescence and enzymatic oxidation and preserves colour and texture; helps to retain volatile compounds contributing to produce a natural aroma and restrict foreign odours; maintains fresh produce structural integrity, and protects against mechanical damages; and serves as carriers of functional or active compounds, such as nutraceuticals, flavouring and colouring agents, antioxidants and antimicrobials, that will maintain/improve product quality and safety (Dhall, 2013 and Ghasemnezhad *et al.*, 2013). Edible coatings are composed of hydrophobic groups, such as lipid-based waxes; hydrocolloid/hydrophilic groups, such as polysaccharide or protein-based materials; or an integration of both groups in order to improve the functionality of the coating (Gol *et al.*, 2013). Within the last decade, there has been a considerable amount of research and innovations focused on the development of edible coatings from natural or synthetic sources in order to control physiological and pathological challenges of fresh produce.

Several edible coatings including chitosan, *Aloe vera*, polyvinyl acetate, mineral oils, cellulose and protein based have shown desirable attributes on fresh produce with good barrier properties, without residual odour or taste and efficient antimicrobial activity (Dhall, 2013). However, more research is required to enhance moisture barrier properties of hydrophilic edible coatings, improve coating adhesion and durability during storage. To maximize the benefits of edible coatings for fresh produce, it is important to understand the effect of storage conditions on the desired functions and the adverse effect on fresh produce quality. The main limitation for the application of edible coating at the industrial level is the cost of scaling up research concepts or investment for new installation of film production and coating equipment, the lack of edible materials with desired physical and functional properties as well as the challenges of regulatory status for the different coating materials. Furthermore, process parameters, such as the method of coating and the amount of additives, can affect the film barrier properties and overall quality of the food product. One of the commercial coating products is Natureseal, which maintains colour, texture and shelf life of a number of fresh-cut fruits *e.g.* apples, pears, carrots, celery, *etc.*, has recorded a good success.

6. Cold Storage

Cold Storage is a special kind of room, the temperature of which is kept very low with the help of machines and precision instruments. India is having a unique geographical position and a wide range of soil thus producing variety of fruits and vegetables like apples, grapes, oranges, potatoes, chillies, ginger, *etc.* Marine products are also being produced in large quantities due to large coastal areas. The present production level of fruits and vegetables is more than 200 million MT and keeping in view the growth rate of population and demand, the production of perishable commodities is increasing every year. The cold storage facilities are the prime infrastructural component for such perishable commodities. Besides the role of stabilizing market prices and evenly distributing both on demand basis and time basis, the cold storage industry renders other advantages and benefits to both the farmers and the consumers. The farmers get opportunity of producing cash crops to get remunerative prices. The consumers get the supply of perishable commodities with lower fluctuation of prices. Commercially apples, potatoes, oranges are stored on large scale in the cold storages.

Cold storages are essential for extending the shelf life, period of marketing, avoiding glut, reducing transport bottlnecks during peak period of production and maintenance of quality of produce. The development of cold storage industry has therefore an important role to play in reducing the wastages of the perishable commodities and thus providing remunerative prices to the growers. Deterioration of fruits and vegetables during storage depends largely on temperature. One way to slow down this change and so increase the length of time fruits and vegetables can be stored, is by lowering the temperature to an appropriate level. It must be remembered that if the temperature is too low the produce will be damaged and also that as soon as the produce leaves the cold store, deterioration starts again and often at a faster rate.

7. Ethylene Scavenging Technology

Ethylene acts as a plant hormone that has different physiological effects on fresh fruits. It accelerates respiration, leading to maturity and senescence and also softening and ripening of many kinds of fruits. Furthermore, ethylene accumulation responsible for a number of specific postharvest disorders in fresh fruits. Although some effects of ethylene are positive, such as degreening of citrus fruits, ethylene is often detrimental to the quality and shelf-life of fruits. To prolong the shelf-life and maintain an accepatble visual and oraganoleptic quality, accumulation of ethylene in the packaging of fruits should be avoided. A number of ethylene sorbing substances are available.

Ethylene is a naturally occuring, chemically simple molecule of the alkene type that regulates numerous aspects of growth, development and senescence of many fruits. As it is effective at part-per million or part-per-billion concentration and its effects are very dose-dependent, it is considered as a plant hormone (Saltveit, 1999). Environmental ethylene can be produced both biologically and non-biologically. Biological source of ethylene includes higher plant tissues, several species of bacteria and fungi, some algae (Zagory, 1995).

In recent years, however, there has been a scarcity of research on developing new and more efficacious ethylene scrubbing materials. In contrast, there has been an exponential increase in research using the ethylene binding inhibitor 1-methylcyclopropene (1-MCP;Watkins, 2006). Thus, research activity has drifted away from ethylene removal to preventing the actions of ethylene through using 1-MCP. Despite various ethylene scrubbing technologies being available (*e.g.* high temperature catalytic degradation, activated carbon, *etc.*) most commercial ethylene control systems rely on both adequate ventilation (often periodic) and oxidation of ethylene using potassium permanganate ($KMnO_4$)-based mechanisms. Ventilation, however, is not appropriate in sealed environments (*e.g.* controlled atmosphere or some packaging formats) or where precise ethylene control is required. $KMnO_4$ supported on activated alumina spheres has limited long-term efficacy in environments with high relative humidity (*e.g.* cold stores). While in activated carbon based technique, various metal catalysts on activated carbon will effectively remove ethylene from air passing over the bed of carbon. Commercial units, known as swingtherm ethylene converters, are based on such a system. However, they require heat and movement of gases and so are not applicable to packaged produce. Activated charcoal impregnated with a palladium catalyst placed in paper sachets effectively removed ethylene in an experiment on maintaining quality of lightly processed kiwifruit, banana, broccoli and spinach (Abe and Watada, 1991).

Treatment to Prolong the Shelf Life of Fruits

1. 1-Methylcyclopropene Technique

The discovery and patenting of cyclopropenes as inhibitors of ethylene perception represents a major breakthrough in controlling ethylene responses of horticultural products (Blankenshi and, Dole, 2003). The process of discovery of the effects of cyclopropenes, and their proposed method of action, has been described (Sisler, 2006 and Sisler and Serek, 2003). Of the cyclopropenes, 1-MCP proved to be extremely active, but unstable in the liquid phase. However, 1-MCP can be complexed with *dinna*-cyclodextrin to maintain its stability; this development represented a major step towards its commercialization as it was then possible to release 1-MCP from the complex to expose to the horticultural products. Regulatory approval for use of 1-MCP has been obtained in more than 50 countries, and approval for use of the technology continues to occur around the world. 1-MCP is registered for use on a wide variety of fresh fruits and vegetables including apple, avocado, banana, broccoli, cucumber, date, kiwifruit, mango, melon, nectarine, papaya, peach, pear, pepper, pineapple, plantain, plum, squash and tomato. 1-MCP affects many ripening and senescence processes (Watkins, 2006 and Serek *et al.*,

2006), including pigment changes, softening and cell wall metabolism, flavour and aroma, and nutritional properties, but to varying degrees in both non-climacteric and climacteric products. The range of responses reflects the enormous diversity of these crops in terms of both inherent diversity and morphological derivation (Huber, 2008). Several generalizations can be made about responses of crops to 1-MCP (Huber, 2008 and Watkins, 2008).

1. Genotype, cultivar and maturity effects can be highly variable
2. Climacteric fruit are affected by 1-MCP treatment, but the capacity to interrupt ripening once initiated varies by fruit and attributes studied;
3. Non-climacteric fruit also respond to 1-MCP, providing insights into ethylene-dependent and -independent events during ripening
4. Treated fruits are firmer, slower to soften, slower to change peel colour, and develop aroma and flavour. If 1-MCP concentrations and exposure periods for each fruit must be appropriate to allow ripening to occur, so the final quality of the treated fruit is similar to that of the untreated product and
5. The effect of 1-MCP on physiological disorders is dependent on the role of ethylene. 1- MCP decreases senescent-related and ethylene-induced disorders (senescent breakdown of apples and water soaking of watermelons). 1-MCP can increase susceptibility of apple fruit to CA storage-related disorders (CO_2 injury). Some chilling-related disorders (woolliness and internal breakdown of peaches and nectarines, chilling injury of citrus and bananas) can be increased by inhibition of ethylene production.Other chilling-related disorders (superficial scald of apples and pears and internal flesh browning of avocados and pineapples) are decreased by inhibition of ethylene production.

2. Nitric Oxide

Nitric oxide (NO) is a highly reactive free radical gas and acts as a multifunctional signalling molecule in various plant physiological processes, such as fruit ripening and senescence of fresh fruits and vegetables (Wendehenne *et al.*, 2004). Optimum NO levels delay the climacteric phase of many tropical fruits, prolong the postharvest storage life by impeding ripening and senescence, suppress biosynthesis of ethylene, reduce ethylene production and, consequently, delay in fruit ripening (Singh *et al.*, 2013). NO gas is applied as a fumigant or released from compounds such as sodium nitroprusside, S-nitrosothiols and also diazeniumdiolates used as a dipping treatment. Reduced ethylene production during ripening in NO-fumigated fruit has been claimed owing to binding of NO with 1-aminocyclopropane-1-carboxylic acid (ACC) and ACC oxidase to form a stable ternary complex, thus limiting ethylene production (Zaharah and Singh, 2011). Other mechanism of NO action include the inhibition of ethylene biosynthesis, cross-communication with other phytohormones, regulation of gene expression (Manjunatha *et al.*, 2012) and amelioration of oxidative postharvest stress (Singh *et al.*, 2009).

Successful application of NO has been reported for apple, banana, kiwifruit, mango, peach, pear, plum, strawberry, tomato, papaya, loquat, Chinese winter jujube fruit and Chinese bayberry (Manjunath *et al.*, 2012). NO treatments reduce the rate of respiration, water loss and inhibit browning, as well as reducing the incidence of postharvest diseases. NO fumigation in combination with cold storage has a synergistic effect in extending storage life of fruit such as plum and mango (Zaharah and Singh, 2011 and Singh *et al.*, 2009). Similarly, NO combined with modified atmosphere conditions extended the postharvest life of green beans, broccoli and bok choy (Soegiarto and Wills, 2004). Commercial application of NO in fresh fruits and vegetables depends upon the development of a smart carrier/ controlled release system for NO.

3. Sulfur Dioxide

Sulfur dioxide (SO_2) is widely used on table grapes to prevent decay during storage, by either initial fumigation of fruit from the field followed by weekly fumigation of storage rooms or slow release from in-package pads containing sodium metabisulfite (Palou *et al.*, 2010). SO_2 technology has also been tested for control of postharvest decay on other fruits such as litchi, fig, banana, lemon or apple (Sivakumar *et al.*, 2010). Cantín *et al.* (2012) reported that SO_2 fumigation followed by CA storage (3 per cent O_2 + 6 or 12 per cent CO_2) is a promising postharvest strategy for fresh blueberries to reduce decay, extend market life and maintain high nutritional value. Rivera *et al.* (2013) demonstrated that SO_2 is an effective and practical technology for reducing the risk of blueberry grey mould decay during storage and it could be used for the export market. There are disadvantages to SO_2 use; the SO_2 concentration necessary to inhibit fungal growth may induce injuries in grape fruits and stems, and sulfite residues pose a health risk for some individuals (Palou *et al.*, 2010 and Palou *et al.*, 2012), as well as firming of the texture of some fruit species (pomaces), incomplete de-sulphiting and incomplete re-colouring of red fruits. Nevertheless, SO_2 treatment is a widespread process because of its advantages of universal antiseptic action and economic application.

References

Abe, K. and Watada, A.E. 1991. Ethylene absorbent to maintain quality of lightly processed fruits and vegetables. *J. Food Sci.*, 56(6): 1589–1592.

Anon., 2004. Ultra Low Oxygen Storage.

Anon., 2015. Ultra Low Oxygen Technology.

Beaudry, R.M., 1999. Effect of O_2 and CO_2 partial pressure on selected phenomenanaffecting fruit and vegetable quality. *Postharvest Biol. Technol.*, 15: 293–303.

Blankenship, S.M., Dole. J.M., 2003. 1-Methylcyclopropene: a review. *Postharvest Biol. Technol.*, 28: 1–25.

Cantín, C.M., Minasa, I.S., Goulas, V., Jiménez, M., Manganaris, G.A., Michailides, T.J., Crisosto, C.H., 2012. Sulfur dioxide fumigation alone or in combination with CO_2-enriched atmosphere extends the market life of high bush blueberry fruit. *Postharvest Biol. Technol.*, 67: 84–91.

Dhall, R.K., 2013. Advances in edible coatings for fresh fruits and vegetables: a review. *Crit. Rev. Food Sci. Nutr.*, 53: 435–450.

Ghasemnezhad, M., Zareh, S., Rassa, M., Sajedi, R.H., 2013. Effect of chitosan coating on maintenance of aril quality, microbial population and PPO activity of pomegranate (*Punica granatum* L. cv. Tarom) at cold storage temperature. *J. Sci. Food Agric.*, 93: 368–374.

Gol, N.B., Patel, P.R., Rao, T.V.R., 2013. Improvement of quality and shelf life of strawberries with edible coatings enriched with chitosan. *Postharvest Biol. Technol.*, 85: 185–195.

Huber, D.J., 2008. Suppression of ethylene responses through application of 1-methylcyclopropene: a powerful tool for elucidating ripening and senescence mechanisms in climacteric and non-climacteric fruits and vegetables. *Hort Science*, 43: 106–111.

Kader, A., 2004. Controlled atmosphere storage. In USDA Handbook, 66.

Kader, A.A., Zagory, D., Kerbel, E.L., 1989. Modified Atmosphere Packaging of Fruits and Vegetables, Crit. Rev. Food Sci. Nutr., 28(1): 1–30.

Manjunatha, G., Gupta, K.J., Lokesh, V., Mur, L.A.J., Neelwarne, B., 2012. Nitric oxide counters ethylene effects on ripening fruits. *Plant Signal. Behav.* 7: 476–483.

Palou, L., Serrano, M., Martinez-Romero, D., Valero, D., 2010. New approaches for postharvest quality retention of table grapes. *Fresh Produce* 4: 103–110.

Rao, C. G., 2015. Engineering for Storage of Fruits and Vegetables. Copy right Book, BSP Book Pvt. Ltd. Published by Elsevier Inc. All right reserved. *Biotechnol. Adv.*, 24: 357–367.

Saltveit, M.E., 1999. Effect of ethylene on quality of fresh fruits and vegetables. Postharvest biology and technology, 15: 279-292.

Saltveit, M.E., 2003. Is it possible to find an optimal controlled atmosphere? Postharvest Biol. Technol., 27: 3–13.

Serek, M., Woltering, E.J., Sisler, E.C., Frello, S., Sriskandarajah, S., 2006. Controlling ethylene responses in flowers at the receptor level. *Biotechnol. Adv.*, 24: 368–381.

Singh, S.P., Singh, Z., Swinny, E.E., 2009. Postharvest nitric oxide fumigation delays fruit ripening and alleviates chilling injury during cold storage of Japanese plums (*Prunus salicina* Lindell). *Postharvest Biol. Technol.*, 53: 101–108.

Singh, Z., Khan, A.S., Zhu, S., Payne, A.D., 2013. Nitric oxide in the regulation of fruit ripening: challenges and thrusts. *Stewart Postharvest Rev.*, 4: 3.

Sisler, E.C., 2006. The discovery and development of compounds counteracting ethylene at the receptor level. *Biotechnol. Adv.*, 24: 357–367.

Sisler, E.C., Serek, M., 2003. Compounds interacting with the ethylene receptor in plants. *Plant Biol.*, 5: 473–480.

Sivakumar, D., Terry, L.A., Korsten, L., 2010. An overview on litchi fruit quality and alternative postharvest treatments to replace sulfur dioxide fumigation. *Food Rev. Int.*, 26: 162–188.

Soegiarto, L., Wills, R.B.H., 2004. Short term fumigation with nitric oxide gas in air to extend the postharvest life of broccoli, green bean, and bok choy. *Hortic. Tech.*, 14: 538–540.

Thomsons, A. K., 2010. Controlled Atmosphere Storage of Fruits and Vegetables, 1–12.

Vargas, M. Pastor, C., Chiralt, A., Mcclements, D. J. and Nez, C.L., 2008. Recent Advances in Edible Coatings for Fresh and Minimally Processed Fruits. *Critical Reviews in Food Science and Nutrition*, 48: 496–511.

Watkins, C.B., 2006. The use of 1-methylcyclopropene (1-MCP) on fruits and vegetables. *Biotech. Adv.*, 24: 389–409.

Watkins, C.B., 2008. Overview of 1-methylcyclopropene trials and uses for edible horticultural crops. *Hort Science*, 43: 86–94.

Wendehenne, D., Durner, J., Klessig, D.F., 2004. Nitric oxide: a new player in plant signalling and defence responses. *Curr. Opin. Plant Biol.*, 7: 449–455.

Zagory, D., 1995. Ethylenen removing packaging in Roony ML, Active Food packaging, London, Blackie Academic and Professional, 38–54.

Zagory, D. and Kader, A.A., 1988. Modified atmosphere packaging of fresh produce, Food Tech., 42(9): 70–77.

Zaharah, S.S., Singh, Z., 2011. Mode of action of nitric oxide in inhibiting ethylene biosynthesis and fruit softening during ripening and cool storage of 'Kensington Pride' mango. *Postharvest Biol. Technol.*, 62: 258–266.

Zhang, Z.K., Huber, D.J., Rao, J.P., 2013. Antioxidant systems of ripening avocado (*Persea americana Mill.*) fruit following treatment at the preclimacteric stage with aqueous 1- methylcyclopropene. *Postharvest Biol. Technol.*, 76: 58–64.

Chapter 22

Low Cost Processing Technologies for Dryland Fruit Crops

V.K. Garande

Professor of Horticulture
College of Agriculture, Kolhapur, M.S.

India is blessed with diverse agro-climatic conditions which enable to produce varieties of fruits. Large numbers of minor fruits accounting for about 5.53 million tones are produced in our country. Many of these fruits are known for their nutritional, therapeutic and medicinal properties. Some of them are not suitable for consumption in their fresh state, but can be processed into excellent products *e.g.* Kokum, Aonla, Bael, Wood apple *etc.*

Fruits and vegetables are fastest growing sectors within horticulture. India produces around 133.54 million MTs of vegetables and 72.28 million MTs of fruits, which respectively accounts for nearly 14.00 per cent and 12.00 per cent of country's share in the world production of vegetables and fruits (Anon, 20014). The food processing industry sector in India is one of the largest in terms of production, consumption, export and growth prospects. The government has accorded it a high priority, with a number of fiscal reliefs and incentives, to encourage commercialization and value addition to agricultural produce, for minimizing pre/postharvest wastage, generating employment and export growth. Due the importance of this subject, the Government of India has created a separate ministry *i.e.* 'Ministry of Food Processing Industry' during 1990.

Traditionally, the main aim of food processing is preservation - to maintain the supply during the year and in particular to preserve for periods when supplies are low. However, the generation of income for the producer and seller is becoming more and more important. A large proportion of fruits produced in developing countries are at risk of loss through spoilage. To overcome this, producers sell their marketable surplus within a short time after harvest at thrown away prices. The opportunities of setting up a fruit processing business are reduce postharvest losses, increase the shelf-life of the fruit, preserve fruit when it is out of season, transform the raw material into edible products, increase food security, add variety to the diet, improve nutrition and health, add value to the fruit and generate increased income, create employment opportunities in producing areas.

The fruit products prepared from these indigenous under-exploited fruits have export potential if carried out in a scientific and sophisticated way. The following products have great potential for export:

- Preserves, jams, candies, glazed and crystallized fruits from ber, karonda, bael, jackfruit, aonla and wood apple.
- Fruit leather: mango, jack fruit, jamun, fig *etc.*
- Fruit juices, pulp syrups, concentrates and fermented beverages from aonla, jamun, phalsa, bael, custard apple, wood apple, karonda *etc.*
- Pickles, chutneys, sauces from mango, cashew apple, karonda, ber, jackfruit, aonla, tamarind, wood apple *etc.*

Low Cost Processing Technologies for Important Dryland Fruits

1. Tamarind

The commonly available market samples of tamarind pulp usually contain extraneous matter such as seed, dust, fibre, *etc.*, and hence can be unfit for various culinary preparations. Value addition is of immense benefit for traders and consumers. Therefore, processed products from tamarind can meet the requirements of consumers for convenience and ready to use. Tamarind is used for the preparation of various processed products. Tamarind fruit pulp is used for the preparation of beverages in different regions. Good quality ready to serve beverage, syrup and concentrate can be prepared with a shelf life of six months at ambient storage (Kotecha and Kadam, 2003). Tamarind pulp was treated with pectiolytic enzymes and the extract obtained was used for the preparation of flavoured RTS tamarind beverage. The carbonated tamarind beverage having 12.5 per cent juice, 16°Brix and 0.4 per cent acidity was found to be highly acceptable for up to two months of storage at room temperature (Lakshmi *et al.*, 2005. A process for the preparation of tamarind juice concentrate was developed by Central Food Technological Research Institute, Mysore, India. The process involves extraction of tamarind juice and concentrating the juice in a vacuum evaporator to total soluble solids of 70°Brix. The tamarind concentrate is shelf stable at room temperature and can be diluted and used in various food preparations. Tamarind kernel powder is used as a gelling, thickening or stabilizing agent in foods in Japan. Tamarind kernel powder (TKP) is now extensively used for sizing fabrics in textile industries (Garande, 2014). There is wider interest over the past few years in the use of tamarind seed polysaccharide extracts in the food industry (Salazar *et al.*, 2002). The antioxidants were extracted by supercritical fluid extraction and by solvent extraction using ether or ethyl acetate. The mixture obtained by solvent extraction with ethanol showed epicatechin yield of 150mg/100g and the highest antioxidant activity in terms of peroxide value as compared to the supercritical fluid extraction (Luengthanaphol *et al.*, 2004).

2. Aonla

The fruit is highly nutritious and is a rich source of pectin and polyphenols apart from ascorbic acid. Aonla fruits are well known for their medicinal properties. The fruits are used for curing chronic dysentery, bronchitis, and diabetes. Aonla fruit is not consumed fresh but the fruit is valued highly in the Ayurvedic system of medicine. In Ayurvedic preparation like 'Chyavanprash' and triphala, Aonla is one of the main ingredients. Aonla fruits are highly perishable in nature and hence its storage in atmospheric conditions after harvesting is very limited, which is accompanied by browning of the skin, loss of glossiness and vitamin C content. Due to its highly acidic and astringent nature, the fruit in fresh form or as a table fruit is not popular and consequently, it is used in the preparation of various ayurvedic tonics like chayvanprash, triphala chrna *etc.* However, aonla fruits are processed into a number of food products like preserve, jam, jelly, candy, toffee, pickle, sauce, squash, juice, RTS beverage, shreds, dried powder, *etc.* (Masalkar and Garande, 2003, Jain *et al.*, 2006; Goyal *et al.*, 2008; Garande, 2014).

3. Bael Fruit

Beal fruit because of its hard shell, mucilaginous texture and numerous seed is difficult to eat out of hand and therefore it is not popular as a fresh fruit. This fruit is used for processing in the mature green from to prepare preserves, squash and jam, syrup and fruit powder. The pulp can be preserved by canning, freezing and addition of SO_2. Preserved pulp can find an export market since it can be used in ice-cream and confectionery preparation. Bael fruit flavour is entirely unknown in the export market and a good range of products like fruit preserve, pulp, nectar, squash, slab (leather), toffee, powder, clear juice *etc.* have been developed formulated by Singh and Roy (1984) and Khurdiya (2001b).

4. Jack Fruit

Raw jackfruit is popularly used as vegetable. Ripe jackfruit is consumed as a dessert fruit. Jackfruit chips are prepared by frying ripe or semi-ripe fruits. Jackfruit leather is also prepared from the ripe or semi-ripe fruit. A palatable beverage concentrate can be made from jackfruit pulp by adding sugar, citric acid and water and even a good quality nectar, preserves, biscuits, chiwda, jam, pickle, minimally processed bulb lets can be prepared from the ripe jack fruit pulp (Garande and Patil, 2013). The wastes (skins, peels and cores), which constitute about 45 per cent of the total fruit weight, have been found to be a fairly good source of pectin. The wastes (skins, peels and cores), which constitute about 45 per cent of the total fruit weight, have been found to be a fairly good source of pectin (Khurdiya, 2001 b and Rao, 2003).

5. Jamun

The ripe jamun fruits have good value in curing the diabetes and they can be utilized for making products such as jam, jelly, burfi, toffee, RTS, squash, syrup, wine, jamun seed powder *etc.* It has been found that maximum yield of jamun juice with a high level of anthocyanins and other soluble constituents can be obtained by pulping; heating to 60°C for 15 minutes and basket press method of juice extraction recorded the highest juice and anthocyanin recovery (Garande, *et al.*, 2007). The juice being highly acidic is not consumed as such. A ready to serve beverage (nectar) is prepared with 25 per cent juice, 18° Brix and 0.6 per cent acidity. Jamun seed powder can be used for curing the diabetes, diarrhea and dysentery. A good quality of wine can be prepared from the ripe jamun fruit (Sonar, *et al.*, 2004. The processed products such as juice, RTS, squash and syrup, wine and carbonated RTS, jam, jelly, burffee, seed powder (Garande and Joshi, 1995 and Khurdiya, 2001 a) can be prepared from jamun fruits while vinegar from unripe fruits (Virmani, 1950). The pomace left after the extraction of jamun juice can be nicely utilized for preparation of burfi seed powder for diabetes (Garande, 2013)

6. Phalsa

The fruits ripen in summer and have an extremely poor keeping quality (about 48hrs). Investigation has indicated that a number of acceptable products *viz.* phalsa juice, nectar, concentrate, and crush can be prepared from phalsa fruit pulp. Such preparations have been standardized by Waskar and Khurdiya (1987). Being highly perishable, the fruit must be utilized within 24 hours after picking. The popularity of phalsa fruit is due to its attractive colour ranging from crimson red to dark purple and its pleasing taste. The juice when extracted gives a deep crimson red to dark purple colour and is very popular. It is rated very high in indigenous system of medicine. The juice is extremely refreshing and is considered to have a cooling effect especially in hot summer. Heating the crushed phalsa fruit to 50° C gives the highest recovery of the juice with an appropriate quantity of anthocyanin and other soluble and insoluble materials. Studies have shown that addition of cane sugar to the juice has a protective effect on colour stabilization (Khurdiya, 2001b).

7. Karonda

Karonda has remained neglected as far as its postharvest handling is concerned. Because of its soft flesh, with high moisture content, its shelf life is less. The products such as squash, syrup, jam, jelly, pickle, sweet chutney (from mature green fruits) and dried karonda fruits could be successfully prepared from mature green and ripe karonda fruits (Pawar, 1988, Jadhav *et al.*, 2003). The karonda fruits after destining are boiled for about 3 to 4 minutes in 0.4 per cent NaOH and washed in water. The washed fruits are steeped in plain water for about 30 minutes. Thus bleached fruits are boiled in red colours and kept overnight for frying into cherry red colour. Such fruits are syruped into sugar syrup and then dried. This candied karonda fruits (cherry) look like red cherries and are used as toping for ice-creams (Garande, 2009).

8. Wood Apple

Wood apple is most important but under exploited indigenous edible wild fruit. This fruit is nutritionally and medicinally valuable but has hard shell, sticky texture and it contain numerous seed. It is difficult to eat by hand as fresh and most of the fruits are wasted. Wood apple can be converted into value added products like jam and fruit bar in order to avoid glut and utilize the surplus during the season, it is necessary to employ methods to extend storage life, for better distribution, to preserve them for utilization in the offseason both in large scale and home scale. Jam is more or less a concentrated fruit processing which has fairly thick consistency and body. Fruit bar is a nutritious product, has a chewy texture, similar to dried raisins and is a good source of fibre and natural sugar. Wood apple juice when mixed with other fruits can serve as an excellent beverage. The extraction of pulp from ripe wood-apple fruit is the main hindrance to processing. This is mainly due to mucilaginous texture and tendency of pulp to develop off flavor and colour. The wood-apple pulp can be successfully extracted by addition of water equal to the weight of pulp (with seed and fibre) adjusting pH to 4.3 with addition of citric acid and heating at 80° C for one minute. The applications of heat not only inactivates enzyme but also helps in dissolving mucilage uniformly to provide homogeneous pulp. The pulp is passed through 20 mesh stainless steel sieve or muslin cloth (Vidhya and Narain, 2011). This pulp is used for preparation of processed products such as RTS, squash, syrup, carbonated RTS, jelly chutney, leather *etc.* (Khurdiya, 2001a. and Garande, 2013).

9. Custard Apple

Custard apple is one of the important fruit crops of drier parts of India, which ripen within four days after harvest. Considering the fast increasing area under custard apple cultivation, methods of its preservation and processing technology needs to be developed to regulate the prices of produce during glut period. Pulp was extracted from custard apples and stored for a period of six months by addition of 1500 ppm of potassium meta-bisulphite. After six months, various products like squash and nectar were prepared. The products were stored at room temperature and cold storage for a period of four months to study the stability and consumer acceptability of the products. All the products stored at cold storage were good physico-chemically when compared to the products stored at room temperature (Khurdiya, 2001 b). In order to remove the seed, hand rubbing the mixture on steel sieve to separate the seed and pulp has been found suitable; cheaper and easy to handle than pulper. The pulp can be mixed with milk to form a delicious drink called custard apple shake. The pulp can be utilized for preparation of good quality products like jam, freeze dried powder, rabdi, milk shake, ice cream *etc.* (Garande, 2014). Sensory and microbiological properties of the ascorbic acid-treated custard apple pulp were of acceptable quality throughout the storage period of 4 weeks (Dhumal, 1994).

10. Ber

Ber fruits are consumed as such or can be processed into different fruit products. Juicy varieties are better suited for pulp and juice extraction. The fully ripe, well-developed fruits are washed de-stoned and juicer extracts juice. Ber juice can be used for the preparation of ready to serve beverage. Carbonated beverage of ber is highly acceptable and has excellent keeping quality. Dehydrated ber is prepared by treating ber fruits with sulphur dioxide at 3.5-10 g/kg for 3 hours followed by sun drying, or carbinet drying below 15 per cent moisture. Ber can be utilized for candy and chiwada (Chavan *et al.*, 1992 and Garande, 2014) and ber pulp can be processed into wine. The steps include diluting the pulp, adding pectinase enzymes adjusting proper Brix with sugar, addition of yeast, fermentation, stabilization and clarification. (Adsule, *et al.*, 1992).

11. Kokum

Versatile medicinal fruit like kokum opens us to a variety of value added products ranging from simple culinary supplements to highly sophisticated anti obesity, anti-allergic, dermatological products. The range also offers opportunities for cottage industry to small and medium scale industry. The beverages like syrup, RTS, carbonated RTS, Kokum butter, amsul and agal *etc.* can be prepared from the kokum. The more modern and sophisticated products dealing in Hydroxy Citric Acid, Garcinol, Anthocyanins and even wines will require further research and development efforts to bring them to level of desired status of international standards before they contribute handsome returns to kokum family. Keeping the focus on ultimate benefit to the farmer, efforts should be to convince the entrepreneurs to work out their manufacturing flow charts in a way that the preliminary steps as much as feasible are passed on to the farmers concerned (Shirodkar, 2013 and Garande, 2014).

12. Fig

Figs can be eaten fresh or dried, used in jam-making and pickling. Most commercial production is in dried or otherwise processed, since the ripe fruit does not transport well, and once picked does not keep well. The natural flavor of figs can be preserved in fig jam, preserves and paste. To make fig concentrate, which replaces sucrose and corn syrup, the water is extracted from the figs. Chopped, diced and sliced figs are incorporated into food products. When dried, figs are added to cookies, bars and snacks. The drying technique for fig was standardized by Gawade and Waskar (2005). The natural humectants in figs make them a beneficial ingredient in such health and beauty products as soap, moisturizers, fragrance and candles. The processed products like dried fig, jam, burfi, leather ice-cream *etc.* can be prepared from the different varieties of fig (Garande, 2015).

References

Adusle, R. N., P. M. Kotecha and S. S. Kadam. 1992. Preparation of wine from ber fruit. Beverage and Food World. 19(4):13-14.

Anonymous. 2014. National Horticulture Data Base, Gurgaon, New Delhi.

Chavan, U. D., R. N. Adsule and S. S. Kadam. 1992. Processing of ber into different products. Beverage and Food World. 20(4):28-29.

Dhumal, N.S. 1994. Studies on processing of custard apple (*Annona squamosa* L.). M. Sc (Agri.) thesis, Mahatma Phule Krishi Vidyapeeth, Rahuri (MS).

Gamag, T. V. E., C. M. C. Yuen and R. B. H. Wills. 1997. Minimal processing of custard apple (*annona atemoya*) pulp. Journal Food Processing and Preservation. 21(4): 289–301.

Garande V. K. and G. D. Joshi. 1995. Storage of jamun fruit products. Asean Food Journal, 10 (2): 54-56.

Garande V. K., S. D. Masalkar and R. S. Patil. 2007. Studies on extraction and quality evaluation of jamun juice, Beverage and Food World, 53(1):85- 86.

Garande, V. K. 2009. Processed products from karonda. Agro-one, 09th May, pp.2.

Garande, V. K. 2009. Value added products of jack fruit. Agro-one, 19th May, pp12.

Garande, V. K. 2013. Preparation of processed products from wood apple. Agro-one, 24h and 25th January, pp.12.

Garande, V. K. *and* Patil, J. B. 2013. Studies on physico-chemical composition and qqality of jackfruit clones of Kolhapur district. Proceeding of National Seminar on Ergonomics for Enhanced Productivity, Department of Family Resource Management, Home Science College and Research Institute, Tamil Nadu Agricultural University, Madurai, 18-19 February, 2013, pp. 243-245.

Garande, V. K. 2013. Processed products from jamun. Agro-one, 21st and 24th December, pp.12.

Garande, V. K. 2013. Value addition in aonla through processing. Agro-one, 8th December, pp.12.

Garande, V. K. 2014. Processed products of custard apple. Agro-one, 30th September, 12.

Garande, V. K. 2014. Processing of ber into different value added products. Agro-one, 21st th and 22nd November, 12.

Garande, V. K. 2014. Processed products of kokum. Shekari Magazine, November, 21-23.

Garande, V. K. 2014. Processed products of tamarind. Agro-one, 3rd November, pp.10.

Garande, V. K. 2015. Processed products of fig. Agro-one, 3rd November, pp.10.

Gawade, M. H. and D. P. Waskar. 2005. Effect of different varieties and pretreatments on yield and quality of dried fig fruits. Indian Journal of Agricultural Research, 39 (2):138-141.

Goyal, R. K., R. T. Patil, P. R. P. Kingsly, H. Wali and P. Kumar. 2008. Status of Postharvest Technology of aonla in India – A Review. American Journal of Technology, 3(1):3-23.

Jadhav, S. B. and G. D. Joshi and V. K. Garande. 2003. Studies on preparation and storage of processed products karonda fruit products. Beverage and Food World, 31(5):46-47.

Jain, S.P., Tripathi, V.K., Ram, H.B. and Surjeet Singh. 2003. Effect of storages condition on storage life of some important squashes. Indian Food Packers, 40(2):36-41.

Kotecha, P. M. and Kadam, S. S. 2003. Preparation of ready-to-serve beverage, syrup and concentrate from tamarind. Journal of Food Science and Technology (Mysore), 40(1): 76-79.

Khurdiya, D. S. 2001a. Postharvest management of underutilized for fresh marketing. Winter school on exploitation of underutilized fruits, pp. 266–274.

Khurdiya, D. S. 2001b. Postharvest management of underutilized fruits for processed products. Winter school on exploitation of underutilized fruits, pp. 291–298.

Lakshmi. K., Vasanth Kumar, A. K., Jaganmohan Rao, L., and Madhava Naidu, M. 2005. Quality evaluation of flavoured RTS beverage and beverage concentrate from tamarind pulp. Journal of Food Science Technology 42 (5), 411-415.

Luengthanaphol, S., Mongkholkhajornsilp, D., Douglas, S., Douglas, P.L., Pengsopa, L. and Pongamphai, S. 2004. Extraction of antioxidants from sweet Thai tamarind seed coat: Preliminary experiments. Journal of Food Engineering 63(3): 247-252.

Masalkar S. D. and V. K. Garande. 2003. Studies on preparation and storage of aonla products. National seminar on 'Production and utility of aonla held at Salem during 8 to10 th August, 2003, pp. 35.

Pawar, C.D. 1988. Studies on postharvest handling and preparation of different products from karonda fruits. M. Sc. (Agri.) Thesis submitted to KKV, Dapoli.

Rao, V. S. 2003. Postharvest Technology and Value Addition in Fruits, Central Research Institute for Dryland Agriculture, Hyderabad - 500 059, pp. 103-110.

Salazar - Montoya, J.A., Ramos-Ramirez, *e.g.* and Delgado-Reyes, V. A. 2002. Changes of the dynamic properties of tamarind gel with different saccharose and polysaccharide concentrations. Carbohydrate Polymers, 49 (4): 387-391.

Shirodkar, Ajit. 2013. Postharvest and processing - status report of Kokum. National symposium on spices and Aromatic crops (SYMSAC- VII) held at Madikeri during 27-29th, November, 2013, pp.159-164.

Sing, R. N. and Roy, S. K. 1984. The bael cultivation and processing, ICAR Publication.

Sonar R. P., S.D. Masalkar, V.K. Garande and R.S. Gaikwad. 2004. Influence of inoculum levels and pH of must on quality of jamun wine. Beverage and Food World, 69 – 70.

R. Vidhya and A. Narain. 2011. Development of preserved products using under exploited fruit wood apple (*Limonia acidissima*). American Journal of Food Technology, 6: 279-288.

Vernon-Carter, E. J., Espinosa-Paredes, G., Beristain, C. I. and Romero-Tehuitzil, H. 2001. Effect of foaming agents on the stability, rheological properties, drying kinetics and flavour retention of tamarind foam-mats. Food Research International, 34(7): 587-598.

Vidhya, R. and Narain, A. 2011. Development of processed products using under-exploited fruits: Wood apple. American Journal of Food technology, 6: 279-288.

Virmani, R. S.(1950). Jaman vinegar and its Composition. Indian Food Packer, 4(1): 13-14.

Waskar, D. P. and Khurdiya, D. S. 1987. Processing and storage of phalsa beverages. Indian Food Packer, 41(5):7-16.

Chapter 23

Blending, Carbonation and Storage of Fruit Based Beverages

A.P. Patil, V.P. Kad, J.K. Dhemre and D.G. Kadam

Postharvest Technology Centre,
Mahatma Phule Krishi Vidyapeeth, Rahuri – 413 722, Dist. Ahmednagar, M.S.

People of all age groups consume beverages to quench their thirst. These beverages are nonalcoholic and include fruit based drinks, synthetic drinks, sweetened aerated water or carbonated drinks. The cold drinks are in demand for greater parts of the year in the country. India produces more than four thousand million bottles of aerated water annually. This aerated water contains synthetic colouring and flavoring additives, which may potentially be allergenic, It is hugs business in comparing to fruit juices and beverages. These aerated waters and would prove a boon to the consumer as well as grower.

Thus, the fruit based carbonated beverages can provide a new product having natural colour and flavour along with nutrition we can:

- ☆ Utilize fruit juices for new type of beverages;
- ☆ Produce fruit based carbonated drinks;
- ☆ Avoid use of synthetic colouring and flavouring gents; and
- ☆ Utilize indigenous fruits.

Precautions for Production of Carbonated Beverages

- ☆ Select only microbiologically safe fruits.
- ☆ Remove all spoiled, rotten fruits.
- ☆ Thick glass bottles should be used, which can sustain CO_2 gas pressure.
- ☆ Maintain clean, hygienic working atmosphere.

☆ Workers should ware apron and head gear.

☆ Proper CO_2 gas pressure should be maintained in the cylinder containing chilled water.

Principle

The fruit based carbonated drinks can be produced and preserved by hurdle technology, *i.e.* chemical preservative, carbon dioxide and heat treatment without addition of synthetic colouring and flavouring additives.

Carbonation is the process of mixing sufficient CO_2 with water or beverage so that when served, the product gives off the gas in fine bubbles and has the characteristic pungent taste suitable to the beverage carbonated. Carbonation adds to the life of a beverage and contributes in some measure to its tang.

Fruit juice beverages are generally bottled with CO_2 content varying from 1 to 8 g per liter. Though this concentration is much lower than that required for complete inhibition of microbial activity (14.6 g/litre), it is significant in supplementing the lethal effect of acidity on pathogenic bacteria. Another advantage of carbonation is the removal of air which results not only in anaerobic condition, but also reduces the oxidation of ascorbic acid.

Preparation and Preservation of Fruit Juices

Fully ripe fruits are selected; injured and spoiled fruits are sorted out. Fruits are cleaned and washed thoroughly with clean water. Fruits are crushed in a fruit mill and the crushed mass is pressed into a hydraulic press to get juice. The juice is filtered and preserved either by chemical preservative or by heat application (*i.e.*, pasteurization or sterilization). The processed fruit juice could be stored at refrigerated temperature for better quality during storage.

Preparation of Carbonated Fruit Drinks

Preparation of fruit syrup: Carbonated drinks are made by post-mixed method. Fresh or preserved fruit juice is used to prepare syrup base. Syrup base is prepared by mixing fruit juice, sugar, citric acid and water and preserved by sodium benzoate. These ingredients are mixed well so that the syrup base has 50° Brix total soluble solids and 1.0 per cent acidity.

The general formulations for production of fruit syrup base are given below:

Recipes	*Aonla*	*Grape*	*Jamun*	*Lime*
Juice (kg)	1.0	1.0	1.0	1.0
Sugar (kg)	1.17	1.066	1.16	3.254
Citric acid (g)	11.43	12.00	11.11	--
Water (kg)	0.322	0.443	0.341	2.254
Sodium benzoate (g)	1.25	1.25	1.25	3.25
Total (kg)	2.50	2.50	2.50	6.50

The amount of above ingredients may be altered as per requirement.

The sugar and citric acid are dissolved in water by heating, filter through a four-fold muslin cloth and then cooled. After cooling, it is mixed with the juice thoroughly. The sodium benzoate is mixed with a small quantity of clean water and then mixed thoroughly in the syrup base and filled into the clean and pre-sterilized glass bottles.

The desired amount (50g) of a fruit juice syrup is dosed into the previously sterilized and clean glass bottles (200 ml capacity) and then filled with the carbonated chilled water and sealed by crown corking.

The fruit based carbonated drinks can be stored for a week if prepared without preservative. But if preserved with sodium benzoate and heat processing then it can be stored for a period of 3 months at ambient temperature (30° C) under dark conditions and 6 months if stored at low temperature (6-7°C)

Heat Processing

Bottles are placed in the water heated to a temperature of 70°C in a stainless steel or aluminium vessel so as to maintain a temperature of 66°C for 30 minutes in the contents of the bottle. After heat processing, the bottles are removed and cooled. After cooling, bottles are washed, labelled, stored, and marketed as and when required.

Calculations

CO_2 gas volume: For measuring gas volume in a carbonated beverage, a gas volume tester is required. It is screwed on the top of the neck of the bottle. By pressing the handle crown cork is punctured and gas pressure is observed in the pressure gauge which is released by unscrewing the side valve until the first bubble rise in the liquid. Then shut the valve and shake the bottle until gauge reaches maximum pressure. Note the pressure on the gauge. Remove the crown cork and note the temperature of the drink. With these values, gas volume is seen through a Table 5.1 Gas volume denotes how many gas volumes are dissolved in one volume of liquid.

Carbon Dioxide Volume

At atmospheric pressure, the amount of CO_2 dissolved by water (W:V) depends solely on the temperature. Solubility is greater at lower temperature than at higher temperature. The amount of CO_2 dissolved in water at a given temperature is proportional to the pressure the former exerts on the latter (Table 28 - 12). Carbonation, therefore, depends on pressure and temperature.

Degree of Carbonation

The level of carbonation required varies according to the type of fruit juice and type of flavour.

Sample Calculation for Carbonation

The carbonation pressure required to carbonate a beverage at 40′ F sp as to maintain gas volume of 4 is calculated as give below:

Solubility of CO_2 at 60′ F = 1 volume

Solubility of CO_2 at 32′ F = 1.7 volume

$$\text{Solubility of } CO_2 = 1 + \frac{(1.7-1)\times(60-40)}{(60-32)} = 1.5$$

At 40′ F

Gas volume required in the beverage = 4

It is therefore, necessary to raise the volume by 2.5 (*i.e.*, 4-1.5) over the atmospheric solubility of 1.5 volumes of CO_2 at 40′.

$$\text{The carbonation pressurerequired} = \frac{15\times 2.5}{1.5} = 25 \text{ lbs.}$$

Unit of Measurement

The unit of measurement accepted by the beverage industry as standard is the 'volume'. This is defined as the amount of gas in milliliters that a given volume of water will absorb at atmospheric pressure, *i.e.*, at 760 mm of Hg at a temperature of 15.5° C (60′ F). These are arbitrary points set by agreement. This condition registers as zero on the gauge commonly used to measure the volume of CO_2 absorbed in carbonated beverages.

Thus 15.5′C, beverage water will absorb one volume of carbon dioxide represented as zero on the gauge. When the pressure is increased to 15 lbs (actually 14.7 lbs) by applying a pressure of one additional atmosphere, the water will absorb two volumes of the gas, and for each additional 15 Ibs or atmosphere of pressure an additional volume of CO_2 will be absorbed. Reduction of the temperature will permit the water to dissolve greater amounts of CO_2. When the temperature is reduced to 0′C(32′F), 1.7 volumes of CO_2 will be absorbed. There will consequently be an additional absorption of 1.7 volumes of CO_2. Thus, if a bottle is filled at 32′ F under 30 lbs of pressure, the volume of CO_2 dissolved in the water will be 1.7 x 3 = 5.1. For intermediate pressure and temperatures, intermediate volumes of the gas will be absorbed.

Solubility of CO_2 in Sugar Solutions

As the concentration of sugar increases the solubility in terms of volume of gas decreases, but this is not very highly significant for practical purpose (1 per cent sugar solution at 60 per cent F dissolves 0.995 volume of CO_2 while 13 per cent sugar solution dissolves 0.902 volume).

Measurement of Carbon Dioxide in Bottled Beverages

Use a device consisting of a gauge with a hollow spike having holes on the side with a slot in the neck to introduce the bottle into the slot provided in the neck of the tester. Secure the bottle in place by tightening with a threaded stem. Insert the pressure gauge until the needle point touches the crown cork. Close the sniff valve on the gauge stem until the needle point of the pressure gauge is forced through the crown cork. Note the reading on the gauge. Open the sniff valve and allow the

Analysis of Fruit and Vegetable Products

Pounds per sq inch Pressure	*32*	*36*	*40*	*44*	*48*	*55*	*60*	*65*	*70*	*75*	*80*	*85*	*90*
15	3.46	3.19	2.93	2.70	2.50	2.20	2.02	1.86	1.71	1.58	1.44	1.37	1.27
20	4.05	3.73	3.42	3.15	2.92	2.57	2.36	3.17	2.00	1.84	1.69	1.58	1.48
25	4.58	4.27	3.92	3.61	3.35	3.04	2.69	2.48	2.29	2.10	1.93	1.80	1.70
30	5.21	4.81	4.41	4.06	3.77	3.31	3.03	2.80	2.58	2.37	2.18	2.03	1.91
35	5.80	5.35	4.91	4.52	4.19	3.69	3.37	3.11	2.86	2.63	2.42	2.26	2.13
40	6.37	5.89	5.39	4.97	4.61	4.05	3.71	3.42	3.15	2.89	2.67	2.49	2.34

Composition of Carbonated Beverages

Beverage	*Sugar (° Brix)*	*Gas Volume*	*Citric Acid (per cent)*	*pH*
Lime	9.17	4.0	0.14	3.02
Lime	11.10	3.7	0.23	2.90
Lemon	11.18	3.2	0.12	3.07
Lemon and Lime	11.04	3.2	0.18	3.01
Orange	13.40	2.3	0.19	3.39
Grape	14.50	1.3	0.12	3.10

pressure to escape until the first bubble rises in the liquid. Shut the sniff valve and shake the bottle until the gauge reaches maximum pressure. Note the pressure gauge. Sometimes, the final pressure noted may be lower than the pressure noted before sniffing the head pressure in the bottle. Usually it is the same as the final pressure when the sample sample is cold (40-50°F). Remove the gas volume tester and crown cork from the bottle and note the temperature with the help of a thermometer.

Equipments/Materials

Preparation and Preservation of Fruit Juice

- ☆ Fruit mill
- ☆ Hydraulic press
- ☆ Stainless steel (S.S.) Tanks
- ☆ Heating vessels (S.S. Patiala)
- ☆ Stainless steel sieve
- ☆ Glass bottles
- ☆ Crown corking machine
- ☆ Crown Corks
- ☆ Heating equipment a) Cooking gas cylinder; b) Gas burner
- ☆ Thermometer
- ☆ Weighing balance

Preparation of Carbonated Fruit Drinks

- ☆ Carbonating machine including CO_2 gas cylinder
- ☆ Water chiller
- ☆ Water filtering unit

Chemical/Materials

- ☆ Fruits juice, (Fruits-Jamun, Amla, Lime and Coloured Grapes)
- ☆ Sugar
- ☆ Citric acid
- ☆ Sodium benzoate
- ☆ Labels
- ☆ Carbon dioxide gas
- ☆ Potable water
- ☆ Filtration cloth

Aseptic Processing Comprises the following

a) Sterilization of the product before filling.

b) Sterilization of the packaging material or container before filling operation.

c) Filling the product into the sterile package by maintaining the aseptic condition.

d) Sterilization of Product

The product sterilization is carried out by in-process or on-line sterilization which is popularly known as the ultra high temperature (UHT) or high temperature short time (HTST) depending upon the product treatment. The essential difference between the sterilization through autoclave, *i.e.* conventional canning method, and on-line UHT treatment is that in canning, the sterilization is done after the product is packed, whereas in UHT method, the sterilization is done prior to packaging. Normally, the sterilization temperature is maintained at 133 – 135° C for 3 to 5 seconds under HTST/UHT method to make the product sterile. However, there may be slight variation in temperature and duration depending on the type of product.

Sterilization of Packaging Materials and Equipment

Different kind of sterilizing agents like heat, chemicals and radiation are used alone or in combination for sterilization of aseptic equipment and packaging materials.

i) Heat

Initially, product supply lines and fillers are commonly sterilized by moist heat in the form of hot water or saturated stream under pressure, 'Dry heat ' in the form of superheated steam or hot air may also be used to sterilize equipment. However, due to the relatively high dry heat resistance of bacterial endospores, the time – temperature requirements for dry heat sterilization is high. System enjoying moist heat are frequently sterilized at temperatures ranging from 121 – 129°C, while 176-232°C is used for sterilization by dry heat.

ii) Hydrogen Peroxide

Hydrogen peroxide (H_2O_2) is not an efficient sporicide when used at room temperature. Therefore, most aseptic packaging systems use (H_2O_2 at concentrations of 30 to 50 per cent) as a sterilant for packing materials followed by hot air (60-125° C) to dissipate residual hydrogen peroxide. Other chemicals which have been used as sterilents for acid foods include various acids, ethanol, ethylene oxide and peracetic acid.

iii) Radiation

A dose of approximately 1.5 megaradians (MGRAD) gamma radiation is commonly used to decontaminate containers for acid and acidified foods. Doses required to sterilize containers for use with low acid foods are considerably higher than those required for acid and acidified foods.

c) Filling of the Product under Aseptic Condition

Once the product has been brought to the sterilization temperature, it flows into holding tube. The tube provides the required residence time at the sterilization temperature. No external heating of holding tube should take place.

A deaerator is used to remove air from the product processed and packed aseptically. The deaerator generally consists of a vessel in which the product is exposed to a vacuum on a continuous flow.

The sterilized product is a accumulated in an aseptc surge tank prior to packaging. The product is pumped into the surge tank and is removed by maintaining a positive pressure in the tank with sterile air or other sterile gas.

Aseptic packaging system has been introduced in India by Tetrapack Company and marketing the different process food in tetra packs. The most commonly system is followed by Tetra pack, called as Tetra classic aseptic (TCA) system. The packages produced have a tetra hedronal shape. The ratio of pack material to packaged product is a very favourable one, *i.e.*the amount of packaging material needed to package the volume of food is small.

Similarly, the TBA/3 (Tetrabrik Aseptic) system was introduced into the Indian market in the beginning of 1970. The containers produced are 'brick shaped' and easier and more efficient to handle during storage and distribution.

The tetrabrik packages are brik like shape and made of six layers of packaging materials. The 6 layers in the brick provide total protection to the packaged product.

A number of processed food products which are liquid or semi in nature, are available in the aseptic packages. The common product aseptically packed are fresh milk and flavoured milk, fruit juices, yougurt, vegetable oil, chocolate milk, milk curd and cream.

Types of Aseptic Packages

A variety of consumer packages may be filled aseptically in Caton box, Bags and pouches, Cups and Trays – Polypropylene base multilayered materials with EVOH barrier material are used, Glass bottles and jars.

The Aseptic Process for Fruit Juices

The aseptic process is a technique for processing liquid food products by exposing them to brief intense heating. Sterility is the key word in the process and involves the same principles as discussed with respect to sterilization. Since very high temperature are employed (125°C to 145°C), the process is also referred to as Ultra High Temperature (UHT). The holding time of the product at high temperature is very short (3 to 15 seconds) and therefore, thermal stress on the product is less while ensuring safety. This rapid heating and cooling process also reduces the energy use and nutrient losses associated with conventional sterilization. Processing at high temperature usually requires that the process be based on enzyme inactivation rather than microbial destruction. The process involves:

i) Sterilization of the product in sterile equipment:
ii) Pre-sterilized containers:
iii) Aseptic environment for filling of the sterile product into the sterile packages and
iv) Sterile packaging.

The advantages of the UHT processing are (i) higher quality (ii) increased shelf life (iii) easy adaptability to many size containers from single serving (200 ml.) to thousands of litres of bulk pack, and (iv) cheaper packaging material. The disadvantages are (i) higher cost (ii) sophisticated control and maintenance to plant, and (iii) higher skill level on the part of operators and maintenance personnel.

Process Equipment

UHT plants are often designed with great product flexibility in order to enable processing of a wide range of products in the same plant. Both low acid products (with pH values below 4.5) can be treated in UHT plant. However, only low acid products required UHT treatment in order to become commercially sterile as sores can grow in a low acid environment. Spores cannot develop in high acid products such as juice and the heat treatment is, therefore, only intended to kill yeasts and moulds. Normal high- temperature pasteurization is sufficient to make high end products commercially sterile. The UHT plant are classified according to the method of heating.

Types of Plant	*Option Available*
Direct Heating (Product and the heating media are in direct contact)	Steam injection (steam into product), Steam infusion (product into steam)
Indirect Heating (Heat transfer surface between the product and the heating media)	Plate Tubular and Scraped surface heat exchangers

I) Direct Heating

There are two main methods for aseptic processing by direct methods: by injecting pressurized steam into product or by injecting product into steam (Steam infusion). Both system work on the principle that the steam comes into contact with the product, it will condense and give up some latent heat causing the product to heat up very quickly. The steam has to be of satisfactory quality, *i.e.* derived from drinking water. The basic principle for both systems is to pass the product form a balance tank to a pre-heating and final heating system, usually by plate heat exchanger at 75-80°C. The product then passes through the main product pump to the steam injection or infusion system. The mixture then flows through a holding tube (2-4 seconds) which enters tangentially into the vacuum chamber. The product passes on an aseptic path-pump, homogenizer cooler and storage tank – to aseptic package. Steam injection is most suitable for low viscosity homogeneous products such as milk and juice. Steam infusion is for more gentler and sensitive products.

II) Indirect UHT Plant

UHT heating plants using indirect heating method relay on having a heat transfer surface between the product and the heating media. There are three main types of indirect heating system, *viz.*, plat heat exchanger, tubular heat exchanger and scraped surface heat exchanger. Each system has benefits as well as drawbacks depending on the product process requirement.

There are several different forms of packages that are used in aseptic processing such as cans, laminates, flexible pouches, thermoformed plastic containers, tetra pack, *etc*. Tetrapack, a revolutionary package, is a laminate of three materials: high-quality paperboard, polyethylene and aluminium. Each material plays a critical role in achieving the unique benefits of the aseptic package. Polyethylene (24 per cent) on the innermost layer forms the seal that make the package liquid-tight. A protective coating on the exterior keeps the package dry Aluminium (6 per cent) is the silver material we see on the inside of the aseptic package. This ultrathin layer of foil forms a barrier against light and oxygen, eliminating the need for refrigeration and preventing spoilage. The processing steps involved in formation of a package are: sterilization of roll and, formation of package, filling and sealing in sterile environment.

Hot Pack of Hot Fill

The terms Hot Pack or Hot Fill Hold-Cool Process or Flash pasteurization refer to the filling of previously pasteurized or sterilized foods, while still hot, into clean but not necessarily sterile containers, under clean but not necessarily aseptic conditions. The hermetically sealed containers are then inverted and held for an adequate time (1-3 minutes) to sterilize the lid as well as the containers. The hot-filled containers are then cooled either in a water – spray tunnel or by immersing in tank containing chlorinated cold water. Rapid cooling is essential to retain the colour, flavour and nutrients. In yet another practice followed in some western countries, the head space is flushed with steam or nitrogen to minimize the residual oxygen.

Hot pack is more effective with acid foods as at pH below 4.5, clostridium botulinum will not grow or produce toxin. Hot pack with low acid foods (above pH 4.5) is not possible unless the product is recognized as being only pasteurized and will be stored under refrigeration. However, hot pack treatment could be combined with some additional means of preservation such as very high sugar content. The temperature and time for pasteurization and hot filling depend upon the specific product's pH and other food characteristics. It is essential that the definite temperature and time must be adhered for hot pack processing to be effective.

References

Balaswamy, K., P. P. Rao, A. Nagender and A. Satyanarayana. 2011. Preparation of Sour Grape (Vitis Vinifera) Beverages and Evaluation of their Storage Stability. *J. Food Process. Technol.* 2:3.

Ilamaran M. and S. Amutha. 2008. Studies on the biochemical and sensory qualities of enzymatic clarified carbonated sapota beverages. *Asian Journal of Home Science*. 3 (2): 230-233.

Patil, S. R., S. P. Kurhekar and R.R. Patil. Study on development of custard apple carbonated beverage. *International Journal of Processing and Postharvest Technology*. 2(1): 56-58.

Sandhan, V. S., D. R. Nandre and B.M. Kushare. 2009. Storage studies of carbonated beverage from pomegranate juice. *International Journal of Agricultural Sciences,* 5 (1): 217-220.

Tompe, A. and A. P. Sonone. 2015. Development of Fruit Juice Blended Carbonated Beverages. *Journal of Management Research and Analysis*. 2(3):185-189.

Chapter 24

Milk Enriched Processed Fruit Products from Dryland Fruit Crops

R.J. Desale, R.V. Karche and M.S. Gavhane

Department of Animal Husbandry and Dairy Science,
Mahatma Phule Krishi Vidyapeeth, Rahuri – 413 722, Dist. Ahmednagar, M.S.

Milk

India is emerging as a highest milk producing country in the world with an annual growth rate of 5 per cent. The level of milk production has risen gradually from 17.00 million tonnes in 1950-51 to a 63.50 million tonnes in 1995, now the milk production in India is 146.31 million tonnes (2014-15). Per capita availability of milk has also increased to 302 gm more than the minimum quantity recommended by the WHO. The enhancement of milk production in India at 6.3 per cent whereas on international scenario there is only an increment of 2.2 per cent. Out of the total milk produced in India, 46.00 per cent is consumed as liquid milk and 54.00 per cent is utilized for conversion into different products (Aneja *et al.*, 2002).

Year	*Milk Production (million tonnes)*		*Per Capita Availability (gms/day)*	
	India	*Maharashtra*	*India*	*Maharashtra*
2010-11	121.8	8044	281	197
2011-12	127.9	8469	290	206
2012-13	132.4	8734	299	213
2013-14	137.7	9089	307	219
2014-15	146.3	9542	322	296

Source: NDDB, 2015.

Milk is considered to be an adequate source of valuable macronutrients (fat, protein, lactose), vitamins and micronutrients (minerals), making it a „wholesome food. It can serve as an excellent carrier product for extra nutrients, and if enriched or fortified it can satisfy the nutritional needs of the population. Non-dairy ingredients find a critical role in synergy of the chemical constituents of dairy foods to enhance their sensory, nutritional profile, at the same time influencing the cost of the resultant product (Wang, 1980). Fruits are invariably used for flavoring of several dairy products. However, due to the presence of phytochemicals in most of the fruits, its involvement has increased looking at the wellness of the product containing it. Fruits are rich sources of various important phytonutrients namely, vitamins, minerals, antioxidants and dietary fibers. Incorporation of fruits in milk products not only aids in „value-addition and „product diversification, but also helps in checking the postharvest losses. (Hirpara *et al.*, 2011). The use of synthetic antioxidants in food has been decreased due to their suspected action as promoters of carcinogenesis, as well for the general consumer rejection of synthetic food additives (Namiki, 1990). Several studies indicated that the use of synthetic antioxidants has begun to be restricted because of their health risks and toxicity (Hirose *et al.*, 1998).

Horticulture

Horticultural crops play a unique role in India's economy by improving the income of rural people. Fruits and vegetables are also rich source of vitamins, minerals, proteins and carbohydrates *etc.* which are essential in human nutrition. Hence, these are referred as protective foods and assumed great importance in nutritional security of the people.

Dryland horticulture has a considerable scope, cultivation of certain fruits and vegetables crops can be restored in the vast rainfed areas of the state. Fruit crops such as pomegranate, annonas, aonla, ber, karonda, tamarind, jambhul, phalssa, custard apple *etc.* can be cultivated in dryland/rainfed areas.these fruits are highly perishable can not be stored for long time. Hence the development of agro based industries is important. Maharashtra state has varied agro-climatic zones. It is ideal state for production of tropical and subtropical fruits.

India has witnessed voluminous increase in horticulture production over the last few years. We are second largest producer of vegetables and fruits and our presence in global market is significant.

Table 24.1: India's Position in World Agriculture 2013

Item	*Production (million tonnes)*		*India's*		*Near to*
	India	*World*	*Per cent share*	*Rank*	
Fruits excluding melons	83	677	12.2	Second	China
Banana	28	107	26.2	First	
Mango, mangosteen and guava	18	43	41.9	First	
Lemon and lime	2.5	15.2	16.4	First	
Papaya	5.5	12.4	44.4	First	

Significant progress has been made in area expansion resulting in higher production. Over the last decade, the area under horticulture grew by about 2.7 per cent per annum and annual production increased by 7.0 per cent. During 2013-14, the production of horticultural crops was about 283.5 million tonnes from an area of 24.2 million hectares (ha). Out of the six categories, that is, fruits, vegetables, flowers, aromatic, spices and plantation crops, the highest annual growth of 9.5 per cent is seen in fruit production during 2013-14. In 2013-14, the total fruit production was highest in the case of Maharashtra which is about 15 per cent (134.6 lakh tonnes) followed by Andhra Pradesh 12 per cent (105.11 lakh tonnes), Gujarat 9 per cent, Tamilnadu 8 per cent, Uttar Pradesh 8 per cent and Karnataka 8 per cent.

Pomegranate

Pomegranate is commercially grown for its sweet-acidic taste. The fruits are mainly used for dessert purposes. El-Nemar *et al.* (1990) analysed the chemical composition of pomegranate fruits and recorded that the edible parts represented 52 per cent of the total fruit weight, comprising 78 per cent juice and 22 per cent seeds. The fresh juice contained 85.4 per cent moisture, 10.67 per cent total sugar, 1.4 per cent pectin, 0.1 g/100 ml total acidity (as citric acid), 19.6 mg/100 ml free amino nitrogen and 0.05 g/100 ml ash. The seeds were rich source of total lipids, protein, crude fibre and ash, representing 27.2, 13.2, 35.3 and 2.0 per cent respectively, and also contained 6.0 per cent pectin and 4.7 per cent total sugar.

Leena borah (1990) reported the following juice composition of pomegranate cv. Ganesh:

Moisture (per cent)	87.20
TSS (°Brix)	12.00
Acidity (per cent)	0.31
pH	2.95
Reducing sugar (per cent)	9.93
Total sugar (per cent)	10.66
Ascorbic acid (mg/100 g)	22.00
Tannins (per cent)	0.092
Ash (per cent)	0.52
Calcium (mg/100 g)	12.07
Phosphorus (mg/100 g)	4.91
Iron (mg/100 g)	0.34

The juice of wild pomegranates is used in the manufacture of citric acid and sodium citrate for medicinal purposes.

The pomegranate antioxidant properties are the most well known. Some studies shown that the juice of the pomegranate has about three times the antioxidant power of green tea and red wine. Common antioxidants in the pomegranate are vitamin C and vitamin E.

The nutrients in pomegranate reduce the risk of heart disease; help to prevent plaque build-up in arteries by reducing the oxidation of LDL (bad cholesterol), reduce inflammation, be effective against osteoarthritis, help to prevent prostate cancer and other cancers, enhance immune function, promote production of red blood cells, expel tapeworms, strengthens the bladder, strengthen the gums, reduce diarrhoea.

Date Palm

Ahmed *et al.* (1995a) studied the composition and properties of proteins from 9 date cultivars (Bunaringa, Barneh, Fardh, Khalas, Khasab, Khuneizy, Handhal, Hilali and Zabad) and found that the most date proteins were water soluble albumins. These proteins were rich source of amino acids and poor in sulphur-containing amino acids such as methionine and cysteine. Results of elemental analyses indicated that dates could be an important source of K when consumed regularly (Ahmed *et al.*, 1995b).

Date is a staple food in many countries in the middle east of Asia and North Africa and an important commercial crop.

Ber

The ripe ber fruits are rich in nutritive value. The fruits are mostly eaten fresh but the other forms, such as dried, candied, pickled and other products like squash or juice and ber butter can also be prepared and used. Various Ayurvedic and Yunani medicines containing ber extract are said to be a blood purifier and also help in digestion. The powder and decoction prepared from the roots are effective in case of fever, ulcers and old wounds. The stem bark is considered to be remedy for diarrhoea. It is believed to nourish the heart yin, augment the liver blood, and calm the spirit. It is used to treat irritability, insomnia and heart palpitation.

The medicinal value of various parts of the tree and fruits of ziziphus spp. are many and yet not fully exploited.

Custard Apple

The fruit is good source of carbohydrate (23.5 per cent) minerals (0.9 per cent), protein (1.6 per cent), moisture (73.5 per cent), fat (0.3 per cent), calcium (0.02 per cent), phosphorus (0.04 per cent) and iron (0.01 per cent). It is good source of vitamin A and C. The fruit yields about 40 per cent pulp having 26.4 brix (TSS) 5.5 pH and (0.5 per cent) tannins (Gopal and Venkatraman, 1990).

The edible portion of fruit varies from 28.55 per cent and the calorific value per kilogram ranges from 822 to 1050 as compared to 741 in mango. It has mild acidic flavour and sweet taste (Salunkhe and Kadam, 1995) custard apple pulp has great demand. They are also used in ice-creams and other milk products and preserved as jam, jelly or other products on a limited scale.

Aonla

The fruit is very nutritive and it is richest source of vitamin C among fruits except Barbados cherry (Asenjo, 1953). The ascorbic acid and other constituents are well retained in dried aonla fruits (Srivastava and Srivastava, 1964). The aonla powder is superior to synthetic vitamin C in treating deficiencies. The stability of ascorbic acid and presence of astringency in aonla fruits may be assigned to the presence of polyphenols or leucoanthocynins (Sastry *et al.*, 1956). Hanif *et al.* (1966) noted marked antioxidant effect of gallic acid present in aonla fruit.

The fruits are valued as an antisorbutic, diuretic, laxative, alterative (Nadkarni, 1927) and antibiotic (Ray and Majumder, 1976).

Wood Apple

The ripe fruits contain sour-sweet, aromatic pulp which is about 70 per cent of total weight. Wood apple pulp contains significant amount of proteins. The fruit is a rich source of riboflavin, acidity is about 2.3 per cent and sugars 7.2 per cent.

The fruit has a great demand from native system of medicines such as Ayurvedic. Fruit is used has astringent effect, stomachic and stimulant while leaves are carminative. Bark is prescribed for biliousness.

Karonda

Karonda fruit is a rich source of iron and contains a fair amount of vitamin C, and therefore, very useful for curing of anaemia and has antiscorbutic properties. Enthomedically, fruits are used as an astringent, antisorbutic and as a remedy for biliousness.

Dairy Products Prepared from Various Dryland Fruit Crops

Adani *et al.* (2014) utilized date paste in the preparation of milk burfi. The levels of date paste used for *burfi* preparation were 2, 4, 6 and 8 per cent by the weight of *khoa* and a Control *burfi* without date paste. Sugar was added @ 25 per cent by the weight of *khoa*. They reported that *burfi* prepared by incorporation of 6 per cent date paste was superior over rest of the treatments on the basis of cost and overall acceptability scores.

Deshmukh *et al.* (2010) studied the chemical and sensory quality of ice-cream incorporated with karonda pulp. The levels of karonda pulp used for ice-cream preparation were 0, 10, 20, 30 per cent and concluded that the ice-cream with 20 per cent Karonda pulp had the superior general appearance and good body and texture and desirable flavour and overall acceptability as compared to rest of the treatments.

Ice-Cream Incorporated with Karonda Pulp

Selection of ingredients

↓

Figuring of mix (Standardization)

↓

Pasteurizing the mix (72° C for 20 min)

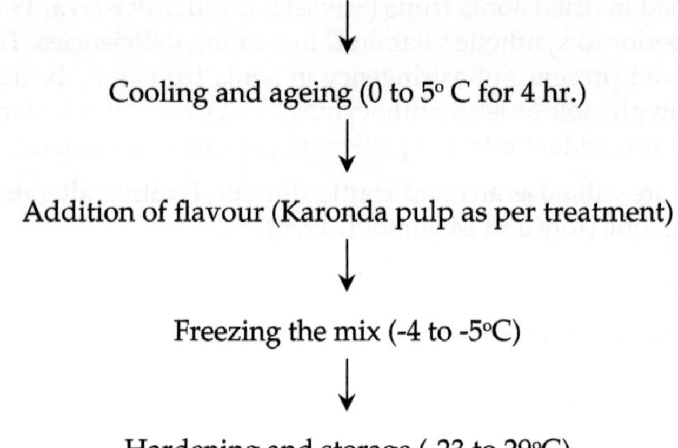

Hanwate *et al.* (2005) Utilized karonda (*Carissa carandas L.*) juice in the manufacture of flavoured milk. On the basis of overall acceptability and chemical composition of the product (Karonda flavoured milk) they concluded to use 10 per cent karonda juice and 7.5 per cent sugar alongwith 0.5 per cent gelatin for preparation of flavoured milk.

Navale *et al.* (2014) prepared herbal *burfi*, with blending of wood apple (*Limonia acidissima*) pulp at 0 per cent, 5 per cent, 10 per cent, and 15 per cent concentrations with buffalo milk khoa and revealed that Indian customers would prefer 10 per cent wood apple *burfi* over the normal *burfi*, and other combinations of wood apple *burfi*. They also concluded that 10 per cent wood apple *burfi* could be projected as an herbal confectionery for persons with gastric ailments like peptic ulcer and/or renal calculi.

Wood Apple *Burfi*

Procurement and filtration of buffalo milk

↓

Standardization of milk (Fat-6 per cent, SNF-9 per cent)

↓

Pan-heating of milk to obtain *khoa*

↓

Addition of sugar

↓

Addition of wood apple pulp

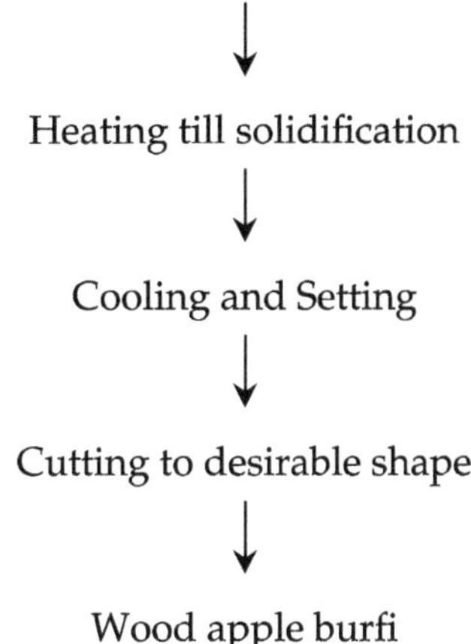

Bhagyashri (2015) prepared custard apple kalakand. The kalakand was prepared from fresh cow milk with addition of 0.02 per cent citric acid and 5 per cent sugar and this was common for all treatments. The levels of custard apple pulp used for kalakand preparation were 0, 4, 6, 8 per cent. She revealed that the overall acceptability scores for all the samples were above 7.0 out of 9.0; hence all of them were acceptable. However, custard apple kalakad prepared with addition of four per cent custard apple pulp was preferred most and secured maximum score (8.75) for its overall acceptability.

Custard Apple Kalakand

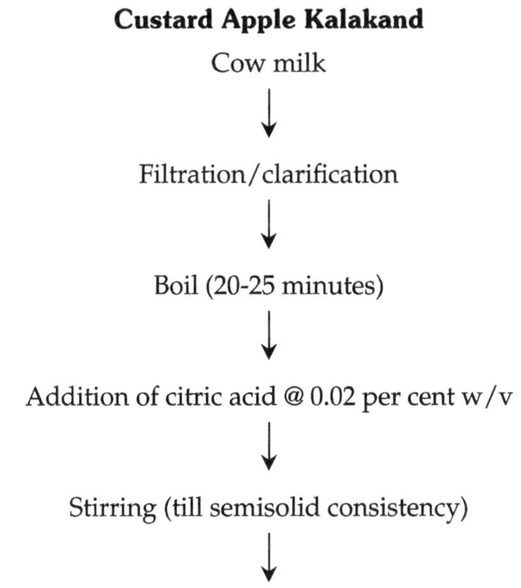

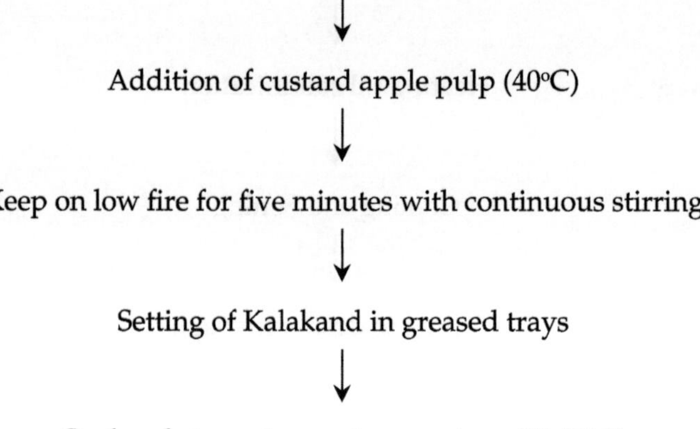

Kranti (2005) prepared pomegranate yoghurt from cow milk and reported that addition of fruit arils (grains) at 10- 15 per cent before incubation in unstirred form with addition of 3- 6 per cent sugar was more acceptable.

Pomegranate Yogurt from Cow Milk

Cow milk

↓

Filtration/clarification

↓

Standardization (fat=3.5 per cent add SMP = 5 per cent)

↓

Stirring

↓

Homogenization

↓

Addition of sugar @ 3 – 6 per cent

↓

Stirring

↓

Pasteurization (85 °C for 30 min)

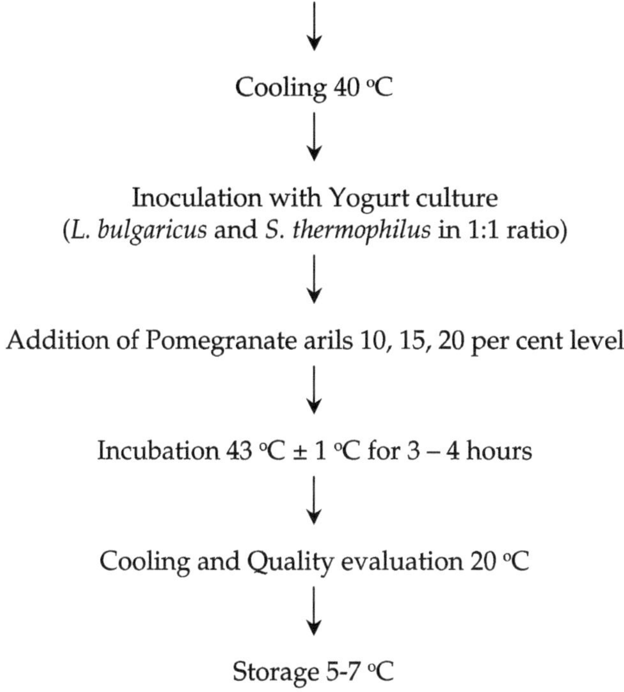

Shital (2014) studied on processing of custard apple for preparation of ice cream and milk shake. She reported that ice cream prepared by using 12 per cent custard apple pulp, 13 per cent sugar and 0.15 per cent sodium alginate and milk shake prepared with addition of 15 per cent custard apple pulp and 7 per cent sugar were more acceptable on the basis of sensory evaluation.

Custard Apple Ice cream

Whole milk

↓

Figuring of mix

↓

Pasteurization (68 °C/30 min)

↓

Homogenization

↓

Cooling mix (at room temperature)

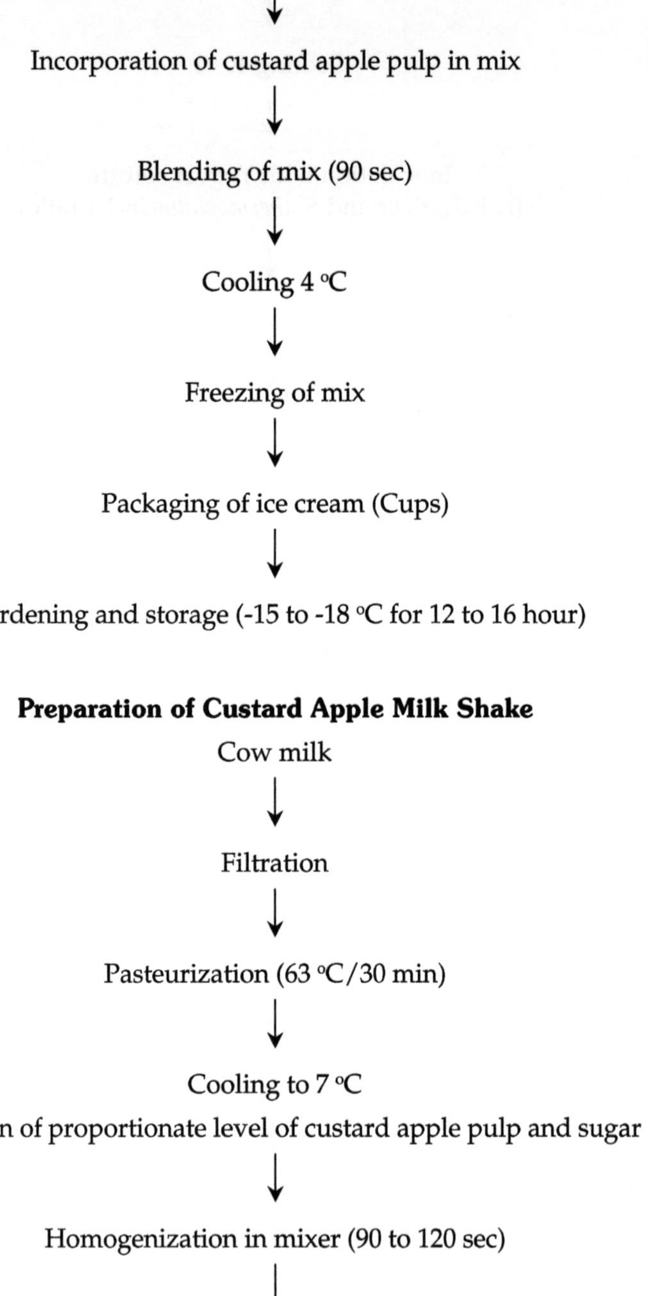

↓

Incorporation of custard apple pulp in mix

↓

Blending of mix (90 sec)

↓

Cooling 4 °C

↓

Freezing of mix

↓

Packaging of ice cream (Cups)

↓

Hardening and storage (-15 to -18 °C for 12 to 16 hour)

Preparation of Custard Apple Milk Shake

Cow milk

↓

Filtration

↓

Pasteurization (63 °C/30 min)

↓

Cooling to 7 °C

Addition of proportionate level of custard apple pulp and sugar

↓

Homogenization in mixer (90 to 120 sec)

↓

Filling in glass bottles

↓

Cooling and storage at refrigeration temperature (7 ± 0.5 °C)

Abd El-Aziz *et al.* (2013) investigated Pomegranate (*Punica granatum*) fruit peels for its phenolic compounds, antioxidant and antoxidative activity on L.A.B in yoghurt during cold storage. Pomegranate peel extract explored considerable inhibitory effect of each of lactic acid bacteria, coliform group and *Staph. Sp.* The increase aqueous extract of pomegranate peel from 1to 5 per cent lead to an inhibitory effect on LAB and other tested pathogenic bacteria. 1 per cent of extract was appropriate for the viability of Lactic acid bacteria which was above 30×10^6 c.f.u D ml. In addition, using of 1, 0.75, 0.5 and 0.25 per cent on yoghurt processing was acceptable for the chemical, microbiological and organoleptic properties of the resultant yoghurt. Moreover, the ratio of 0.75 per cent was more acceptable, comparing with other added doses. The use of Pomegranate peels with yoghurt was advantageous due to inhibitory effect on *staphylococcus sp.*, Coliform, moulds and yeasts. In addition a Pomegranate peel is safe for public health. The results highlighted the possibility of the treatment (0.75 per cent). The developed treatment was evaluated and proved to be of good quality, long shelf life and could be kept at 4°C for 14 days without significant microbial growth or loss of the product colour and texture during manufacture and storage.

Kanchan (2011) standardized the technique for production of wood apple burfi and concluded that the best quality of wood apple *burfi* can be prepared by using 20 per cent wood apple pulp and 25 per cent (sugar + *jaggery*) as a sweetener which can be stored in good condition and oranoleptic acceptable up to 15 days at ambient temperature and 40 days at refrigerated temperature.

Flow Diagram for Preparation of Wood Apple *Burfi*

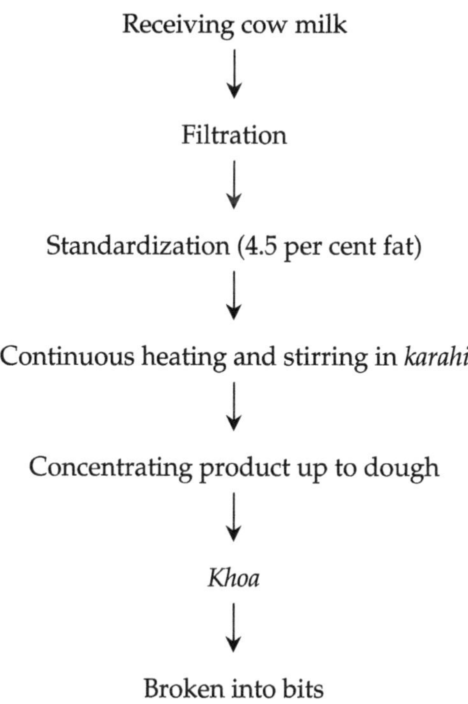

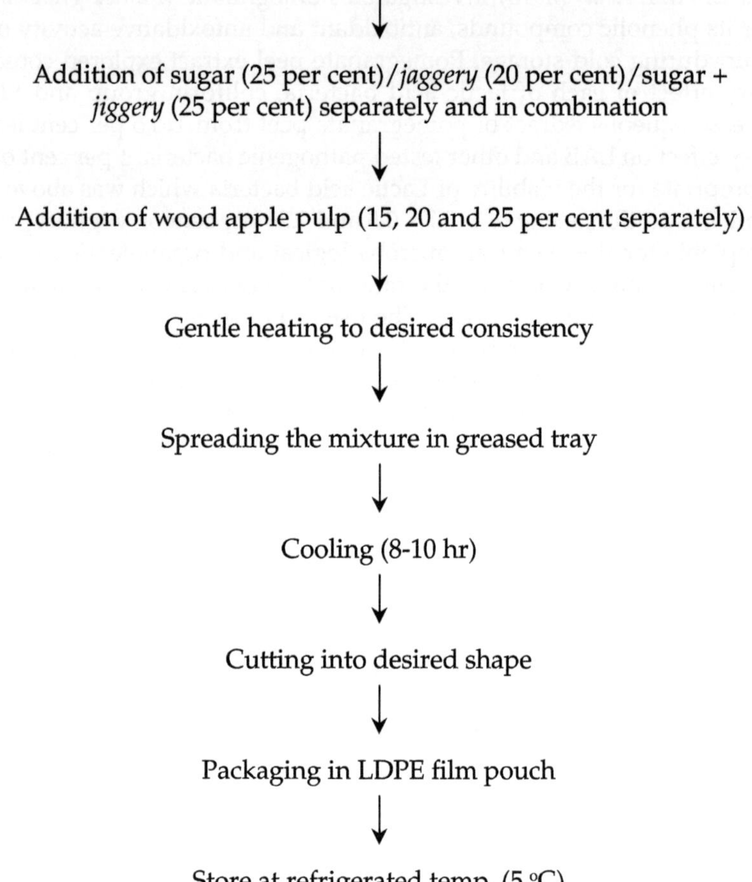

Conclusion

Most of the dryland fruits are enriched with nutritional and medicinal value. Also, milk is considered to be an adequate source of valuable macronutrients (fat, protein, lactose), vitamins and micronutrients (minerals), making it a "wholesome food". There is always demand from consumers for new, delicious, nutritious and attractive products. To satisfy his demand, there is a constant effort to develop products from diverse sources. The potentiality of processed milk enriched products from some dryland fruits is still untapped. However, efforts have been made by the various researchers' for the development of value added products from these dryland fruit crops as explained above.

References

Abd El – Aziz, M. E., Rania E. El – Gammal, Abo-srea, M. M. and Faten I. Youssf (2013). Antioxidant and antimicrobial activity of pomegranate (punica granatum l.) fruit peels extract on some chemical, microbiological and organoleptical properties of yoghurt during storage, J. Food and Dairy Sci., Mansoura Univ., Vol. 4 (7): 387 – 400.

Adani, R. J., Zinjarde, R. M., Atkare, V. G., Gajarlawar, S. and Pooja Mule (2014) utilization of date (Phoenix dactylifera) in the preparation of cow milk burfi. 5th International Conference on Advances in Food Technology and Health Sciences (ICFTHS- 2014), page 19.

Ahmed, I. A., Ahmed, A. W. K. and Robinson, R. K. (1995) Food Chem., 54: 305-309.

Ahmed, I. S. A., Al-Gharibi, K. N., Darr, A. S. and Kabir, S. (1995b) Food Chem., 53: 441-446.

Asenjo, C. F. (1953) Bull. De Colegio de Quimicos de, Puerto Rico, 10: 8-17.

Bhagyashri S. Thakur (2015). Preparation of custard apple kalakand, M.Sc. (Agri.) Thesis (unpub.) submitted to M.P.K.V., Rahuri, (M.S.), India.

Deshmukh, A.R., Gaikwad, R.P., Adangale, S.B. and Deokar D.K. (2010) Study of the sensory quality and determine the optimum level of Karonda pulp for ice-cream preparation, Res. J. Animal Hus. and Dairy Sci., 1 (2): 55-57.

El-Nemr, S. E., Ismail, I. A. and Ragab, M. (1990) Nahrung, 34: 601-606.

Gopal and Venkatraman (1990). The Custard Apple and Fruits in Fruit Culture of India. Indian council of Agriculture Research, New Delhi.pp.217-224.

Hanif, M. (1966) Pakistan J. Sci. Res., 18: 61-63.

Hanwate, B. D., Kadam, R.M., Joshi S.V. and Yadav D.N. (2005) Utilization of Karonda (Carissa carandas L.) Juice in the Manufacture of Flavoured Milk. Souvenir, National Seminar on Value Added Dairy Products, NDRI, karnal, pp: 164.

Hirose, M., Y. Takesada, H. Tanaka, S. Tamano, T. Kato and T. Shirai (1998). Carcinogenicity of antioxidants BHA, caffeic acid, sesamol, 4- methoxyphenol and catechol at low doses, either alone or in combination, and modulation of their effects in a rat medium-term multi-organ carcinogenesis model. Carcinogenesis 19 (1): 207–212.

Hirpara Krupa, H. Jana Atanu, and H. G. Patel (2011). Synergy of dairy with non-dairy Ingredients or product: A review. African Journal of Food Science Vol: 5 817-832.

Kanchan S. Bhingardive (2011) Standardization of technique for production of wood apple burfi, Ph.D. (Agri.) Thesis (unpub.) submitted to M.P.K.V., Rahuri, (M.S.), India.

Kranti G. Kale (2005) Studies on preparation of pomegranate yoghurt from cow milk, M.Sc. (Agri.) Thesis (unpub.) submitted to M.P.K.V., Rahuri, (M.S.), India.

Leena Borah (1990) M. Sc. (Food Technol) thesis, University of Mysore,India.

Nadkarni, A. K. (1927) Indian Materia Medica, S. K. Bijur, Bombay.

Namiki, M. 1990. Antioxidants/antimutagens in foods. Crit. Rev. Food Sci. Nutr., 29: 273-300.

Navale, A.S., Deshmukh, B.R., Korake, R.L., Narwade, S.G. and Mule P.R. (2014) Production profile, proximate composition, sensory evaluation and cost configuration of wood apple burfi, Animal Science Reporter, 8 (3):114- 120.

Ray, P. K. and Majumder, S. K. (1976) Econ. Bot., 30:317-320.

Salunkhe, D. K. and Kadam, S. S. (1995). Handbook of Fruit Science and Technology: Production Composition Storage and Processing, First Edn., Markel Dekkar, Inc. New York, Basel, Hongkong, pp.382.

Sastry, L. W., Satyanarayana, M. N., Srinivasan, M., Subramanian, N. and Subrahmanyan, V. (1956) J. Sci. Indus. Res. 15: 70-80.

Shital V. Deshmukh (2014) Studies on processing of custard apple for preparation of ice cream and milk shake, M.Sc. (Agri.) Thesis (unpub.) submitted to M.P.K.V., Rahuri, (M.S.), India.

Srivastava, R. P. and Srivastava R. K. (1964) Sci. and Cult., 30: 446-447.

Wang H. (1980). The importance of traditional fermented foods. Bio Sci., 30: 402-404.

Chapter 25

Postharvest Handling and Value Addition of Pomegranate

V.P. Kad, J.K. Dhemre, S.D. Shelar and S.D. Magar

Postharvest Technology Centre,
Mahatma Phule Krishi Vidyapeeth, Rahuri – 413 722, Dist. Ahmednagar, M.S.

Pomegranate

Most consumers are unfamiliar with pomegranates and even fewer know that only the juice and seeds are edible. However, they are gaining popularity in food service due to their sweet-tart flavour, the multitude of uses and the fact that they are high in antioxidants. The fruit is nearly round with a crown at the base. Pomegranates have a tough, leathery skin with membranous walls and spongy tissue holding fleshy, juicy pulp. The seeds embody more than half of the fruits weight.

It is native from Iran to the Himalayas in northern India and has been cultivated since ancient times throughout the Mediterranean region of Asia, Africa and Europe. Its name in Latin means "apple with many seeds," but it actually looks somewhat like a petrified tomato. The pomegranate plant is well adapted to the hot, dry climate of the Southwest and grows in a wide range of soils. The pomegranate plant can be used for its aesthetic beauty in hedges, as a shrub, a tree or for its fruit.

Pomegranate Uses

Health benefits of pomegranate are innumerable. Apart from being healthy, pomegranate is delicious too. Pomegranate consists of antioxidant, antiviral and anti-tumour properties. It is said to be a good source of vitamins as it includes vitamin A, C and E as well as folic acid. This fruit consists of three times the antioxidants of wine or green tea. It is said to be the powerhouse of health. Pomegranates are known mostly for curing the problems related to heart and for maintaining effective and healthy blood circulation. Other health benefits include cure of stomach disorders, cancer, dental care, osteoarthritis, anemia and diabetes.

Health benefits of pomegranates can be extracted by consuming it in several ways. You can take it in the form of a juice or can eat its seeds, syrup, paste, nectar or concentrate. Pomegranates are being traditionally used for clearing the skin and reducing inflammation. Pomegranate's juice can participate in treatment of a sore throat. According to a research study stated in American Journal of Clinical Nutrition, pomegranate being rich in antioxidants can prevent the oxidization of 'bad' cholesterol. Pomegranate syrups are useful in treating the cartilage damage and protest the babies from the brain damage during their birth. This is the reason that various advantages of pomegranates have been recorded since years.

Several advantages of pomegranates in curing various diseases can be listed below:

Stomach Disorder

Pomegranates peel, bark and leaves are used to calm the disorder of the stomach or diarrhea caused due to any kind of digestive problems. Drinking tea made from the leaves of this fruit helps in curing digestive problems. Pomegranate juice is also used for handling problems of dysentery and cholera.

Heart Problems

Regular intake of pomegranate juice can maintain good flow of the blood in the body. Along with this, it decreases the risk of heart attack and heart strokes. The antioxidant properties that are consisted in this fruit helps to keep the bad cholesterol in the forming position and thus keeps the arteries clear of any clots. The clots are clear because pomegranate has the ability to make the blood thinner.

Cancer

Pomegranates consist of high level of antioxidants called flavenoids. These flavenoids are thought to be effective in counteracting various cancer radials. The people that face high risk of prostate and breast cancer should start drinking the juice of this fruit, as this will help them to reduce further risk of developing cancer. Regular consumption of pomegranates can fight the existing cancer cells in the body.

Dental Care

The best benefit of pomegranate is that its juice, along with its antibacterial and antiviral properties, helps to reduce the effects of dental plaque.

Other Advantageous Osteoarthritis

Pomegranate helps to reduce the illness caused in various forms, like atherosclerosis and osteoarthritis. The damages that are caused due to the thickening and hardening of the arterial walls and in cartilage and joints can be cured by eating this fruit. Also, pomegranate is capable of preventing the creation of enzymes that are liable for breaking down the connective tissues.

Anemia

Healthy blood flow can be maintained in the body by taking this fruit in any form. Pomegranate supplies iron to blood and thus, help to reduce the anemic symptoms including exhaustion, dizziness and weakness and hear loss.

Diabetes

Drinking of pomegranate juice by a diabetic patient can reduce the risk of coronary diseases. Along with this, there is a slowdown in hardening of the arteries, which can fuel non-occurrence of various heart diseases.

Other Advantageous Forms of Pomegranate

With the passage of time, more and more people have started realizing the importance of eating pomegranates. There are other benefits too, like pomegranate reduces the possibility of having premature babies and it is also beneficial for the expected mothers to avoid having low weight babies during birth. Pomegranate also reduces the possibility of developing Alzheimer's disease among the older generation. It helps to control aging problems like wrinkles and thus, facilitates youthful and glowing skin. Other than this, it helps a woman to overcome from her depression period, especially form the menopause period. Thus, the advantages of pomegranates are uncountable and profitable.

However, pomegranates are a super food full of healthy antioxidants and vitamins that may help to prevent heart diseases and cancer. It's unfortunate that so many people are discouraged from eating pomegranates because it takes a lot of patience to get the small seeds or arils out.

World Scenario

- ✰ Popular in Eastern as well as Western parts of the world.
- ✰ The crop is also cultivated in the Mediterranean region. Pomegranates thrive well in regions with semi-arid to sub-tropical climatic conditions.
- ✰ The crop is cultivated in Spain, Iran, India, Turkey, South East Asia, Afghanistan, tropical Africa, parts of USA *etc.*
- ✰ It is grown for its fully luscious grains called 'Arils'; the fruits are very attractive with sweet acidic taste.
- ✰ The fruits are mainly used for dessert purposes.
- ✰ The fruits are also processed to make juice, syrup, jam, jelly, wine, to flavour cakes, baked apples, *etc.*
- ✰ Of late, its nutritional and medicinal values are given ample importance.

Pomegranate Cultivation in India

Comparative Advantage

- ✰ Production of Pomegranate almost throughout the year.
- ✰ Supply of Pomegranates even during off - season to the European Countries.

- ☆ Cultivation in Arid and Semi-arid tracts of the states Karnataka, Maharashtra, Andhra Pradesh, Gujarat, Rajasthan, Uttar Pradesh, Haryana, Tamil Nadu
- ☆ Maharashtra is the largest producer.

Farmers Cultivate the Crop

- ☆ Carefully selecting varieties.
- ☆ Adopting advanced Technologies.
- ☆ Spurt in cultivation and trading of Pomegranate in India from the last two decades.
- ☆ Important varieties cultivated are Ganesh, Dholka, Seedless(Bedana), Bhagawa, Phule Araktha, Mridula, Bhagawa Super
- ☆ Seedless verities such as Paper Shell, Spanish Ruby, Mascut Red and Velladu shows great potential
- ☆ Peak harvest: January to April

Area and Production

India

Area: 131 thousand ha (2013-14)

Production: 1346 thousand MT (2013-14)

Productivity: 10.3 MT/ha (2013-14)

Maharashtra:

Area: 90 thousand ha (2013-14)

Production: 945 thousand MT (2013-14)

Productivity: 10.3 MT/ha (2013-14)

Salient Features of Pomegranate Production in Maharashtra

- ☆ High productivity at progressives farmers field around 30-35 tons/ha.
- ☆ Well defined agro techniques.
- ☆ Availability of fresh pomegranate almost 10 months of the year.
- ☆ Quality production due to cultivation in arid region.

Consumer Preference

- ☆ Consumer preference - changing from time to time and from country to country.
- ☆ Earlier, Ganesh variety with big sized fruits was the preferred one, and the fruits were exported to the Gulf countries.
- ☆ Now, in Europe and other parts of the world, varieties such as Bhagawa (Shendrya), Mridula are the suitable and accepted ones.

- Soft seeded, coloured varieties with high per cent of juice with easy to remove arils are preferred.
- Fruits weighing more than 500 gms with superior qualities - immediate and ready acceptance in the international markets.

The Desirable Fruit Characters of Fresh Pomegranate for Export Purpose

- Dark rose pink colour of the fruit.
- Fruit weight around 500 gms.
- Round shape of the fruit.
- Uniform size and shape of the fruit in a pack or box.
- Dark rose pink arils.
- Softness of the seeds.
- Higher sugars near about 16-17° Brix.
- Free from scars, rusting, disease spots, insect injury, scratches, *etc.*
- Smooth cutting at the stem end.
- Bracts/calyx without any damage and having freshness.
- Pleasant flavour and aroma.

Marketing and Export Potential

Marketing

- No organized marketing systems for pomegranate.
- Domestic marketing for pomegranate is fairly balanced as of now.
- Farmers normally dispose their produce to contractors who take the responsibility of transport to far off markets.
- Some progressive farmers who have developed contacts in urban areas despatch them on their own and are able to get a reasonable price for their produce.

Export Potential

- The pomegranate fruit is considered an exotic in European and Middle East countries.
- There is scope for exporting Indian pomegranates to Bangladesh, Bahrain, Canada, Germany, United Kingdom, Japan, Kuwait, Sri Lanka, Omen, Pakistan, Qatar, Saudi Arabia, Singapore, Switzerland, U.A.E. and U.S.A.
- The other potential markets are, Belgaum, Spain, France, Italy, Netherlands, Philippines, South Africa, Brazil and Mexico.
- India's contribution to the export is less than 5 per cent, although we are in first position in production of Pomegranate.

- ☆ Scope for diversified use of pomegranate and hence value added products can be promoted. Consumer preference, colour,TSS and demand for crop are important aspects to be focused upon.
- ☆ Time of export, choice of variety, quality maintenance and regulation of fruit size are important aspects in the production of pomegranate for export.

Postharvest Handling

- ☆ Thorns need to be removed along with dead branches.
- ☆ Thin out fruits in clusters.
- ☆ Clippers need to be developed for harvesting.
- ☆ Grading rings.
- ☆ Individual fruit wrapping with partial permeable films.

Harvesting

Bahar	*Month of maturity*
Ambe bahar	June-August
Mrig bahar	November-December
Hasta bahar	February-April

Fruit maturity required 135-150 days in Ganesh, Mridula and Phule Arakta, 180-190 days in Bhagawa

Maturity and Maturity Standards

- ☆ Distinctive fruit colour characteristic of the variety.
- ☆ Metallic sound on tapping
- ☆ Ganesh cultivar: TSS reaches 14.5°B or acidity 0.3 per cent to 0.35 per cent
- ☆ Mridula cultivar: TSS 16°B, Acidity 0.45 per cent.

Methods and Time of Harvesting

- ☆ Harvest pomegranates early in the morning or in evening hours
- ☆ Use clippers and bags for harvesting
- ☆ Do not pull the fruits
- ☆ Cut the stem end close to the base
- ☆ Do not bruise the base while cutting the fruits. Handle carefully at harvest
- ☆ Do not keep longer stem end, as it can cause damage to other fruits
- ☆ Keep the fruits in plastic crates for emptying the bags
- ☆ Do not keep fruits on the ground
- ☆ Do not keep empty crates in the sun but in shade

- ☆ Do not overfill the crates with fruits and keep them in shade and not in open sun
- ☆ Do not damage calyx while harvesting or while keeping in crates or during field storage.

Transporting Fruits from Field to Pack-house

- ☆ Transporting immediate to pack-house without any loss of time
- ☆ For higher value realization shiny appearance, bright colour of rind
- ☆ Fruit should not be spoiled during loading, unloading and transportation
- ☆ Vehicle should be clean to avoid infection
- ☆ No jerks or bumps in the vehicle
- ☆ No other product should be transported.

Receipt of Pomegranates at Pack-house

- ☆ Unloading should be done gently to avoid brushing and tumbling
- ☆ Unloading should be preferably in sorting bay having cool temperature
- ☆ Avoid multiple handling of fruits
- ☆ Do not stack too many crates over each other
- ☆ Note down both tare and net weight of crates
- ☆ Grower's code is to be maintained to avoid mixing.

Sorting, Washing and Treatment with Fungicide and Wax on a Packing Line

A. Sorting

- ☆ Inspection roller-cum-conveyer are used for sorting
- ☆ Fruits showing greenish colour, fruits with sunspots, affected by fruit borer, disease spot, rusting, rough surface be eliminated
- ☆ Having cuts, blemishes, damages must be rejected.

B. Washing

- ☆ Use of disinfectant sodium hypochlorite @ 100-200 ppm.
- ☆ Rinsing with clean water
- ☆ Potable water should be used.

C. Elimination of Excess Water on the Fruit Surface

- ☆ Use of absorbent roller.

D. Waxing and Fungicide Application

- ☆ Thiabendazole or Benomyl @ 1000 ppm concentration

- ✰ Only good quality edible wax must be used
- ✰ Monitor edible wax quantity
- ✰ Monitor concentration of the fungicide to avoid crossing of international residue limit
- ✰ Effective wax coating is very important.

E. Drying the Wax Coating

- ✰ Temperature of drying tunnel must not to beyond 30-35°C
- ✰ Bulbs, heaters, hot air circulation along with fans, air knives must work properly for proper drying.

Sizing of Fruit

Sl.No.	*Different Size Ranges*	*No. of Fruits in*	
		4.0 kg box	*5.0 kg box*
1.	450 to 500	Not exported	
2.	400 to 450	(Indian Domestic market)	
3.	350 to 400	10 - 12	12 - 15
4.	300 to 350	12 - 15	15 - 18
5.	250 to 300	12 - 15	15 - 18
6.	200 to 250	15 - 18	18 - 20

- ✰ For export purposes mostly 200 - 400g fruit weight are used
- ✰ Variation in a size grade should be minimum
- ✰ Bad quality of fruits must not go into good quality fruits

Grading of Fruits

The fruits after harvesting are graded as per their colour, size and weight. The different grades are as follows:

1. Super size fruits are free from spots and individual fruit weight is more than 750grams.
2. King size fruits are attractive and individual fruit weight is 500-700 grams.
3. Queen size fruits are attractive, red and individual fruit weight is 400-500 grams.
4. Prince size fruits are attractive, red and individual fruit weight is 300-400 grams.

Nose Cleaning

Before packing a box, fruits at stylar end or calyx side are wiped with a cloth soaked in a fungicide solution (Benomyl or Thiabendazole @ 1000 ppm) to avoid cross contamination in a packed box.

Packing

All the graded fruits are wrapped in paper and packed in corrugated fiberboard (CFB) boxes. Brown coloured 3 fold CFB boxes are used for local market while; white coloured 5 fold CFB boxes are used for distant markets. Paper shreds are used as padding material. For 'Super' and 'King' grades, boxes of size 32.5 x 22.5 x 10 cm are used. 'Queen' grade fruits are packed in 37.5 x 27.5 x 10 cm size boxes while 'Prince' grades are packed in 35 x 25 x 10 cm size boxes.

Packing and Coding

- ✰ 4.0 to 5.0 kg box is used with 10-20 number of fruits per box
- ✰ 5.0 kg box containing 10-15 fruits of Ganesh and Mridula or Arakata 18-20 number
- ✰ Dimensions of 4.0 kg box: 375 x 275x100 mm and 5.0 kg box: 480 x300 x100 mm.

Characteristics of Boxes

- ✰ Cardboard boxes must have ventilation holes, hand holes *etc.*
- ✰ Fruits should not press each other to avoid bruise on the skin
- ✰ No movement of fruits inside the pack by putting accurate size of fruits
- ✰ Pallets should be 1000 mm x 1200 mm. For export to Europe, dimension of 800 mm x 1200 mm should be used
- ✰ Only plastic material should be used for strapping and clipping for securing the cartons to the pallets.

Information on Boxes

- ✰ Name of the produce
- ✰ Variety
- ✰ Origin of produce
- ✰ Commercial specifications of class
- ✰ Date of packing
- ✰ Name of packing organization
- ✰ Gross weight/net weight
- ✰ Number of fruits/fruit size
- ✰ Recommended storage temperature
- ✰ Additional labeling requirement of importer should be included.

Pre-cooling

- ✰ Before storing, the fruits needs to be pre-cooled by forced air.
- ✰ The ideal pre-cooling temperatures must be 5 -7°C at least within 10 hours after harvesting.

- ☆ Chilling damage of fruit is possible if stored below 5°C.
- ☆ Humidity should be around 90 - 95 per cent during pre-cooling.
- ☆ Prompt pre-cooling is very important for avoiding postharvest losses.
- ☆ Filled cartons should be shifted to pre-cooling chamber without any delay.
- ☆ On transfer of product, outflow temperature and humidity should be properly recorded.
- ☆ Product should remain in the pre-cooling chamber till the temperature of the fruits reaches to required stage.
- ☆ Do not allow to fall the temperature otherwise chilling injury will make the fruits unfit for consumption
- ☆ Temperature, humidity and air circulation of the room shall be properly maintained
- ☆ After pre-cooling and palletisation, strapping should be done. This needs to be done only in cold room to maintain the cool chain.

Cold Storage

Fruits can be stored upto two months at temperatures 5 - 7°C with relative humidity 90-95 per cent throughout transportation and storage. In case of storage beyond two months, temperature should be maintained at 10°C to avoid chilling injury. Chilling also varies with maturity and variety. Pomegranates subject to low temperature develop brown discolouration of rind, pale colour of arils and brown discolouration of white segments separating the arils. Pomegranates are very susceptible to water loss resulting in shriveling of the skins. Storing fruit in plastic liners and waxing can reduce water loss, especially under conditions of lower relative humidity.

Cool Chain

Cool chain is essential during the transport of export quality commodity all the way from the farm to the customer. This helps in maintaining the temperature inside the box at the same low level as in the cold storage.

The various stages of the cool chain are:

1. Cold store at the farm.
2. Refrigerated truck from farm to the airport
3. Cold store at the airport.
4. Building up of the pallet in a cold store at the airport.
5. Loading the aircrafts directly from the cold store in a short time.
6. Cargo aircraft maintains cold store temperature in hold.
7. Off loading direct into a cold store in the receiving country.
8. Refrigerated truck to the customers.

Quality Appraisal

Quality checking should be done with regard to following parameters:

1. Cutting stem end to appropriate size
2. Fruit size/uniformity
3. Colour of the fruit/uniformity
4. Condition of calyx
5. Freedom from sunscald, disease spots, fruit borer, rusting and rough surface
6. Fungicide residue in no case should exceed prescribed limits
7. T.S.S. (°Brix)
8. Spoilage at the port due to diseases before dispatch to importing countries.

Level for Quality Checking

- ☆ Pack-house before dispatch to the port
- ☆ At the port before dispatch to the importing country
- ☆ Feedback for further improvement. Acceptability and grading given in the importing country is very important for managing future export.

Equipment Required for Quality Checking

- ☆ Refractometer - T.S.S.
- ☆ Burette, pipette, volumetric flask, glass rod, NaOH, Phenolphthalein - acidity analysis
- ☆ Scales for recording weight
- ☆ Pressure tester to test the firmness of the fruit
- ☆ Temperature probe to record temperature inside the fruit
- ☆ Spectrophotometer/HPLC to determine residue levels of fungicides in the fruit.

Container Loading

- ☆ Maintain the cool chain temperature of the loading bay be controlled preferably to 10°C
- ☆ Transfer of pomegranates should be smooth and quick so that it does not result in temperature increase of pomegranates before shifting of pomegranates to containers or Reefer vans at recommended temperature and humidity inside the container.
- ☆ The doors of containers must be closed before and after loading pomegranates
- ☆ Ensure the temperature and humidity recorders are placed inside the container

- ☆ Arrange the cartons in such a manner that there is no movement within the container.
- ☆ Ensure that only single type of product is loaded to one container

Implementation of Code of Hygienic Practices at Postharvest Handling Stage of Fresh Pomegranates

- ☆ Pack-house should not ingress of vermin from birds *etc.* It should be protected by fence and flooring *etc.* to enable proper cleaning.
- ☆ Potable quality of water for washing to minimize introduction of pathogens, *etc.*
- ☆ Disinfectants should be monitored and residues should be as per CODEX.
- ☆ Recycled water should not become a risk to the safety of fresh pomegranates.
- ☆ Use only waxes, fungicides which are safe and approved by CODEX.
- ☆ Properly cleaned sprayers and automatic nozzles be calibrated to control the accuracy to avoid contamination.
- ☆ Air cooling system be maintained in a manner to prevent contamination.
- ☆ Cold storage should be temperature controlled and monitored to minimize microbial growth.
- ☆ Cooling system be maintained in a clean and sanitary condition.
- ☆ Cleaning schedule of equipment and personnel hygiene in the pack-house be strictly implemented.
- ☆ All hygienic aspects during transportation must be looked into.

Fruit Use

The edible portion of the fruit includes the seed and the juicy translucent flesh covering the seed. The tart flavor of the pulp and juice can add interest and variety to other foods. The flesh-covered seeds can be used as a garnish in fruit cups, compotes, salads, desserts, and as a snack. The juice is used to make jellies, puddings, desserts, and drinks. Grenadine, made from pomegranate juice, is a flavoring in some beverages. Grenadine is also a delicious topping for ice cream and chilled fruits such as peaches, pears, and bananas.

Removing the Seeds

When need a large number of seeds, cut off the crown and score the pomegranate lengthwise in three or four places. Seeds can also be removed by cutting the pomegranate in half, placing the cut face down, and rapping the shell firmly with a blunt instrument such as a hammer handle. The juice sacs are somewhat easier to remove. While holding the fruit under water, split the sections and separate the seeds from the section walls. The seeds will sink to the bottom and the peel and membranes will float. Skim off membranes and drain seeds in a colander.

Pomegranate arils extractor developed by Dr. A. S. College of Agricultural Engineering and Technology, Mahatma Phule Krishi Vidyapeeth, Rahuri, Dist. Ahmednagar (MS) is also used to remove the seeds (Arils) from pomegranate.

Extracting the Juice

To get the juice, crush or press the seeds. Seeds can be pressed in a sieve, basket press or food mill. Juice can also be extracted by whirling seeds, in a blender or food processor until liquefied, then straining them. Strain crushed pulp through a double thickness of cheese cloth or nylon netting to remove seeds. Another method of obtaining juice is to cut the fruit in half and use a juice press or juicer.

Freezing

The juice can be frozen or canned. Freezing is recommended because juice will hold its flavor and color better. To freeze, fill freezer containers, leaving 1/2-inch headspace for expansion. Label with the date, then store in an upright position until the juice is frozen. Freeze the seeds for use in salads and appetizers by spreading them in a single layer on a cookie sheet. Quickly freeze, then transfer to and store in moisture, vapor-proof freezer bags or containers. For other uses, pack seeds in containers and cover with cold, 30 per cent syrup. Make the syrup by heating two cups of sugar in four cups of water until sugar is dissolved. Cool syrup, then pour over seed. Leave 112 inch headspace, seal, and freeze.

Canning

Pomegranate Juice

Canned juice may turn brown and lose its flavor. Adding lemon juice to pomegranate juice helps retain color and flavor. Bring juice quickly to a boil, pour into hot jars, adjust lids, and process. Jars that will be processed for less than 10 minutes should be sterilized before filling. Sterilize by heating jars in boiling water for 10 minutes.

Pomegranate juice makes a delicious drink. The juice can be extracted by using a spiral-type screw press without crushing the seeds. The juice is clarified by heating in a flash pasteurizer at 79-82°C, cooling, setting for 24 hour, racking up and filtering or decanting. The clear juice can be preserved by heat treatment or by using chemicals (600 ppm sodium benzoate). The use of sulphur dioxide is banned for pomegranate juice due to loss of colour by bleaching action of SO_2. After heating at 80°C, it is filled into bottles while still hot. The bottles are crown corked and pasteurized at 80°C for 30 minutes.

Concentrate

Fruit juice concentrate are products with great potential on account of their wide applicability in fruit-based beverages. The beverages prepared from pomegranate are highly refreshing and nutritious. For concentrating the pomegranate juice, vacuum concentrator is used to maintain their nutritional and sensory quality.

Syrup

Syrup is prepared by mixing 25 per cent juice with sugar and citric acid to adjust the TSS (65°B) and acidity (1.2 per cent) and stir. It was preserved by pasteurization or by addition of sodium benzoate. Use as topping for ice cream or chilled fruits or in beverages.

Jelly

Combine pomegranate juice, lemon juice, and pectin. Bring to a boil over medium-high heat, stirring constantly. Stir in sugar until well blended; return to a boil and continue boiling, uncovered, stirring occasionally for 2 minutes. Add all ingredients as per adjustment of TSS (67°B), acidity (0.8 -0.9 per cent). Remove jelly from heat. Let stand a minute to allow foam to form; then carefully skim off the foam. Pour hot jelly quickly into hot half pint jars, filling jars to within about 1/4 inch of tops. Carefully wipe off rim of jar. Put lid on each jar as it is filled, screwing ring band on hand tight. Process in hot water bath for 10 minutes, cool and store in a cool, dark area. For maintaining the proper texture of jelly, pectin is added having quantity more than 1 per cent.

Pomegranate Jam

A product is also known as anar-rub with fairly good keeping quality can be made by concentrating pomegranate juice, adding sugar and heating the mixture. The jam is prepared by adding 45 per cent juice, sugar and citric acid to maintain the TSS more than 68.5°B and acidity 0.8 -1.2 per cent. The finished product has a thick consistency and contains 68.5°B or more total soluble solids. Anar rub can be stored for one year and it is utilized as a jam, which can be enjoyed with bread, *etc.*

Anardana

Pomegranate arils are dried and sold as anardana. This is used as an acidulant in place of tamarind or dried green mango for cooking purposes in North India, in Indian style curries, chutney and other culinary preparations. For preparation of anardana, high acidity varieties are used. Mahtma Phule Krishi Vidyapeeth, Rahuri has developed high acidity variety for preparation of anardana. If acidity of arils is less, add citric acid externally in the arils and dried to a moisture content less than eight per cent using solar or mechanical dryer.

Frozen Arils

One of the major problems in pomegranate fruits is cracking at fruit maturity leading to huge economic loss to farmers. The traditional utilization of these fruits lies in drying a value added by-product known as Anardana used as acidulent and condiment in Indian curries and also used in Ayurvedic and Unani medicines. For preparation of Anardana, fresh and pure pomegranate arils that are free from any kind of bruises are separated from pomegranate. By using blast freezer or individual quick freezing technique, the arils are frozen having temperature is reduced to -20 to -18°C. They are packed in hygienic packets and boxes as per the requirements of market to ensure their taste and freshness. After packaging, the arils are stored at -20°C and below temperature so as to maintain the quality. The dried product Anardana contains more acid (5.8-15.4 per cent), total sugars (9.3-17.5 per cent) and crude fibre as compared to fresh fruit.

Minimally Processed Pomegranate Arils

Pomegranate consumption is limited due to difficulty in peeling to obtain the arils. Thus arils as a minimally processed product (MPP) have a great interest because it facilitates its consumption overcoming the difficulty of extraction by consumers. Therefore, pomegranate arils are processed into ready-to-eat convenient form. The arils are separated from the pomegranate fruit and are washed (to eliminate juice and decontaminate) and disinfected for 1-2 min in waved water at 5°C, with 50-150 ppm free chlorine and pH 6-6.5 (adjusted with citric or ascorbic acids, which are also anti-browning agents at about 5 g l^{-1}). After rinsing, they are air-dried, packed in polythene bags and stored at cold storage. Arils are then rinsed in non-chlorinated water at 5°C, dried by cold (5°C) forced air, weighted and MAP packed, by using right polymers (polyethylene, polypropylene or ethyl-vinyl-acetate, among others), of right thickness in order to passively generate the intended gas composition (usually 5-7 kPa O_2 + 5-10 kPa CO_2).

Pomegranate Wine

Juice is extracted from the pomegranate fruit arils and is mixed with sugar to make 22-23°Brix TSS. Potassium metabisulphite is added to the juice to prevent the growth of undesirable microorganisms. Then, juice is inoculated with wine yeast starter culture (2 per cent v/v) and allowed for fermentation and ageing. If a sweet table wine is desired sugar is added to 8-10°Brix. Finally, the wine is flash pasteurized at 60°C, bottled hot and sealed. Fortifying the sweetened wine to about 20 per cent alcohol can make wine like port.

By Products

Inedible portion of pomegranate also contain antioxidant, phenolic compounds and flavonoids in high quantities. Further it can be a good nutrient source. Pomegranate peel contains substantial amount of polyphenols such as ellagic tannins, ellagic acid and gallic acid. It has potential for use in medicines, dying industries, tooth powder *etc.* The rind of pomegranate is separated, clean and dry in solar or tray dryer to reduce the moisture below 6 per cent. The dried rind is converted into powder form using grinder or pulverizer.

References

Adsule, R.N. and Patil, N.B. 1995. Pomegranate: in Handbook of fruit science and Technology, D.K. Salunke and S.S. Kadam(eds), Marcel Dekkar, New York, pp. 455-464.

Anonymous, 2015. Indian Horticulture Database-2013-14. Pub. National Horticulture Board, Ministry of Agriculture, Govt. of India.

Al-Zoreky, N. S. 2009. Antimicrobial activity of pomegranate (*Punica granatum* L.) fruit peels. International Journal of Food Microbiology, 134, 244-248.

AOAC International. 1995. Official Methods of Analysis of AOAC International, 16th ed.; AOAC International: Arlington, VA.

Artés F., Tudela, J.A. and Gil, M.I. 1998b. Keeping quality of pomegranate fruit by intermittent warming. In: Z. Lebensm. Unters. Forsch, 207: 316-321.

Artés, F. and Tomás-Barberán, F.A. 2001. Postharvest technological treatments of pomegranate and preparation of derived products. In: Production, processing and marketing of pomegranate in the Mediterranean region: advances in research and technology. Options Mediterranées, Series A, 42: 199-204.

Artés, F. and Artés-Hernández, F. 2012. Advances in postharvest and refrigeration techniques in whole and minimally processed pomegranate. II International Symposium on the Pomegranate, Options Méditerranéennes, A, 103: 169-175.

Barman, K., Koley, T.K., Sharma, S. and Patel, V.B. 2013. Pomegranate (Punica granatum L.): Nutraceutical compounds and medicinal properties. Indian Food Industry Mag, Vol 32, Mar-Apr 2013.

Basu, A. and Penugonda, K. 2009. Pomegranate juice: a heart-healthy fruit juice. Nutrition Reviews, 67: 49-56.

Cakmak, B., Alayunt, F.N., Akdeniz, R.C., Aksoy, U. and Can, H.Z. 2010. Assessment of the quality losses of fresh fig fruits during transportation. Journal of Agri. Sci., 16: 180-193.

Cam, M., Içyer, N.C. and Erdogan, F. 2014. Pomegranate peel phenolics: Microencapsulation, storage stability and potential ingredient for functional food development. LWT - Food Science and Technology, 55: 117-123.

Conesa, A., Aguayo, E., López-Rubira, V. and Artés, F. 2005. Combined use of clean room, disinfection and modified atmosphere packaging for improving quality of minimal fresh processing pomegranateNarils. In: *Australasian Postharvest Horticulturae*. Rotorua, New Zealand.

Dhumal, S.S. 2012. Studies on processing of pomegranate (*Punica grantum* L.) into value added products. Ph.D. (Agri.) Thesis submitted to MPKV, Rahuri, (Maharashtra) India.

Domingo, M. R., Maria S., Angel C., Salvador C., Fernando R. and Daniel, V. 2004. Mechanical damage during fruit postharvest handling: Technical and Physiological Implications. *Production Practices and Quality Assessment of Food Crops*, 3: 233–252.

Elyatem, M.S. and Kader, A.A., 1984. Postharvest physiology and storage behavior of pomegranate fruits. In: Scientia Hort., 24: 287-298.

Gil, M., Martínez, J.A. and Artés, F. 1996a. Minimally processed pomegranate seeds. Lebensm. In: Wiss. Technol., 29: 708-713.

Gautam, R., and Sharma, S.C. 2012. Effect of Punica granatum Linn. (peel) on blood glucose level in normal and alloxan-induced diabetic rats. Research Journal of Pharmacy and Technology, 5: 226-227.

Kulkarni, A.P. and Aradhya, S.M. 2005. Chemical changes and antioxidant activity in pomegranate arils during fruit development. Food Chemistry, 93: 319-324.

Kumbhar, S.C., Kotecha, P.M. and Kadam, S.S. 2002. Effect of method of juice extraction on quality of pomegranate wine. Beverage and Food World, 29(11): 35-36.

Marti, N., Perez-Vicente, A. and Garcia-Viguera, C. 2001. Influence of storage temperature and ascorbic acid addition on pomegranate juice. J. Sci. Food Agric., 82: 217-221.

Melgarejo, P., Hernández, F., Legua, P., Tomás-Barberán, F.A. and Artés, F., 2001. Evolution of pomegranate juice antochyanins during the ripening of fruit of three clones: ME16, VA1 and BA1. In: Production, processing and marketing of pomegranate in the Mediterranean region: advances in research and technology. Options Mediterranées, Series A, 42: 123-128.

More, H.G., Unde, P.U. and Chobe, R.S. 2001. Processing of pomegranate and pomegranate juice. Beverage and Food World, 28(1): 37-40.

O'Brien, M., Gentry, J.P. and Gibson, R.C. 1965. Vibrating characteristics of fruit as related to in-transit injury. Transactions of ASAE, 8(2): 241–243.

O'Brien and Guillou, R. 1969. An in-transit vibration simulator for fruit-handling studies. Transactions of ASAE, 12(1): 94-97.

Rania, J., Nejib, H., Messaoud, M., Mohamed, M. and Mokhtar, T. 2007. Characterization of Tunisian pomegranate (*Punica granatum* L.) cultivars using amplified fragment length polymorphism analysis. Scientia Horticulturae, 115(3): 231-237.

Shahbazi, F., Rajabipour, A., Mohtasebi, S. and Rafie, S. 2010. Simulated in-transit vibration damage to watermelon. J. Agr. Sci. Tech., 12: 23-34.

Shulman, Y. Fainberstein, L., and Lavee, S. 1984. Pomegranate fruit development and maturation. In: J. Hort. Sci., 59: 265-274.

Singh, D. and Singh, R.K. 2004. Processed products of pomegranate. Natural Product Radiance, 3(2): 66-68.

Sumner, M.D., Elliott-Eller, M., Weidner, G., Daubenmier, J.J., Chew, M.H., Marlin, R., Raisin, C.J. and Ornish, D. 2005. Effects of pomegranate juice consumption on myoca rdial perfusion in patients with coronary heart disease. Amer. J. Cardiol., 96: 810-814.

Yilmaz, Çelik, Isik. 2007. Mineral composition and total phenolic content of pomegranate molasses. Journal of Food, Agriculture and Environment, 5(3 and 4).

Zhang, M. 2000. Status and development of processing technology of fruit and vegetable in China. Food Mach., 76: 4–6.

Chapter 26

Processing and Value Addition of Fig

J.K. Dhemre, V.P. Kad and V.M. Ingale

Postharvest Technology Centre,
Department of Horticulture
Mahatma Phule Krishi Vidyapeeth, Rahuri – 413 722, Dist. Ahmednagar, M.S.

Fig is one of the oldest delicious and healthy cultivated fruit. The common fig is not much grown inspite of the fact that soil and climatic conditions suitable for fig cultivation are widely found in India. Ficus is one of the largest plant genera of the tropics that enjoy a cosmopolitan distribution. Fig belongs to family Morraceae. It includes an estimated 900-950 species and thought to be a native to southern parts of the Arabian Peninsula, Italy, the Balkan peninsula and the USSR. Greece, is now the leading fig producing country followed by Algeria, Morocco, Syria and Italy. However, Turkey is the largest producer of dried figs. Although wild figs were present in India for thousand of years.

Fresh market figs must be harvested when almost fully ripe and firm to be of good eating quality. Skin, colour and flesh firmness is dependable maturity and ripeness indices. Figs are consumed fresh or dried, preserved, candied or canned. Fresh figs are very delicious, wholesome and nutritious and are called "nature's most nearly perfect fruit". They are often served with cream and sugar. Peeled or impeded, the fruits may be merely stewed or cooked in various ways, as in pies, puddings, cakes, bread or other bakery products, or added to ice cream mix. The whole fruit can be preserved in sugar syrup or used for the preparation of jam, marmalade, or paste. The whole fruits are sometimes candied commercially. In Europe; Western Asia, Northern Africa and California, commercial canning and drying of figs are the industries of great importance. The larger bulk of fruit is consumed as dried fruits.

In Mediterranean countries low-grade figs are converted into alcohol. An alcoholic extract of dried figs has been used as a flavoring for liqueurs and tobacco.

Fig leaves are used as fodder in India. In southern France fig leaves are used as a source for perfume material called "fig-leaf absolute". From dried Latex powder protein-digesting enzyme ficin is obtained which is used for tenderizing meat, rendering fat, and clarifying beverages. In tropical America, the latex is often used for washing dishes, pots and pans. Fig has many medicinal uses. A decoction of the fruits is gargled to relieve sore throat; figs boiled in milk are repeatedly packed against swollen gums. The leaf decoction is taken as a remedy for diabetes and calcifications in the kidneys and liver. The latex is widely applied on skin, ulcers and sores.

Health and Nutritional Benefits of Eating Figs

- The rich potassium content of figs helps to maintain the blood pressure of the body.
- Figs are rich in dietary fiber, which makes them very effective for weight management program.
- Even leaves of the fig plant have healing properties. Eating fig leaves helps diabetic patients reduce the amount of insulin intake.
- Consuming figs helps to reduce the risk of breast cancer.
- Calcium and potassium present in figs prevent bone thinning and help to promote bone density.
- Tryptophan, present in figs, induces good sleep and helps get rid of sleeping disorders like insomnia
- Eating figs relieves fatigue and boosts memory power.
- Drinking fig juice helps tc soothe irritated bronchial passages.
- Consumption of figs with milk cures illness like anemia.
- Due to their laxative effect, figs are helpful for treating chronic constipation.
- Daily consumption of soaked figs (2-3) is an effective remedy for curing hemorrhoids or piles.
- The soluble fiber, called pectin, in figs helps in reducing blood cholesterol.

Dried figs are fast energy giving fruits. Young and adolescents require calcium, Iron and Vitamin Bl, which is available in dried figs. It contains high proportion of fiber and iron magnesium, phosphorous, calcium and does not contain cholesterol, fat or sodium. It also contains easily digestible natural sugars (glucose, fructose) minerals and vitamins.

Harvesting

Harvesting starts from middle of June and continue till June. For best flavour, fruits should be picked when they are soft and wilt at the neck, hanging down from their own weight. Milky latex exuding from the stem when the fruit is pulled off indicates that the fruit is still immature. For distant markets, harvest the fruits when slightly immature.

Fresh figs are highly perishable. At 40° to 43°F (4.44° to 6.11°C) with 75 per cent relative humidity, figs remain in good condition for 8 days but have a shelf life of only 1 to 2 days when removed from storage. At 50°F (10°C) and relative humidity of 85 per cent, figs can be kept no longer than 21 days. They remain in good condition for 30 days when stored at 32° to 35° F (0° to 1.67° C). If frozen whole, they can be maintained for several months. Due to this, they have to be marketed immediately after being harvested. Further more, the fruits are soft and delicate such that they require special care during transportation. Therefore, it is necessary to carryout studies to investigate the feasibility of converting this fruit into a value added product by drying so as to enhance its storage life without loosing its original quality. It is also reported that the degradation of the quality of dried fig is due to contamination of dirt, dust and insect infestation. To overcome these problems there is necessity to provide an appropriate drying technology, which gives quality dried fig and enhance the shelf life of the product.

Different Methods of Drying Figs

1. Open yard sun drying
2. Solar Dryng
3. Microwave drying
4. Vacuum drying
5. Tray drying
6. Osmo dehydration

Drying or Dehydration

Drying or dehydration is the removal of the majority of water contained in the fruit or vegetable and is the primary stage in the production of dehydrated fruits and vegetables. Several drying methods are commercially available and the selection of the optimal method is determined by quality requirements, raw material characteristics, and economic factors. There are three types of drying processes: sun and solar drying; atmospheric dehydration including stationary or batch processes (kiln, tower, and cabinet driers) and continuous processes (tunnel, continuous belt, belt-trough, fluidized-bed, explosion puffing, foam-mat, spray, drum, and microwave-heated driers); and sub atmospheric dehydration (vacuum shelf, vacuum belt, vacuum drum, and freeze driers). Sun drying (used almost exclusively for fruit) and solar drying (used for fruit and vegetables) of foods use the power of the sun to remove the moisture from the product. Sun drying of fruit crops is limited to climates with hot sun and dry atmosphere, and to certain fruits, such as prunes, grapes, dates, figs, apricots, and pears. These crops are processed in substantial quantities without much technical aid by simply spreading the fruit on the ground, racks, trays, or roofs and exposing them to the sun until dry. Advantages of this process are its simplicity and its small capital investment. Disadvantages include complete dependence on the elements and moisture levels no lower than 15 to 20 per cent (corresponding to a limited shelf life). Solar drying utilizes black-painted trays, solar trays, collectors, and mirrors to increase solar energy and accelerate drying.

Atmospheric forced-air driers artificially dry fruits and vegetables by passing heated air with controlled relative humidity over the food to be dried, or by passing the food to be dried through the heated air, and is the most widely used method of fruit and vegetable dehydration. Various devices are used to control air circulation and recirculation. Stationary or batch processes include kiln, tower (or stack), and cabinet driers. Continuous processes are used mainly for vegetable dehydration and include tunnel, continuous belt, belt-trough, fluidized-bed, explosion puffing, foam-mat, spray, drum, and microwave-heated driers. Tunnel driers are the most flexible, efficient, and widely used dehydration system available commercially.

Sub atmospheric (or vacuum) dehydration occurs at low air pressures and includes vacuum shelf, vacuum drum, vacuum belt, and freeze driers. The main purpose of vacuum drying is to enable the removal of moisture at less than the boiling point under ambient conditions. Because of the high installation and operating costs of vacuum driers, this process is used for drying raw material that may deteriorate as a result of oxidation or may be modified chemically as a result of exposure to air at elevated temperatures. There are two categories of vacuum driers. In the first category, moisture in the food is evaporated from the liquid to the vapor stage, and includes vacuum shelf, vacuum drum, and vacuum belt driers. In the second category of vacuum driers, the moisture of the food is removed from the product by sublimination, which is converting ice directly into water vapor. The advantages of freeze drying are high flavor retention, maximum retention of nutritional value,

Osmo-Dehydration of Fig Fruits with Live Sugars

Fig fruits were subjected to osmotic dehydration by means of five different sugar syrups, in order to check the aptitude of the species to this kind of processing. Osmotic dehydration was carried out at 25°C and at atmospheric pressure on whole fig fruits of the local cultivar Verde. Processing lasted 24 h and the following dehydrating syrups were used: sucrose, glucose, fructose and two commercial hydrolyzed corn starch syrups, Frudex 70 and Glicosa. At fixed intervals (0, 1, 2, 3, 6, 12 and 24 h) appropriate determinations were done on fruit pulp, in order to calculate weight loss, water loss, solid gain and normalized solid content. Results obtained with no doubt indicate that whole fig fruits show a low aptitude to be osmo-dehydrated, probably for their morphology and anatomy and for processing conditions used in this trial.

Influence of Industrial Dehydration on Quality of Fig Fruits

The effectiveness of a dehydration process on the quality of "Niedda longa" fig fruits was studied. Drying was carried out either with an industrial two-stage dehydration system, requiring no blanching treatment, or with a simulated sun-drying procedure. During the industrial process fruits were: (a) initially moved for 3 hours in a tunnel with a gradient air temperature from 95°C to 85°C; (b) then moved into a second tunnel at 85°C for 14 h. Sun-drying was simulated keeping fruits for 120 h at 32°C and 35 per cent of relative humidity (RH). Finally dehydrated figs were rehydrated, or equilibrated with environmental RH (sun-dried figs), and both treated with potassium-sorbate solution, then film-packaged into hermetically sealed air tight bags. Moreover, half of the bags were packed in-modified atmosphere (MAP). Fruits were, then, stored at 20°C for 6 months. A 11 h dehydration time was sufficient to obtain microbially stable fruits without chemical treatments, reaching a dry matter content (DM) of 88 per cent and water activity (aw) of 0.60, whereas after 17 h DM and aw were 91 per cent and 0.57, respectively. DM of industrial rehydrated figs and environmental equilibrated sun-dried figs were 71 per cent and 74 per cent, respectively, requiring a potassium-sorbate treatment. After packaging and 6 months of shelf life, no microbial growth was revealed on 'fruits. MAP resulted in no additional improvement on the microbial quality of the product. Based on these results we can state that industrial dehydration performed well, even if fruit calibration is needed before drying.

Effect of Temperature on the Shelf-life of Fig Fruit during Refrigerated Storage

Fig is a very perishable fruit. Storage temperature is a very important parameter to extend the postharvest period of this fruit.The aim of this paper is to find out the conditions to extend the postharvest life of fresh fig cv. white "Cuello de dama". For that fresh fig of this variety was employed. Three storage temperatures (0, 3 and 6 °C) were tested, and the effect of partial modification of the atmosphere at a given temperature (6 °C) was studied. Fruit were analyzed as fresh and at 3, 7 and 10 days of refrigerated storage. The analyses carried out were: dry matter, °Brix, pH and titratable acidity, instrumental texture, color (L*, a* and b*) and also a sensory analysis. Figs stored at 0 and 6°C shown the higher values for soluble solids. At 6°C the samples without film wrapping have a higher °Brix due to a more important water loss. For titratable acidity samples wrapped stored at 6 °C show less variation. As for sensory analysis, the best samples for global score (that include aspect, texture, juiciness and taste) were those samples stored at 6 °C with film wrapping.

Effect of Postharvest Treatments with Sodium Bicarbonate or Active Acid on Storability and Quality of Fig Fruit

The effect of postharvest treatments with sodium bicarbonate (BCS) and acetic acid (AAc) on the quality preservation of breba fig fruit (*Ficus carica* L.) cv. Lampa Preta during storage. Fruits were dipped for 2 min. in SBS and AAc solutions at two concentrations (0.5 and 1 per cent) and left to dry at ambient temperature. No treatment was applied to control fruits. Then, fruits were stored at 2°C and relative humidity at about 85-90 per cent. After 8, 14 and 20 days storage, fruits were analysed for firmness, total soluble solids content, titrable acidity, humidity, and latter, with stored juice, glucose and the organic acids citric, malic, fumaric and piruvic. Organoleptic evaluation was done at harvest, after 14 and 20 days storage. The treatments with BCS gave a better effect on firmness, titrable acidity and moisture content. Total soluble solids content on fruits treated with 1 per cent AAc registered the higher values on this study. Glucose was in major quantity in figs treated with AAc after 20 days. The organic acids identified were in major quantity citric, followed by malic, fumaric and piruvic. Loss of fruits from diseases was higher in control than in the other treatments. AAc treated fruits were preferred by panellists. Results indicated that treatments with 1 per cent sodium bicarbonate gave better results on preserving fruit quality characteristics (higher firmness and moisture and lower °Brix). However, treatments with acetic acid were more efficient on reducing fruit loss and were preferred by consumers. Overall fig fruits could be stored in good conditions up to 3 weeks at 2°C.

Different Technologies for Fig Fruit Processing Process

Flow Chart for Fig Drying

Fresh fruits

↓

Washed in clean water

↓

Dipping for 20 minutes in the solution of 2 per cent potassium Meta bisulphate (KMS),solution (20 gm of KMS per litre of water)

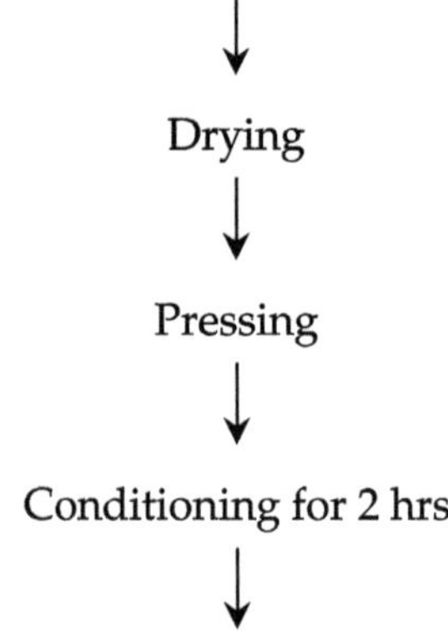

↓

Drying

↓

Pressing

↓

Conditioning for 2 hrs

↓

Packaging in polythene bags

1. Preheat the dryer to 60ºC. It's necessary to dry the figs at a low, even temperature. Drying them at a higher temperature would result in cooked figs.
2. Wash the figs thoroughly with water. Carefully trim away any damaged parts and drain the water.
3. Fumigation with sulphur @ 2g/kg in air tight sulphur chamber.
4. After fumigation, dip the fig in dippening solution for half an hour. Place them in the dryer for drying the fig upto 16 per cent moisture content.
5. During the drying, continuously press the fig as per market requirement.
6. Also turn the figs during drying process for even drying.

Processing and Value Addition of Fig

Canning

Water bath processing is recommended for fig canning. Hot Pack - Prepare light syrup. Wash and drain figs. Do not peel of remove stems. Cover figs with water and boil for 2 minutes. Drain. Gently boil figs in syrup for 5 minutes. Add citric acid to adjust the acidity as 0.6 to 0.7 per cent. Fill jars with hot figs and cooking syrup leaving 1/2 inch headspace. Wipe rims. Adjust lids and process.

Fig Conserve

Cut figs and oranges into small pieces. Combine figs, oranges, sugar and raisins. Cook over low heat until thick and transparent. This will take about one hour. Stir periodically. Add nuts five minutes before removing from heat. Transfer to jars and process in boiling water canner using the times given above for fig preserves.

Storage

Store fresh figs in the refrigerator for up to 2-3 days. Frozen figs that are packaged appropriately and held at 0°F or below arc good for a year. Storage temperature has a major effect on canned figs: at 40°F hold for up to 48 months, at 70°F hold up to 24 months and if at 90ºF don't hold more than 12 months

Fig jam

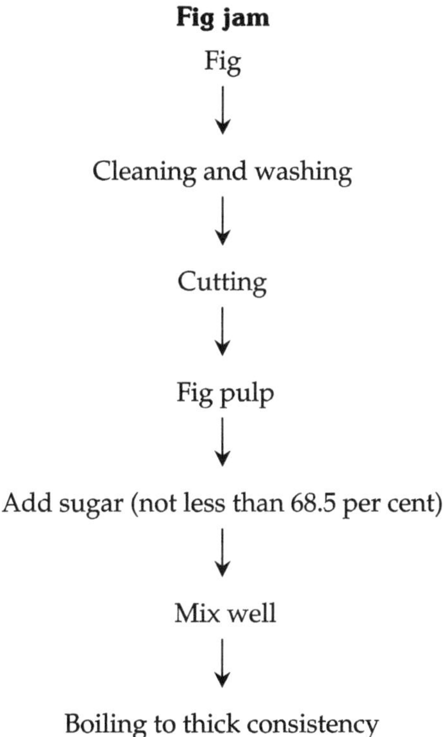

↓

Add preservative (Sodium benzoate and citric acid)

↓

Filling the pouches

↓

Sealing

↓

Sterilization

↓

Jam

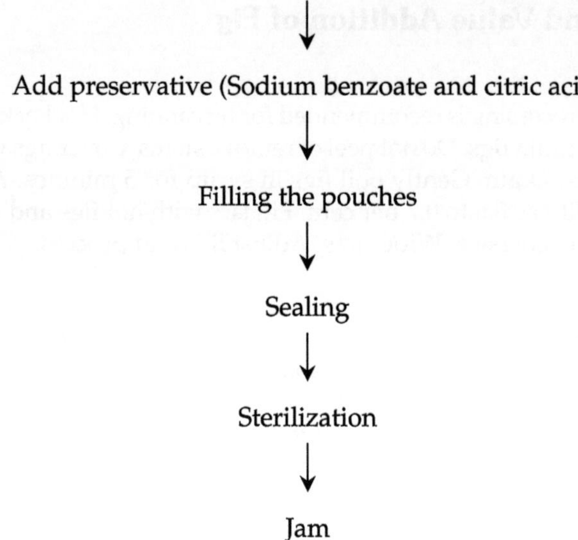

Fig and Pineapple Blended Jam

Fig Fruit	Pineapple
↓	↓
Cleaning and washing	Cleaning and washing
↓	↓
Cutting	Peeling
↓	↓
Fig pulp	Remove the core position
	↓
	Pinapple pulp

↓

Blending

↓

Add sugar (not less than 68.5 per cent)

↓

Mix well

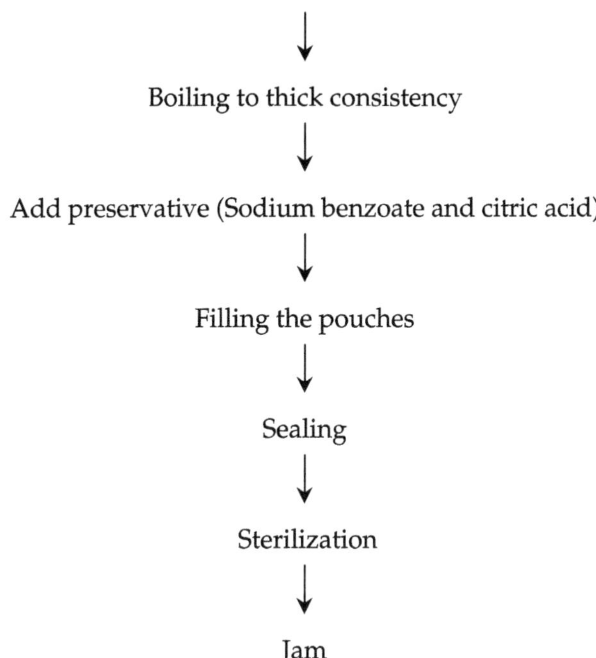

Procedure

1. Prepare pulp with the help of screw type pulper.
2. Add sugar to figs and slowly bring to boiling, stirring occasionally until sugar dissolves.
3. Cook rapidly until thick. Stir frequently to prevent sticking. Add citric acid and cook.
4. Pour hot jam into hot jars, leaving ¼-inch headspace.
5. Sealing
6. Sterilization

Fig Cake recipe

Beat the eggs. Add the sugar and the oil to the eggs. Sift the flour, salt, allspice, cinnamon, nutmeg and cloves into a bowl. Mix the soda and the buttermilk. Add this to the flour mixture. Add the flour mixture to the egg-sugar mixture while beating. Add the vanilla, pecans and fig preserves to the mixture."Bake in a greased and floured pan at 350°F for 35-40 minutes.

Fig Biscotti

Preheat oven to 375°F. Place the sifted flour in a large bowl with the baking powder, cocoa powder and sugar. In another bowl, combine the eggs, chopped chocolate and chopped figs. Gradually add the egg/choc/fig mixture to flour mixture, stirring to form a stiff dough. Transfer the dough to an even work surface, lightly dusted with flour, and knead briefly with your hands until smooth. Divide the dough in half and shape into two long logs. Place logs on a baking tray lined with non-stick baking paper and press along top of log to slightly flatten. Bake for 20 minutes, then remove from the oven and allow to cool for 15 minutes. Reduce the oven temperature to 300°F. Cut logs into half-inch thick slices, then place on baking trays and bake for another 10 minutes or until hard. Allow to cool on a wire rack.

Dried Figs in Syrup Recipe

Mix the water, orange juice, orange zest, cinnamon, brown sugar and vanilla in a small saucepan. Place over medium heat on the hob. Add the figs and cook until reconstituted - usually about 10 minutes.

Chocolate Mint Figs Recipe

Preheat your oven to 350°F (175°C). In a mixing bowl, combine the chocolate chips, chopped mint and almonds, stirring thoroughly. Press a bit of this mixture into each fig half. Lightly spray a baking sheet with vegetable oil. Place the figs on the baking sheet and bake for 15 minutes. Serve warm or at room temperature.

Candied Figs

Wash figs, combine sugar and water and boil it. Place figs in boiling syrup and simmer one hour, uncovered, cool. Repeat the second day. Third day boil slowly one hour, uncovered. Drain it. Place on paper towel and dry in sun for three days.

Chapter 27

Value Added Products from Aonla

V.P. Kad, M.S. Jadhav, J.K. Dhemre and V.S. Sonone

Postharvest Technology Centre,
Mahatma Phule Krishi Vidyapeeth, Rahuri – 413 722, Dist. Ahmednagar, M.S.

The Amla or Aonla (*Emblica officinalis* Gaertn) also known as Indian Gooseberry is a minor sub-tropical deciduous tree indigenous to Indian sub-continent. It can be grown successfully in dry and neglected regions owing to its hardy nature, suitability to various kinds of wasteland. A mature aonla tree can tolerate freezing as well as high temperature of 46°C. In India, aonla is grown in an area of about 50,000 ha with a production of around 2,00,000 metric tonne.

The main varieties of aonla are Banarasi, Chakaiya, Hathijhool, Bansi red, Pink-tinged, NA7, *etc.* The aonla gets ready for harvesting during November-December. However, the fruit may be allowed to remain on the tree till February without much fruit drops. The optimum stage of harvesting falls between the periods extending from the second week of December to the third week of January in Northern plains. However, at some places in Southern India, the fruit is available throughout the year.

The area under aonla (Indian gooseberry) has been expanding rapidly in the last couple of years. From about 3,000 ha in the early 80s the area had stretched to over 25,000 ha in 2000. It has doubled to 50,000 ha in the last two years. Intensive cultivation is being done in the salt-affected districts of Uttar Pradesh including the ravenous area in Agra, Mathura, Etawah, Fatehpur and semi-arid tracks of Bundelkhand. It is rapidly spreading in the semi-arid regions of Maharashtra, Gujrat, Rajasthan Andhra Pradesh, Karnataka, Tamil Nadu and the Arawali ranges in Haryana, Kandi area in Punjab and in Himachal Pradesh.

Harvesting and Yield

The change in seed colour from creamy white to brown is indicative of fruit maturity. Fully developed fruits, which show sign of maturity, are harvested. This helps in size gain of remaining fruit. Delay in harvesting results in heavy dropping of fruits. It also adversely affects the following years bearing. Individual fruits are plucked by climbing on tree with the help of pegged bamboo or ladder. Harvesting should be done in early or in the late hours.

Physical Characteristics

The aonla fruits are round, ribbed and pale green. It is divided into six segments through pale liner grooves. The surface of the fruit is shiny and the size varies from small marble to a large plum. It is quite hard with a thin and translucent skin. The raw fruit, due to its high acidic nature and astringent taste, is unacceptable to consumers. The average fruit weight and seed weight varies from 22.29-25.20 and 1.54-1.82 g, respectively among Krishna, NA7 and Chakaiya cultivars (Goyal *et al.*, 2007). The seed pulp ratio is reported to vary from 1:15 to 1:22 (Kalra, 1988).

Chemical Composition

The chemical composition of aonla fruits is influenced by environmental factors. The chemical composition of fresh fruit in respect to moisture, protein, fat, crude fibre, starch, sugars, minerals and vitamins has been reported by lot a of researchers (Tripathi *et al.*, 1988) and shown in Table 27.1. The fruits are rich in ascorbic acid and tannins.

Table 27.1: Chemical Composition of Fresh Aonla Fruit

Characters	*Composition*
pH	2.50
T.S.S., °B	14.00
Acidity, per cent citric acid	2.24
Ascorbic acid, mg/100 g	571.60
Reducing sugars, per cent	2.37
Non-reducing sugars, per cent	0.74
Total sugars, per cent	3.11
Pectin, as per cent Ca pectate	0.54
Tannin, as per cent gallotannic acid	0.55
Protein, per cent	0.88
Moisture, per cent	84.36

The total sugars content in aonla fruit varies from 7 to 9.6 per cent, reducing sugars from 1.04 to 4.09 per cent and non-reducing sugars from 3.05 to 7.23 per cent among the various varieties. Aonla is particularly rich in vitamin C. The pulp of fresh fruit contains 200-900 mg of vitamin C as reported by several workers (Kalra, 1988).

Nutritive Value/Benefits

Aonla fruit is highly nutritive with a great medicinal use and the richest source of vitamin C. It contains 500-1500 mg of ascorbic acid per 100 g of pulp (Chauhan *et al.*, 2005). This is much more than the vitamin C content of guava, citrus and tomato fruits. The fruit juice contains nearly 20 times as much vitamin C as in orange juice. Its other constituents are phenols and tannins containing gallic acid, elegiac acid and glucose which prevent oxidation of vitamin C. A tablespoonful each of fresh aonla juice and honey mixed together forms a very valuable medicine for the treatment of several ailments like tuberculosis of lungs, asthma, bronchitis, scurvy, diabetes, anemia, weakness of memory, cancer, tension, influenza, cold, loss and grayness of hair *etc.* Because of its highly acidic and astringent nature, the consumers do not relish this fruit in fresh form and consequently it is used in the preparation of various ayurvedic tonics like Chayvanprash, Triphala, *etc.*

Storage

Aonla fruits are highly perishable in nature as it's storage in atmospheric conditions after harvesting is very limited. The objective of the storage of fresh fruit is to extend the period of their availability. Storage facilities such as cold storage and controlled/modified atmospheric storage are very expensive and not in the direct reach of poor farmers (Kumar and Nath, 1993). Singh and Kumar (1997) stored fully mature aonla fruits at room temperature, modified storage condition, zero energy chamber and zero energy + modified storage condition. It was found that decay loss was minimum (26.56 per cent) in modified storage condition on 24th day of storage whereas it was maximum (48.70 per cent) in zero energy chamber. The fruits may be kept in cold storage for 7-8 days at 0-2°C and 85-90 per cent relative humidity.

Due to its perishable nature, it is difficult to store or transport aonla fruits over long distances. In order to have good income from aonla, it must be sold immediately in the market. But, the problem arises when there is glut in the market. To get profit during that period, proper storage facilities should be available to help the farmers get maximum profit. The extension of shelf life during storage should be made possible by checking the rate of respiration, transpiration and microbial infection. Plant growth regulators, certain chemicals and fungicides have played a great part for short period storage.

Nath *et al.* (1992) studied the effect of postharvest treatments on shelf life of aonla fruits with calcium nitrate (1 per cent), GA3 50 ppm and Borax (4 per cent) and found that the physiological loss in weight and pathological loss increased with the length of storage period. Calcium nitrate (1 per cent) minimized the weight loss during the storage period and no pathological loss was observed with Borax upto 9 days of storage.

Traditional/Conventional Methods of Processing and Products

Aonla has been in use for pickle and preserve since ages in India and the methods employed were based on traditional knowledge. Besides, aonla has been an important ingredient for chavanprash, a ayurvedic health tonic. The methods used previously were unhygienic in nature and time consuming. The nutrient loss in these methods was higher. The manual methods are costly, laborious and cannot maintain quality of the products.

Modern Methods of Processing

Aonla fruit is highly nutritive with great medicinal benefits. As aonla fruits are highly perishable in nature, its storage is very limited. Thus, researchers are paying due attention towards processing aspects of aonla fruits so that the demand of its products increases among the consumers. Due to its astringent nature, consumers are hesitant to eat it in raw form. Attempts are being made to produce products which are not only nutritionally delicious but also acceptable among the consumers.

The modern methods for preparation of different aonla products are hygienic, consume lesser time and provide maximum retention of nutrients. The processes for preparation are standardized with proper preservation.

Aonla Preserve

In India preserves or Murrabbas of various kinds are used for taste as well as for medicinal purpose. Process flow chart and aonla preserve as product are shown below:

Flow Chart for Preservation of Aonla Preserve

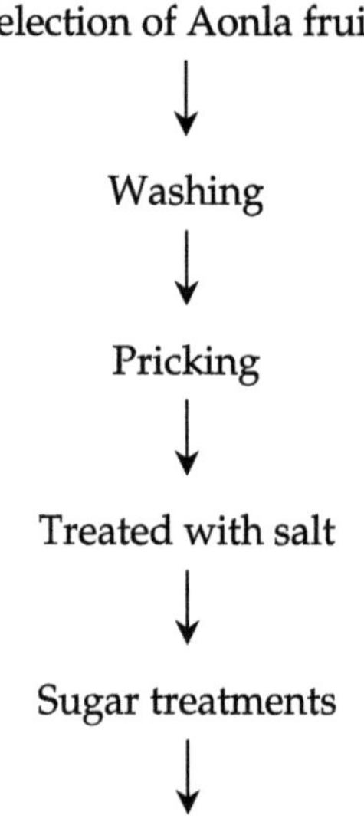

(Dipping in sugar solution of 40,50, 60 and 70°B concentration each per day)

↓

Reaching end point of concentration more than 65°B

↓

Tripathi *et al.* (1988) reported the chemical changes occurring in aonla preserve during processing and storage for 135 days. They observed that most constituents including total soluble solids content, vitamin C, total sugars, pectin, tannin, protein decreased in as compared to the fresh fruits. Further, increased acceptability was observed in aonla preserve up to 135 days of storage. It is rich in sugars and calcium but poor in ether extract, protein, phosphorous and iron.

The effect of sugar concentration at different temperature on the kinetics of total sugar gain, moisture loss, ascorbic acid loss, TSS gain in aonla preserve was studied by Geetha *et al.* (2005). They observed that the rate of constant kinetics for all the constituents increased with the increase in sugar concentration and temperature.

Candy and Jam

Tandon *et al.* (2003) studied the effect of blanching and lye peeling on candy preparation. They found that the candy prepared from lye peeled fruits of aonla showed decreased content of ascorbic acid than blanched fruits. The candy prepared from Lakshmi-52, Kanchan and Chakaiya was found to be the best. However, the effect of the blanching on some of nutritional parameters was less severe than that of lye peeling. General process flow chart for osmotic dehydration of aonla segment for candy and process flow chart for jam preparation is shown below.

Flow Chart for Preparation of Aonla Segment

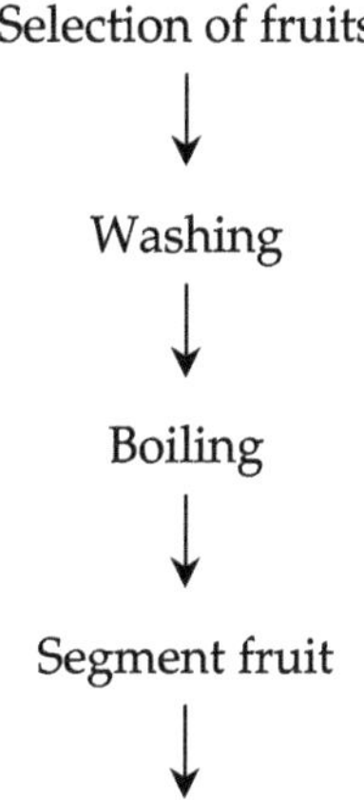

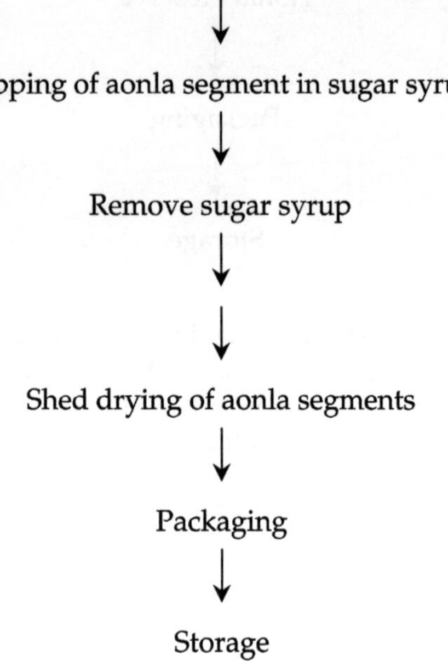

Flow Chart for Preparation of Aonla Jam

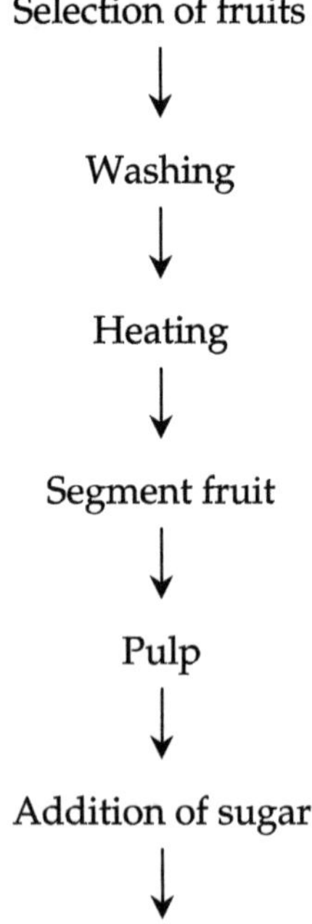

Heating to a thick consistency having TSS not less than 68.5°B

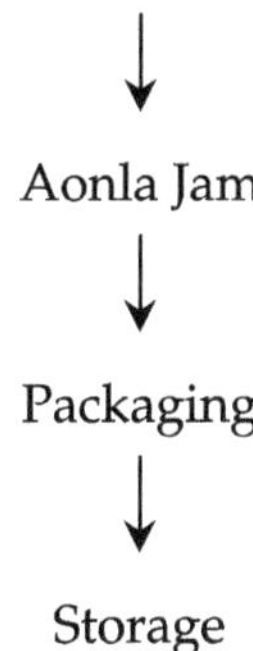

Sauce, Juice and Pulp Preparation

Nath (1999) carried out a study on the extraction of aonla pulp and suggested a method for preparation of aonla pulp from fully matured fruits. In this process, the fruits are blanched in boiling water for about 10 min to separate the segments from stone. Equal quantity of water is added to the segments and in the pulper to make pulp. If the pulp has to be preserved, it should be heated to 75°C and cooled to room temperature. Potassium metabisulphite (600 ppm per kg of pulp) should be mixed thoroughly and the pulp should be filled in clean sterilized bottles and then sealed.

Deka *et al.* (2001) conducted different experiments to study the feasibility of blending juices and pulp from lime, aonla, grapes, pineapple and mango in different proportion for the preparation of ready-to-drink fruit juice beverages. The highest sensory scores were obtained with a formulation comprising of 95 per cent lime and 5 per cent aonla juice/pulp. Process flow chart for juice is shown below:

Flow Chart for Preparation of Aonla Juice

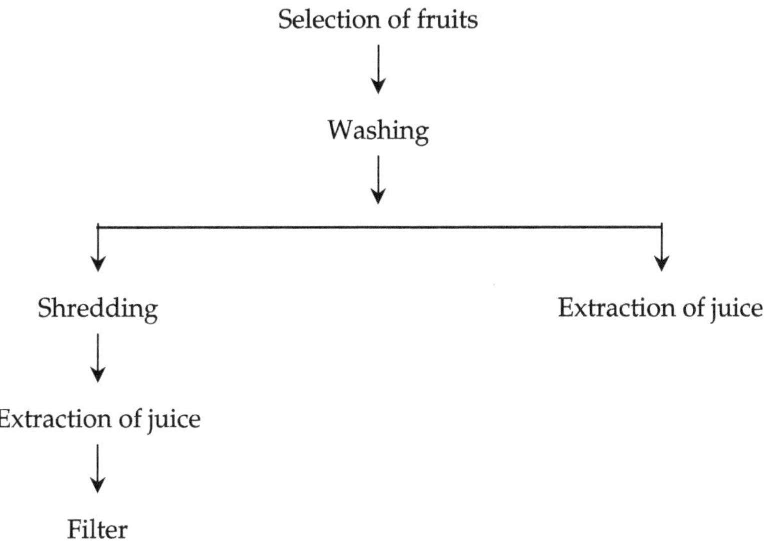

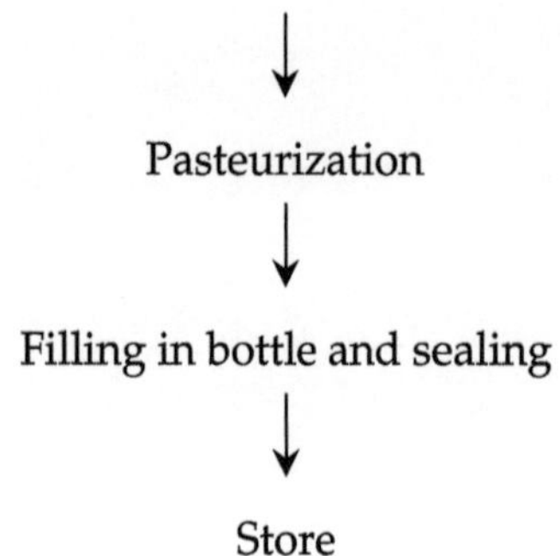

Shred Drying and Powder Preparation

A study carried out by Pragati *et al.* (2003) on aonla fruit drying by 4 different methods *viz.*, osmo-air drying, direct sun drying, indirect solar drying and oven drying. They observed that the osmo-air drying method was best for the drying of aonla fruits, resulting in better retention of nutrition, such as ascorbic acid and sugars. They also found that the level of anti-nutrition, such as tannins were also lower in osmo-air dried aonla, because the leaching and the browning were minimal. The nutrition content in osmo-air dried aonla was satisfactory after 90 day of storage.

Alam and Singh (2005) have conducted a study to prepare aonla powder from its fruits. The pricked aonla fruits were blanched for 5 min in 5 per cent boiling salt solution containing 0.15 per cent $NaHCO_3$ and 0.10 per cent MgO. The blanched aonla fruits were suliphited for 30 min in 0.5 per cent KMS. The treated fruit were sliced manually with knife. For the dehydration of aonla slices, the mechanical dryer (50, 60, 70°C), solar and cabinet dryers were used. They found that the mechanically dried slices contained higher in vitamin C content and were organoleptically superior than slices dried under solar and cabinet dryer.

In a study, the raw and treated (KMS) samples of Banaras and Chakkiya varieties of aonla were dehydrated by sun and mechanically dried (60°C) for the preparation of aonla powder (Sharma *et al.*, 2002). The processed dehydrated powder was packed in 100 gauge packages of high-density polyethylene (HDPE) and low-density polyethylene (LDPE). The packages were stored under ambient (5-18°C and 60 per cent RH) and refrigerated (4±1°C and 90 per cent RH). It was found that the aonla powder prepared from pretreated, Chakkiya variety and mechanically dried can be stored effectively in HDPE packages under refrigerated condition for 3 months without much loss in vitamin C, having better overall acceptability in terms of appearance, taste, flavor and texture.

Kavitha *et al.* (2003) studied the effect of osmotic dehydration on vitamin C content of aonla at different salt concentrations and different temperatures. The overall retention of vitamin C was found better in the un-blanched osmotically dehydrated and air dried samples. Flow charts for shred drying and powder making processes are shown below:

Flow Chart for Preparation of Aonla Shreds

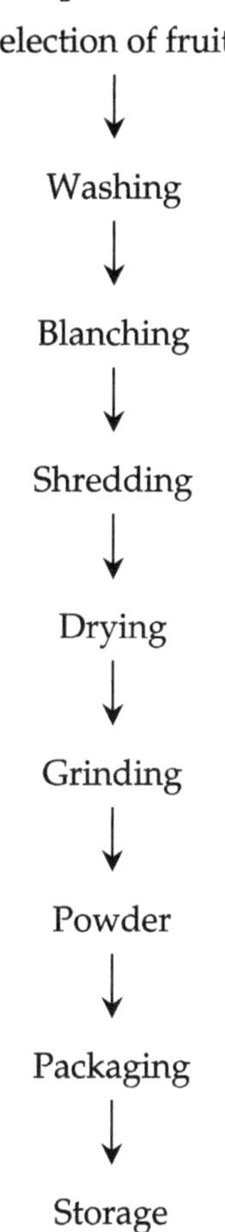

Pickles

Small size aonla fruits, which are not suitable for preparation of preserves and other confectionary items, may be utilized for pickle making. To improve upon the texture of the fruit and also to remove astringency, brining is important in pickling. Add necessary oils, spices *etc.* and leave it for few days in sunshine. When pickle is ready after few days, store it at room temperature.

Flow Chart for Preparation of Aonla Pickle

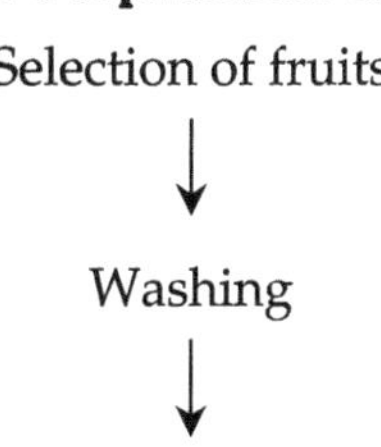

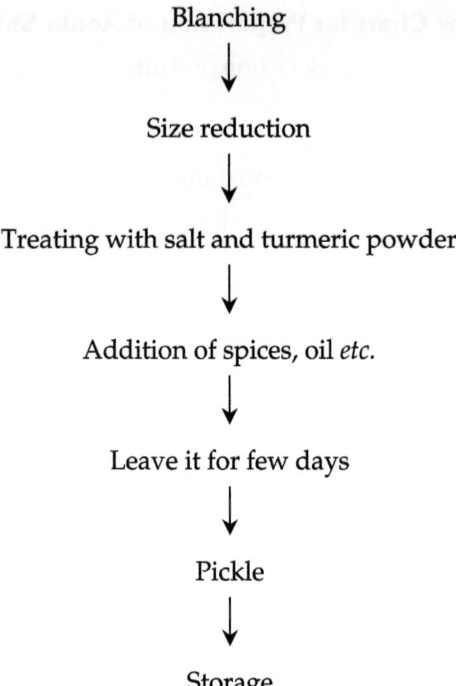

Conclusions

The Amla or Aonla (*Emblica Officinalis* Gaertn) also known as Indian Gooseberry is a minor sub-tropical deciduous tree indigenous to Indian sub-continent and it can be grown successfully in dry and neglected regions owing to its hardy nature, suitability to various kinds of wasteland. The growing popularity for alternate medicines, health foods and herbal products are enhancing the requirement for aonla fruit. The fruits are rich in vitamin C, protein, fat, crude fibre, starch, sugars and minerals. The fruits are rich in ascorbic acid and tannins. Because of its highly acidic and astringent nature, the consumers do not relish this fruit in fresh form. Hence, it is necessary to process this fruit and develop novel innovative products of high value. The modern technologies mentioned in the study for preparation of different aonla products are hygienic, consume lesser time and provide maximum retention of nutrients. The information generated by other researchers will be basis for development of pilot plant for aonla processing which is need of hour to mechanize processing operations to produce quality products of aonla for domestic as well as global market.

References

Agrawal, S. and C.S. Chopra, 2004. Changes in ascorbic acid and total phenols in making aonla product. Beverage Food World, 31: 32-34.

Alam, S. and A. Singh, 2005. Process for dehydration aonla powder. Abstr. Presented convention Indian Soc. Agric. Eng., 39: 222-222.

Chauhan, O.P., S. Srivastava, P. Pandey and G.K. Rao, 2005. A study on the development of aonla blended sauce. Beverage Food World, 32: 31-33.

Deka, B.C., V. Sethi, R. Parsad and P.K. Batra, 2001. Application of mixture methodology for beverage from mix fruit juice/pulp. Food Sci. Technol., 38: 615-618.

Geetha, N.S., S. Kumar and M.K. Garg, 2005. Osmotic concentration kinetic of aonla preserve. Abstr. Presented Annu. Convention Indian Soc. Agric. Eng., 39: 278-278.

Goyal, R.K., A.R.P. Kingsly, P. Kumar and H. Walia, 2007. Physical and mechanical properties of aonla fruits. J. Food Eng., 82: 595-599.

Kalra, C.L., 1988. The chemistry and technology of amla: A resume. Indian Food Packer. July-August, pp: 67-82.

Kavitha, C.V., V.V. Sreenarayanan, C.T. Devadas and S.K. Krishanaven, 2003. Effect of osmotic dehydration on vitamin-C of aonla. Abstract Presented Annu. Convention Indian Soc. Agric. Eng., 37: 134-134.

Kumar, S. and V. Nath, 1993. Storage stability of amla fruits: A comparative study of zero energy cool chamber versus room temperature. J. Food Sci. Technol., 30: 202-203.

Nath, V., 1999. Delicacies of aonla. Indian Hortic., 44: 15-17.

Nath, V., I.S. Singh, S. Kumar and A.K. Pandey, 1992. Effect of postharvest treatments on shelf life of aonla fruits. Prog. Hortic., 24: 79-82.

Pathak, R.K., 2003. The aonla. Central Institute of Sub-Tropical Horticulture, Rehmankhera, Lucknow.

Pragati, S. Dahiya and S.S. Dhawan, 2003. Effect of drying method on nutrition composition of dehydrated aonla fruit. Plant Food Hum. Nutr., 58: 921-928.

Sharma, S.R., S. Alam and S. Gupta, 2002. Storage study on dehydrated aonla powder. Abstr. Presented Convention Indian Soc. Agric. Eng., 35: 155-156.

Singh, P., J.P. Singh and C.S. Chopra, 2003. Techno-economic study on processing of aonla products. Beverage and Food World, January-2003, pp: 68-69.

Singh, R. and S. Kumar, 1997. Studies on the effect of different storage conditions on decay loss of aonla var. Chakaiya. Haryana J. Hortic. Sci., 26: 9-12.

Tandon, D.K., R.C. Yadav, S. Sood, S. Kumar and A. Dikshit, 2003. Effect of blanching and lye peeling on the quality of aonla candy Indian Food Packer, 57: 147-149.

Tripathi, V.K., M.B. Singh and S. Singh, 1988. Studies on comparative compositional changes in different preserved products of amla var. Banarasi. Indian Food Packer, July-August 1988, pp: 60-66.

Alam, S. and A. Singh. 2006. [illegible] for dehydrated aonla powder. Abst. Presented [illegible] convention Indian Soc. Agric. Engg. 39: 222-322.

Chauhan, O.P., S. Srivastava, P. Pandey and G.K. Rao. 2005. A study on the development of aonla blended sauce beverage. Food World, 12: 34-41.

Deka, B.C., V. Sethi, R. Parsad and P.K. Batra. 2001. Application of mixture methodology for beverage from mix fruit juice/pulp. J. Food Sci. Technol. 38: 615-618.

Gupta, N.S., S. Kumar and M.K. Garg. 2006. Osmotic concentration kinetic of aonla preserve. Abstr. Presented Annu. Convention Indian Soc. Agric. Engg. [illegible]

[illegible] aonla fruits. J. Food Engg. [illegible]

[illegible]

Chapter 28

Studies on Preparation of Ready to Serve Beverage from Wood Apple Pulp

S.B. Lande

Senior Research Assistant
Mahatma Phule Krishi Vidyapeeth, Rahuri – 413 722, Dist. Ahmednagar, M.S.

The wood apple (*Feronia limonia*) is an important fruit of family *Rutaceae*. The original home of wood apple is south India and Sri Lanka (Anonymous, 1981). Today, wood apple grows in Thailand, Malaysia, Cambodia, and other parts of Southeast Asia. Wood apple grows in abundance throughout India's drier regions and is cultivated along both peninsulas of the country. Along with arid conditions, the fruit requires monsoons to thrive. Wood apples grow along the plains in the north, south, east and central areas of India. States growing the fruit include Maharashtra, Andhra Pradesh, Tamil Nadu, Kerala, Karnataka, Madhya Pradesh and the western Himalayas. In the cooler northern regions, the fruit grows well up to 450 meters. India does not grow specific cultivars, as it's not a popular commercial crop despite its ubiquity. However, Maharashtra grows some of the larger, sweeter varieties thanks to its small interest in the crop. Wood apple season is October through March. In these months, the fruit is easily found for those seeking its tangy, musky flesh. Shops large and small sell wood apples, including juice shops willing to blend the pulp with a bit of milk and sugar.

Useful Parts of the Plant

Though the rind of the wood apple fruit is so hard that it has to be cracked with a hammer, the sticky pulp is edible and used to make syrups, drinks, jelly, and jams. Other useful parts of the plant are the roots, fruits, and leaves of this tree that are used for medicinal purposes.

The wood apple tree is cultivated in India and Sri Lanka as well as parts of Southeast Asia such as Malaysia, Thailand, and Cambodia. Long considered a 'poor man's food' it was only in the 1950 that various parts of the plant began to be used for medicinal and other uses. The wood apple fruit can be found in two forms the large sweet version and the small acidic ones. Both wood apple types have a number of health benefits.

Wood apple fruit can be eaten plain or made into jams and sweets. In Indonesia, wood apple is mixed with honey and eaten for breakfast. In Thailand, they eat the leaves in salads while in India the pulp is used in savory chutneys.

Nutritional Information and Properties

Hundred grams of wood apple pulp contains 140 kcal. The fruit contains carbohydrates and proteins. It is also rich in beta-carotene, vitamin B, vitamin C, thiamine, and riboflavin. Wood apple fruit that grow in the wild tend to have more tannin than those cultivated for commercial purposes.

Health Benefits of Wood Apple

- ✰ Wood apple is good for digestion. It destroys intestinal worms and cures chronic dysentery.
- ✰ If you have piles or stomach ulcers, the fruit can help cure these conditions
- ✰ Mix the ripe pulp with honey and cumin seeds to treat diarrhea and indigestion
- ✰ The laxative property of the fruit also solves constipation problems
- ✰ Wood apple fruit juice mixed with warm water and sugar can purify the blood and cleanse the digestive system.
- ✰ It can cure scurvy, as it is rich in vitamin C
- ✰ The gum found in tree is used to treat diabetes
- ✰ Wood apple leaves can cure sore throat, chronic cough, and other respiratory conditions
- ✰ The thiamine, riboflavin, and beta-carotene make it an important ingredient in medicines for the heart and liver
- ✰ Roots of the wood apple tree are used to treat earaches
- ✰ A wood apple chutney made with salt and tamarind can cure hiccups
- ✰ Mix the ripe fruit pulp with honey, cumin, and cardamom to prevent breast or uterine cancer. It can also help cure sterility.
- ✰ Wood apple taken after childbirth can prevent post-partum depression
- ✰ Wood apple poultice made from the leaves can relieve joint pain
- ✰ An infusion of the bark and stem can reduce heart palpitations and hypotension
- ✰ Leaf juice mixed with pepper can treat anorexia and jaundice

Other Uses of Wood Apple

- ✰ The pectin of the wood apple plant is first purified and then used for a number of purposes.
- ✰ The hard shell or rind of the fruit can be used as a small container or made into boxes.
- ✰ The gum of the wood apple tree can be used as a substitute for gum Arabic. This gum is also used to make paints such as watercolors, dyes, inks, and varnishes.
- ✰ The wood is used for various construction purposes, as it is both heavy and long lasting.
- ✰ The wood is also used as fuel.
- ✰ Wood apple fruit is used for medicinal purposes in form of a tonic, a powder or a poultice
- ✰ Wood apple leaves, bark, and roots can treat snakebites and other venomous wounds
- ✰ The fruit pulp is also used as a household cleaner
- ✰ The rind of the fruit yields an oil that is used to make fragrances and to produce a dye used on fabrics

Precautions/Side Effects/Warnings

- ✰ Eating too much wood apple can result in constipation, flatulence, and other digestive problems.
- ✰ Patients with thyroid problems should also avoid eating wood apple as should pregnant and nursing women.

The fruit exhibit excellent nutritional and medicinal properties. It is rich in pectin, acid, minerals, vitamins and sugars. The fruit is largely used as remedy to diarrhea, dysentery and other stomach disorders (McCann, 1985). Wood apples are an antiscorbutic, carminative, and digestive stimulant. The pulp mixed with cardamom, honey and cumin seeds remedy liver cirrhosis of malnourished children, piles, and diarrhea. Traditional healers have also prescribed the fruit to fight breast and uterine cancer. Wood apple pulp has potent antioxidants and is a natural antimicrobial agent. The fruit's use as a liver health remedy: the liver of rats improved significantly when treated with wood apple fruit extracts.

The wood apple's traditional use as a male contraceptive may be affirmed based on the pulp's powerful ability to lower sperm count, mobility and viability. These effects were reversed with no side effects after 45 days from treatment. Wood apple helped control hyperlipidaemia and hyperglycaemia in fluoride-exposed rats.

The antimicrobial quality of extracted essential oil, exhibit its effectiveness against 12 human pathogenic bacteria (Geda and Bokadia, 1980). The fruit pulp is also used for preparation of jelly, chutney and gum. The gum is used for preparation of water colour, dyes, varnishes (Anonymous, 1956). The wood apple contains more amount of pectin and is very good for making the jelly (Rao *et al.*, 1989). However, little systematic work has been done on the preparation of processed products from wood apple fruit. Keeping in a view the above facts, the present investigations was undertaken to standardize the technology for processing the wood apple fruit pulp into ready-to-serve beverage and its consumer acceptability.

Nutritional Value of Wood Apple

According to the National Institute of Nutrition's book, "Nutritive Value of Indian Foods," wood apples contain the following values per 100g:

Parameter	Content
Energy	134 kcal
Moisture	64.2g
Protein	7.1g
Fat	3.7g
Minerals	1.9g
Fiber	5.0g
Carbohydrates	18.1g
Carotene	61IU
Vitamin C	3mg
Thiamin	0.04mg
Riboflavin	0.17mg
Niacin	0.8mg
Calcium	130mg
Phosphorous	110mg
Iron	0.48g
Magnesium	41mg
Chromium	0.21mg
Manganese	0.18mg
Zinc	0.10mg

Preparation of RTS

The fresh fruits of local cultivar were obtained from market. The fully ripened, healthy and uniform size fruits were selected and brought to the laboratory for experimentation. The fruits were manually opened and flesh removed. The pulp was extracted by dilution with water at 2.0 times using sieves. The pulp was heated to 60°C for 30 min (Figure 28.1).

Extract the juice of clean muslin cloth. Centrifuge the extracted juice to obtain the clear juice and use this juice for preparation of RTS beverage. The RTS requirements for wood apple juice are 10 per cent. TSS, 15°Brix and acidity 0.3 per cent.

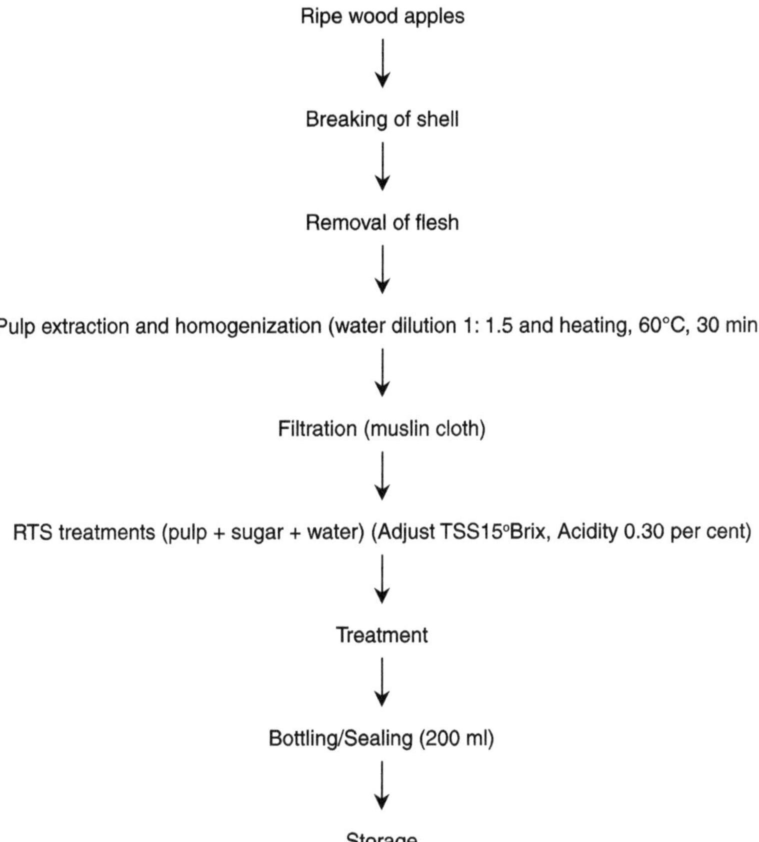

Figure 28.1: Preparation of RTS-beverage from Wood Apple Fruit.

General Recipe of RTS Beverage

Ingredients	Quantity	Quantity
Juice	10 per cent	100ml
Sugar	15 per cent	150g
Citric acid	0.3 per cent	3g
Water	74.70 per cent	747ml
Total quantity of RTS	100 per cent	**1000ml**
Sodium Benzoate	100ppm	100ppm

Considering TSS 5.2°Brix and 1 per cent acidity in wood apple juice, the recipe for preparation of wood apple RTS is

Ingredients	*Quantity*
Juice	100ml
Sugar	144.80g
Citric acid	2g
Water	753.20ml
Total quantity of RTS	1000ml
Sodium Benzoate	100ppm

Procedure

Weigh the water in stainless steel vessel and dissolve the sugar in water. Add the wood apple juice in sugar water mixture and mix it well and add the required quantity of citric acid by dissolving it in small quantity of water and heat the content till it begins to boil. At last add Sodium Benzoate @ 100ppm in similar manner to the RTS and mix well. Fill the sherbet/RTS in clean and previously sterilized glass bottles and store the bottles in refrigerator.

Table 28.1: Chemical Composition of Fresh pulp (1:2) and RTS-beverage

Parameter	*Pulp*	*RTS-beverage*
T.S.S., °Brix	5.2	15
Acidity, per cent	1.0	0.3
Reducing sugars, per cent	2.1	0.4
Total sugars, per cent	2.8	13.8
Ascorbic acid, mg per cent	2.0	0.2

References

Anonymous. 1956. The Wealth of India. A Dictionary of Indian Raw Materials and Industrial Products. Council of Scientific and Industrial Research, New Delhi Vol. IV. F-G, 18-20.

Anonymous. 1981. Silviculture of Indian Trees. III. Edition Forest Research Institute and College, Dehradun. p. 101.

Geda, A. and Bokadia, M.M. 1980. Antimicrobial activity of essential oils on human pathogenic bacteria. Science Culture. 46 (1): 33-35.

MaCann, G. 1985. Trees of India. Periodical Export Book Agency, Delhi, p. 207.

Rao, N.S., Nagarijaah, C. and Raju, B. 1989. Woodapple - suitable tree for farmlands and wasteland. Indian Farmers Digest. 22 (12): 17-18.

Chapter 29

Value Added Products from Ber and Custard Apple

R.S. Gaikwad

Department of Food Science and Technology
Mahatma Phule Krishi Vidyapeeth, Rahuri – 413 722, Dist. Ahmednagar, M.S.

Fruit are the major source of dietary fiber, mineral and vitamins. In addition, they contribute proteins,fats and carbohydrates to the human nutrition. Fruits are rich in calcium, phosphorous, iron and magnesium. These are good source of vitamin C, vitamin A and thiamin, riboflavin, niacin, pantothenic acid and folic acid.fruit and vegetables provide dietary fiber essential for bowel movement and possibly for prevention of several diseases such as colon cancer, diabetes, gallstone and obesity.

Fruit are highly perishable in nature and therefore, suffer heavy losses after their harvest particularly during post handling. These are harvested, stored, graded, packaged and transported to the market. Significant physical and chemical changes occur during postharvest handling, storage and marketing of fruits which result in heavy losses. Hence, some of these fruits are processed into dehydrated, canned and fermented or other types of value added products by employing different method of food preservation.

Ber

Ber fruit can be processed into several products such as jam, jelly pulp, juice, dried ber and powder, candy, slices, tuti-fruity and wine. The byproducts of processing industries can be utilized for extraction of pectin. However, commercialization of all the products will not be an economic proposition. Based on the market demand, few products can be commercialized with a concomitant improvement in technology of preparation of other products.

A. Juice

Juicy varieties of ber are better suited for preparing juice (Khurdia and Singh, 1975). A process for extraction of juice from ber fruits has been developed (Kadam *et al.*, 1991). The juice is extracted in a juicer and is allowed to stand at 4° C for 2hrs. The scum which forms a layer at top surface is removed using stainless steel spoon and the juice is passed through four layer of muslin cloth. It is then filtered through an ordinary filter paper. Alternatively , the scum can be removed by heating the juice at 70°C for 5 min. The recovery of juice is about 40 per cent of the fresh weight of the fruit. The juice can be utilized for preparation from ready-to-serve beverage. A good quality RTS can be prepared from fresh ber juice at relatively low cost. However, the beverage has no appealing flavour (Kadam *et al*,.1991).Whatever aroma exists for fresh juice does not remain after few days. A good quality mixed fruit juice beverage based on ber pomegranate and guava can be prepared (Vaidya *et al.*, 1989). The volatile components of ber juice have not been identified. A carbonated beverage of ber was highly acceptable and had excellent keeping quality (Khurdiya, 1989)

B. Dried Ber and Powder

Dehydrated ber can be prepared by giving treatment of sulphur-dioxide at 2 to 4 g/kg fruits for 2 hours followed by drying in sun for 7 to 10 days or in a cabinet drier at 60°C for 20-35 hour till the moisture content of products is reduced to 15 per cent (Khurdiya and Singh, 1995). The product can be consumed as such or can reconstituents in 10 per cent sugar solution to be consumed as a liquid beverage. The dried fruits can also be soaked in syrup for 20 hours and cocked for 10 min to prepare ber candy. A dry powder has been successfully prepared from ber fruits (Chavan *et al.*, 1992)

C. Candy

Ber candy is becoming increasingly popular product in Maharashtra. Earlier attempts to prepare candy resulted in having dark brown coloured product (Gharate, 1984). Recently, a method of preparing ber candy has been modified to obtain the product with attractive golden yellow colour and better storage life (Khurdiya and Singh, 1975). The candy making process accelerated by syruping the fruit at 40 degree celcious or by vaccum infiltration (Trymbake, 1998). Ber candy can also be prepared from dried ber by soaking them in syrup for 24hrs and cooking for 10 min. The storage of candy at 4°C improves the shelf life and prevents browning of the product.

D. Dried Chips

Fruits are selected, washed thouroughly and chips are prepared with the help of slicer. The chips are blanched in boiling water containing 2 per cent CaCl2 for 3 min. and subjected to sulphur fumigation (2gm/kg) for 1 hr. The Syrup containing 70 per cent sugar and 1 per cent citric acid is prepared and the chips are dipped in it for 3 days. Then, chips are removed from the syrup and dried under sun for 72 hrs. The resultant product is then pressed under basket press. The yield of final product is found to be 40 per cent (Masalkar, 1998). The product looks like dried Fig and exhibits attractive golden yellow colour and appealing taste.

E. Jam and Jelly

An excellent Jelly can be prepared from unripe Ber fruits by adding proper amount of pectin and acid to this fruit or by mixing it with other fruits. The jam prepared from Ber fruits is leathery due to the presence of mucilaginous compound(s) in the fruits.

Custard Apple

Custard apple shows very short storage life due to its fast ripening. It presents climacteric behavior, high respiration rate and ethylene production (Pal and Kumar, 1995; Prasanna *et al.*, 2000). The fast fruit softening, firmness loss is the main characteristic which reduces the fruit quality and its commercialization. The major constraints in custard apple processing are: the appearance, the bitterness develops gradually due to limonin after extraction of pulp and the enzymatic browning reaction catalyzed by polyphenol oxidase (PPO). Control of enzymatic browning during processing and storage is important to preserve the quality of fruit pulp.

Out of the total production of processed products, fruit pulp and juice constitute 27 per cent, followed by the beverages (13 per cent), pickles (12 per cent) and wine (2 per cent). During the past two decades, there has been a renewed emergence of a market for ready-to-serve beverage and wine in India and other developing countries. The peoples are becoming more aware about the health benefits of RTS beverages and wine. The development of soft drink market in developing countries can be directly traced to improved affordability of the consumers. Further, the convenience in the use of snacks and beverages of the liquid as well as refreshment in between regular meals enhances their applicability. Forceful marketing strategy coupled with increasing variety of products by the food industry has helped the growth of the beverage industry. The food beverage and wine industry have been avocations favorable health and effects of specially beverages and wine. The nutritive fortified drinks and these are now being referred as health or sport drinks.

Preparation of Ready-to-Serve Beverages

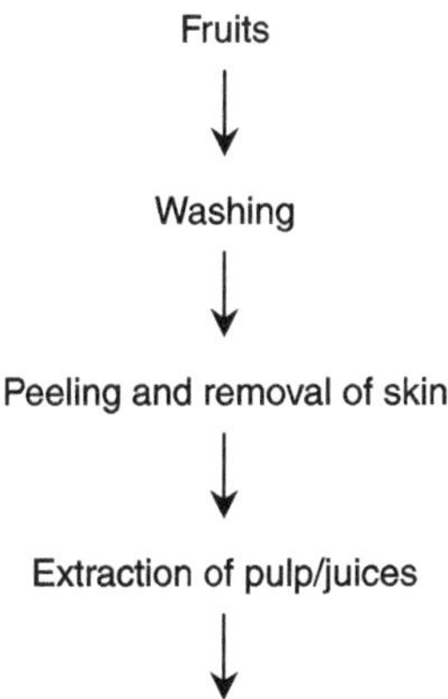

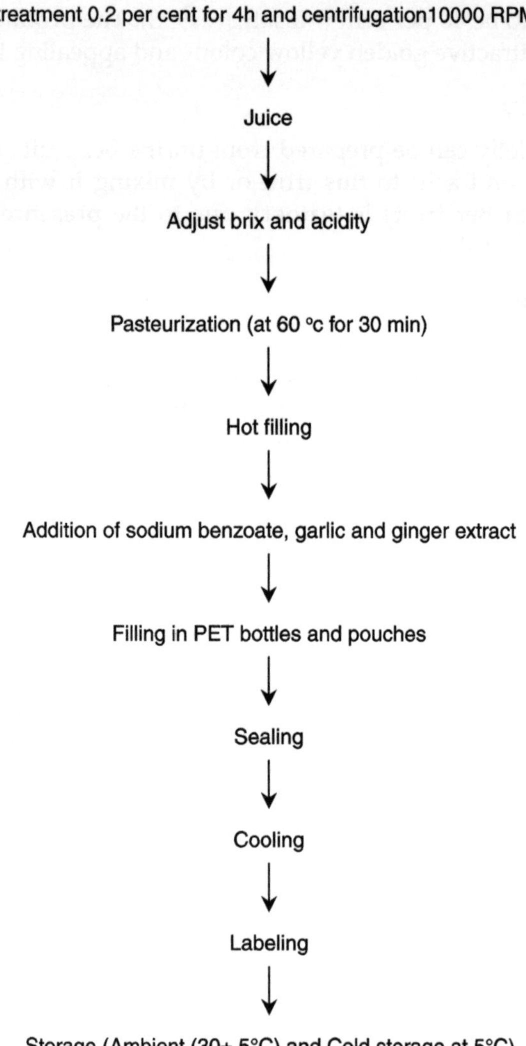

Figure 29.1: Flow Sheet for Preparation of Ready-to-Serve Beverage Wine.

Preparation of Wine

Custard apple fruit juices were supplemented with 0.5 per cent ammonium sulphate and 0.1 per cent potassium dihydrogen phosphate as a source of nitrogen. 400 ml juice of each cultivars was taken in different (six) flasks and pasteurized at 85 °C for 30 min. After cooling, each sample was inoculated with loopful of pure culture of *S.cerevisiae* var. *ellipsoideus* NCIM-3045. All the samples were incubated at 28 ± 2°C for 48h. The wine was prepared using these inoculums @ of 4 per cent using process depicted in Figure 29.2.

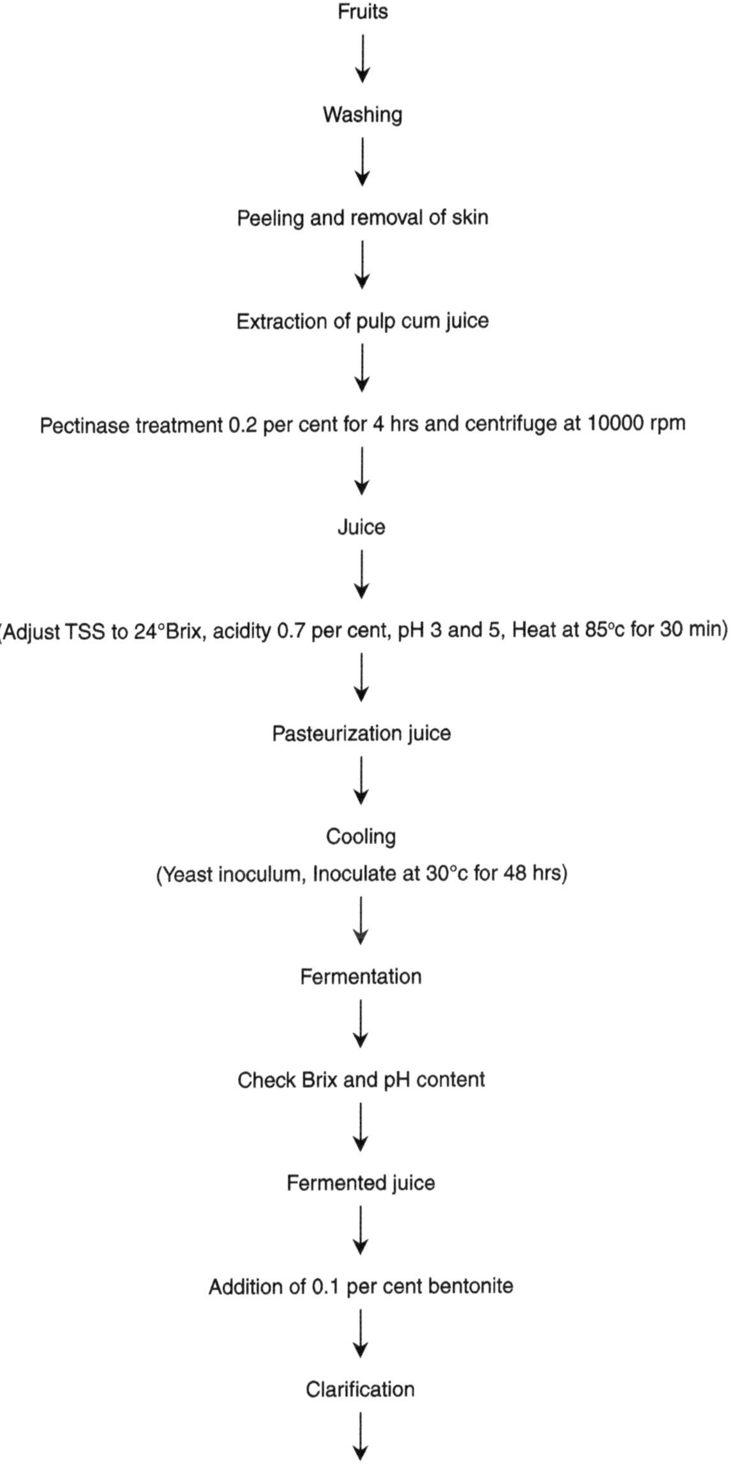
Fruits
Washing
Peeling and removal of skin
Extraction of pulp cum juice
Pectinase treatment 0.2 per cent for 4 hrs and centrifuge at 10000 rpm
Juice
(Adjust TSS to 24°Brix, acidity 0.7 per cent, pH 3 and 5, Heat at 85°c for 30 min)
Pasteurization juice
Cooling
(Yeast inoculum, Inoculate at 30°c for 48 hrs)
Fermentation
Check Brix and pH content
Fermented juice
Addition of 0.1 per cent bentonite
Clarification

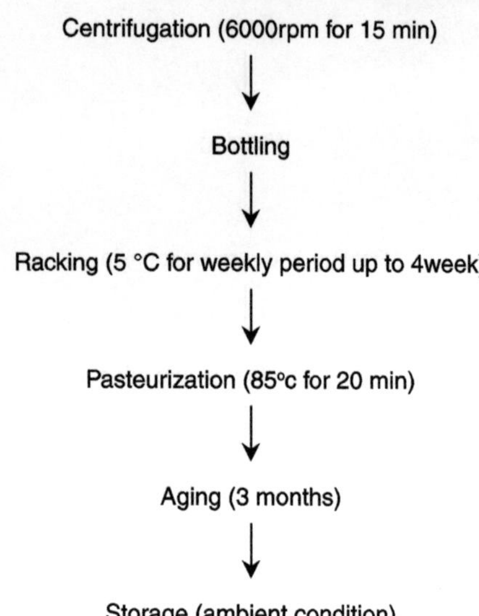

Figure 29.2: Flow Sheet for Preparation of Wine.

Chapter 30

Recent Advances in Mechanization of Fruit Crops

V.D. Deshmukh

Principal Investigator, AICRP on FIM,
Mahatma Phule Krishi Vidyapeeth, Rahuri – 413 722, Dist. Ahmednagar, M.S.

The area under horticultural plantation in Maharashtra State is increasing. The major horticultural and vegetable crops in Maharashtra State are Mango, Grape, Onion, Pomegranate *etc.* There is need to mechanize different farm operations in horticultural and vegetable crops. Hence, mechanization of horticultural and vegetable crops has to be priotized. An overview of improved farm implements and machinery useful for orchard crop management have been taken in this research paper.

Tractor Operated Laser Guided Land Leveler

It is important to level the field for fruit crop plantation particularly at the time of establishment of orchards. At present, tractor drawn levelers are used for leveling operation. But then for irrigating these fields, proper slope is not maintained which results in non-uniform supply of irrigation water to the plants. In view of this, if proper slope of the field is maintained at the time of orchard establishment then it will facilitate proper irrigation to all corners of the field. Tractor operated laser guided land leveler ensures precise and proper slope of the field cross ways also (Figure 30.1).

Tractor Operated Post Hole Digger

It is necessary to dig pits for fruit crop plantations. These pits are used for planting the rooted cuttings of cultivated varieties in the main field. For this,usually square pits of size 1 mtr long x 1 mtr wide x 1mtr deep are dug for planting. Manual

Figure 30.1

digging of these pits is expensive, lanorious and time consuming particularly when required to dig large numbers of pits. These pits can be dug with tractor operated post hole digger. It can be operated by 45 hp and above tractor through its PTO. About 20 to 25 pits of 90cm diameter and 90 cm deep can be dug in one hour of operation. Soil removed from the pit can be placed around the pit which becomes easy to fill the pit back after plantation (Figure 30.2).

Figure 30.2

Power Weeders

Power weeders are useful for inter-row inter-culturing operation in fruit crops. These power weeders are equipped with petrol start kerosene run engines of 3.00 to 6.50 hp. They have small size rotavator which performs the weeding operation. There is saving in man-power by using these machines. These power weeders can

Figure 30.3

also be used for for mending of round shaped check basins around the trees in orchards (Figure 30.3).

Rotavator

Weeds between the rows of vines and orchard trees are also removed mechanically by power tiller rotavator and garden tractor drawn rotavator (Figures 30.4a,b).

Figure 30.4a-b

Tractor Operated Weed Badger

In horticultural plantations particularly grape crop, weeding between plants is difficult to be done by existing machinery., Tractor operated weed badger equipped with sensors are used in the developed countries for weeding between the plants (Figure 30.5).

Figure 30.5

Tractor Operated Air Sleeve Boom Sprayer for Pomegranate

Pomegranate is one of the important fruit crop commercially grown in Maharashtra State. Tractor operated air sleeve boom sprayer is suitable for pomegranate. Better canopy coverage, uniform spray deposition, saving in chemicals, timeliness in spraying operation characterizes the working of the air sleeve boom sprayer (Figures 30.6a,b).

Figure 30.6a-b

Salient Features

- ✰ The sprayer is suitable for spraying in Pomegranate orchards
- ✰ Uniform spray, effective spraying, saving in chemicals characterizes the working of this sprayer
- ✰ Excellent for pomegranate growers
- ✰ Effective field capacity of tractor operated air sleeve boom sprayer has been found to be 1.40 ha h^{-1}.

- ☆ Saving in chemical was observed to be 50 to 60 per cent over conventional method.
- ☆ Timeliness of operation is 40 to 50 per cent as compared to conventional method.

Tractor Operated Air Assisted Boom Sprayer for Pomegranate and Grape

In view of the infestation of pest and diseases on pomegranate and grape vineyards, pesticide application is most important farm operation in pomegranate and grape cultivation. Presently, tractor operated air carrier sprayers/mist blowers are used by the farmers. Their spraying efficacy is better than conventional hydraulic energy sprayers (Figures 30.7a-c).

Figure 30.7a-c

Tractor Operated Tycoon Sprayer with Rotary Nozzle for Tall Trees

Rotary nozzle produces higher air velocity which gives fluttering movement to plant leaves and hence covers both the surfaces of the leaves. The rotary nozzle shows significant results over the hollow cone nozzle for control of mango hoppers. Tractor operated Tycoon sparyer is equipped with rotary nozzles (Figure 30.8).

Figure 30.8

Electrostatic Sprayers

Uniform spray pattern, judicious use of costly insecticides and pesticides, target toxic and environmentally benign characterizes the working of electrostatic sprayers. These sprayers are mostly imported and introduced on limited scale amongst the grape growers. High cost of these sprayers is the constraint in its adoption by the individual farmers. Custom hiring of electrostatic sprayers is being done on limited scale in some parts of the Maharashtra state. It is necessary to conduct feasibility study of these imported sprayers to assess its performance in Indian conditions. Efforts are required for development of indigenous design of efficient and low cost electrostatic sprayers (Figure 30.9).

Figure 30.9

Case Study of Custom Hiring of Tractor operated Electrostatic Sprayer by the farmer

Shri. Bhausaheb Pandharinath Mhaisdhune A/p Mungsara Tq. And Dist. Nasik has purchased 80 SI ESS Electrostatic sprayer in September 2012. Total cost of the unit is Rs. 16.00 Lakh (Figures 30.10a-b)

The Salient Features of the Electrostatic Sprayer

- ☆ 80 cm Width
- ☆ Poly propylene tank of 113 ltrs capacity.
- ☆ Weight- 105 kg
- ☆ Covers 12 ha/day of 8 hours
- ☆ Droplet in the range of 40-50 micron.

Figure 30.10a-b

- ☆ Sprayer is charged positive hence uniform deposition on vine canopy.
- ☆ Useful for GA application.
- ☆ Effective deposition on foliage and grape bunches.
- ☆ 75 to 80 ltrs chemical solution with water per hectare.
- ☆ Design is based on patented technology of max charge.
- ☆ Manufactured by Electrostatic Spraying System Inc., Georgia (USA).
- ☆ **Economics: The farmer charges Rs. 5000/ha for spraying in grape vineyards. He has earned an amount of Rs. 10 lakh in 1 year.**

Tractror Operated Dusters

Tractror operated dusers are useful for dusting operation in horticultural crops (Figure 30.11).

Figure 30.11

Pneumatically Operated Vine Pruners

Pneumatically operated vine pruners are needed to be introduced in the grape crop management (Figure 30.12).

Tractor Operated Banana Clump Remover

Presently, banana clump removing is done by using tractor operated disc harrow followed by rotatilling operation as such it is cost intensive. Tractor operated banana clump remover is having simple mechanical linkage and is very effective in removing banana clumps. The implement with 700 mm working width is compatible

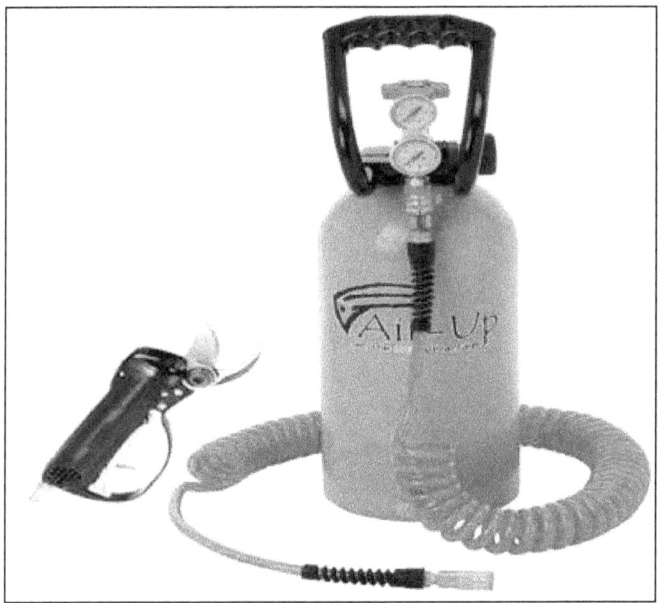

Figure 30.12

with 55 to 60 hp tractor.Effective field capacity of the implement is 0.11 ha h^{-1}. Net saving by using this implement over traditional method is Rs. 2500 $ha^{-1}$for banana clump removing operation. About 30 to 40 per cent timeliness and operational saving up to 50 to 60 per cent is achieved (Figure 30.13).

Figure 30.13

Tractor Operated Banana Stem Shredder

This machine is useful for shredding of banana stems. In one hour of operation, it can shred about 120 to 130 banana stems per day (Figures 30.14a-b).

Figures 30.14a-b

Tractor Operated Chipper Shredder

This machine is useful for shredding crop residues *etc.* in the orchards. The machine can be operated by any 40 hp and above tractor with dual clutch, attached to 3 point linkage and powered by PTO with 540/1000 rpm through drive shaft. The shredded crop can be spread in the field or can be collected into a trolley and becomes rich biomass resource for compost making (Figure 30.15).

Figure 30.15

Mechanical Fruit Harvester

The harvesting of fruits is done manually and is very labour intensive. The manual harvesting is time consuming and tedious, the harvesters will increase the efficiency of the labour and reduce the harvesting time. India is the largest producer of mango and about 23 lakh hectare of land is under cultivation of mango. Tractor operated hydraulic platform has been developed for harvesting mangoes from tall trees. The tractor operated hydraulic platform is having a variable height of upto 6 m. The fruits can be easily harvested from a height of 9-10 m. The platform can also be used for pruning and spraying of the orchards (Figure 30.16).

Figure 30.16

Future Strategies

1. It is necessary to identify the present mechanization gaps and future needs for improved farm equipments on the basis of objective analysis of prevailing agro-socio-economic factors for different fruit crops.
2. To improve the versatility of power tillers and garden tractors by development/adoption of improved implements for different farms operations
3. To test imported farm machinery in the laboratory and field for evaluation of its feasibility in our region.
4. To promote manufacturers by involving them in developmental process of farm machines useful for fruit crop management and their production technology for manufacturing of improved farm implements
5. Imported farm machinery are very costly. Hence, development of indigenous machinery useful for orchard management, should be taken up by using reverse engineering techniques.
6. To conduct feasibility testing of prototypes of proven designs of farm implements and machinery, selected from different regions for adoption under local conditions with a view to bridge the identified mechanization gaps in orchards
7. To educate farmers on the significance of use and maintenance of improved farm equipments through various media including trainings and demonstrations

Chapter 31

Use of different Machineries and Equipments Used in Postharvest Management of Dryland Fruit Crops

A.P. Patil, V.P. Kad, J.K. Dhemre and S.D. Shelar

Postharvest Technology Centre,
Mahatma Phule Krishi Vidyapeeth, Rahuri – 413 722, Dist. Ahmednagar, M.S.

Digital Vernier Caliper

A digital vernier caliper of 0.01 mm least count was used for measurement of linear dimensions of fruits.

Digital Micrometer Screw Gauge

A digital micrometer screw gauge of 0.001 mm least count was used for measurement of linear dimensions of fruits.

Refractometer

A refractometer (Range: 0°-32°B, 32°-64°B, 64°B and above) was used for measurement of total soluble solids of fruits. Sugars are the major soluble solids in fruit juices and therefore soluble solids can be used as an estimate of sweetness. A hand-held refractometer can be used outdoors to measure per cent TSS (equivalent degrees Brix for sugar solutions) in a small sample of fruit juice. Temperature will affect the reading (increasing about 0.5 per cent TSS for every 5°C or 10°F), so should adjust the measurement for the ambient temperature. Clean and standardize the refractometer between each reading with distilled water (should read 0 per cent TSS at 20°C or 68°F) (Figure 31.1).

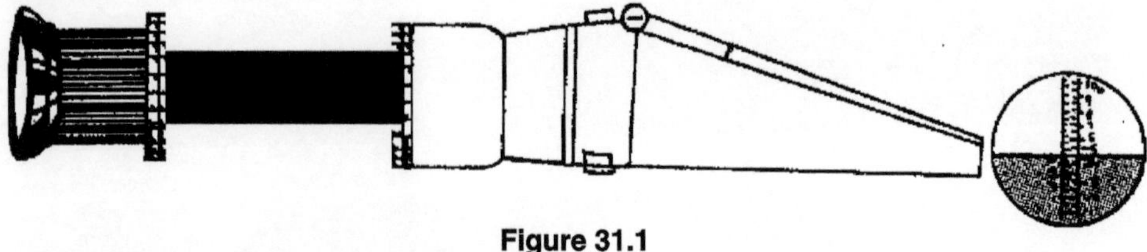

Figure 31.1

Electronic Weighing Balance

A digital display electronic weighing balance having different capacity with different least count was used to measure the weight of fruits.

Firmness Tester

The degree of softness or crispiness can be estimated by squeezing produce, or by taking a bite. Objective measurements can be made with inexpensive penetrometers. The most common way to measure firmness is resistance to compression or pounds-force (lbf). The fruit penetrometer is a hand-held probe with a gauge for pounds-force.

To measure firmness, use fruit that are uniform in temperature, since warm fruit are usually softer than cold fruit. Use fruits that are uniform in size, since large fruit are usually softer than smaller fruit. Make two puncture tests per fruit on larger fruits, once on opposite cheeks, midway between stem and blossom ends. Remove a disc of skin (larger than the tip to be used) and choose the appropriate plunger tip (see below). Hold the fruit against a stationary, hard surface, and force the tip into to fruit at a slow, uniform speed (take 2 seconds) to the scribed line on the tip. Take the reading to the nearest 0.5 lb-force.

Harvesting Containers

Picking baskets, bags and buckets come in many sizes and shapes. These harvesting containers can be made by sewing bags with openings on both ends, fitting fabric over the open bottom of ready-made baskets, fitting bags with adjustable harnesses, or by simply adding some carrying straps to a small basket (Figure 31.2).

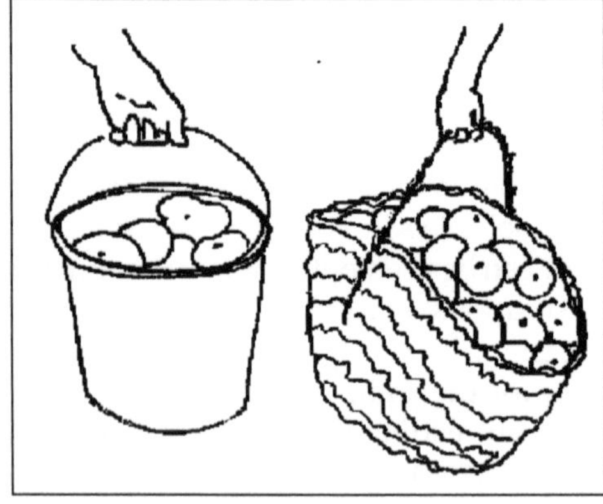

Figure 31.2

Plastic crates are relatively expensive but are durable, reusable and easy to clean. When empty, they can be nested to save space in storage or transport. When filled they can be stacked if every other crate is turned in the direction opposite to the one below. If plastic crates are well vented along the sides and/or bottom, they can also be used to wash and/or cool produce after harvest (Figure 31.3).

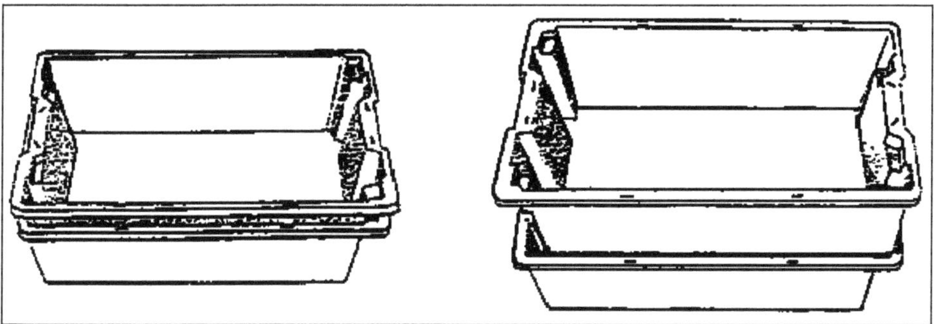

Figure 31.3

Harvesting Tools

Clippers or Knives

Some fruits need to be clipped or cut from the parent plant. Clippers or knives should be kept well sharpened. Penduncles, woody stems or spurs should be trimmed as close as possible to prevent fruit from damaging neighboring fruits during transport (Figure 31.4).

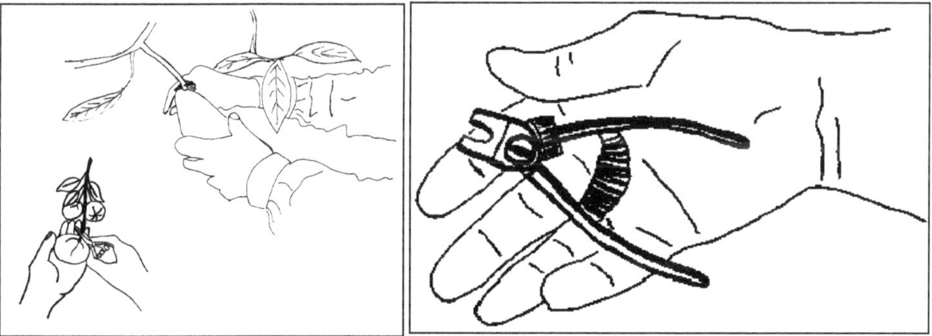

Figure 31.4: Clippers or Knives.

Pole Mounted Cut and Hold Picking Shears

Using a cutting tool attached to a long pole can aid picking of crops such as mangoes and avocados when the fruit is difficult to reach (Figure 31.5). Cutting edges should be kept sharpened and the catching bag should be relatively small The angle of the cutting edge and the shape of the catching bag can affect the quality of the fruit harvested, so it is important to check performance carefully before using any new tools.

Picking Pole (Figure 31.5a-b)

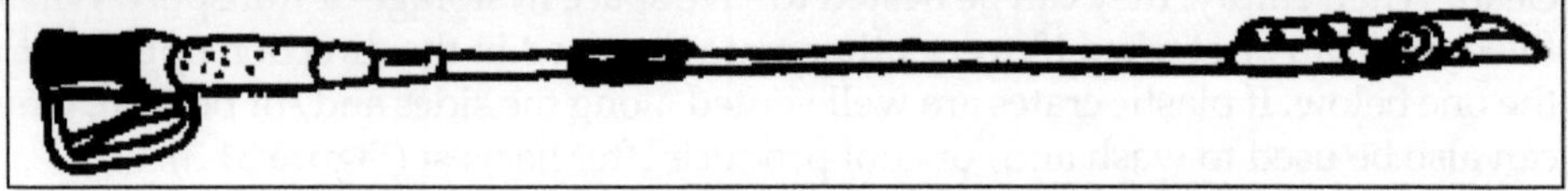

Figure 31.5a-b

The collection bags illustrated below were hand woven from strong cord or sewn from canvas. The hoop used as the basket rim and cutting edges can be fashioned from sheet metal, steel tubing or recycled scrap metal (Figure 31.6).

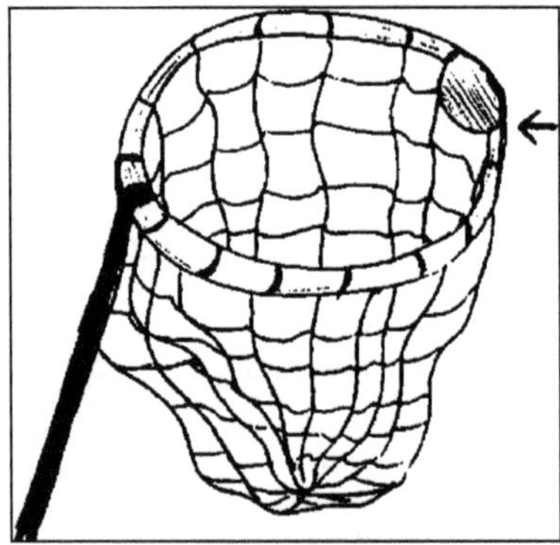

Figure 31.6

Hand Cart

Small hand carts or small mobile field packing stations can be designed to be moved along with the packers and to provide shade for packing operations. Harvesters place an empty carton on the cart, then pick, sort, grade and pack directly into the container. Filled cartons are moved immediately into coolers or keep in a well-shaded area (Figure 31.7).

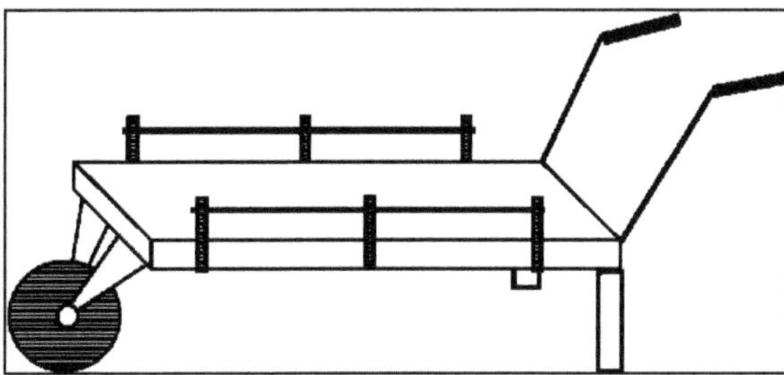

Figure 31.7

Self Propelled Field Pack System

A self-propelled field pack system allows field workers to cut, trim, tie/wrap and pack *in the field,* thus eliminating the expense of operating a packing shed. In the illustration below, a fiat bed truck is moving along-side the field pack system, and packed produce is being loaded for transport (Figure 31.8).

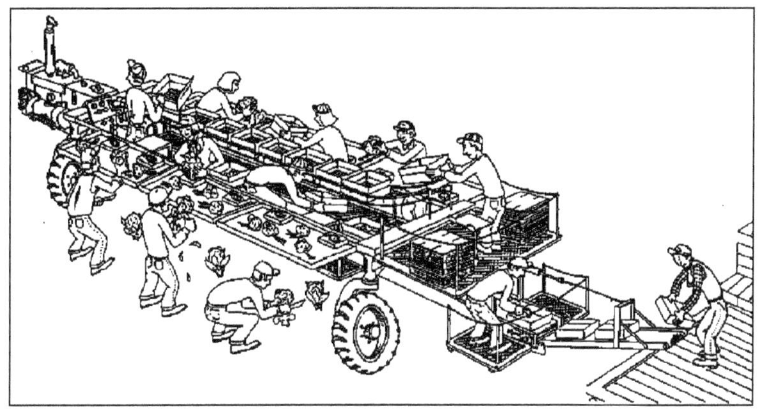

Figure 31.8

Sorting

The table illustrated below is a combination sorting and packing stand. Incoming produce is placed in the sorting bin, sorted by one worker into the packing bin, and finally packed by a second worker. If workers must stand to sort produce, a firm rubber pad for the floor can help reduce fatigue (Figure 31.9).

Improved Zero-Energy Cool Chamber

The low cost cooling chamber illustrated below is constructed from bricks. The cavity between the walls is filled with sand and the bricks and sand are kept saturated with water. Fruits and vegetables are loaded inside, and the entire chamber is covered with a rush mat, which is also kept moist. Since a relatively large amount of materials are required to construct this cold storage chamber, it may be useful only when handling high value products. During the hot summer months in India,

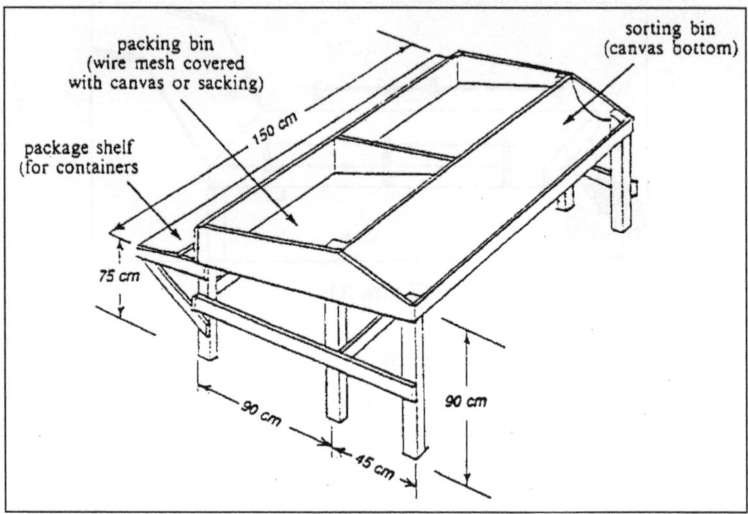

Figure 31.9

this chamber is reported to maintain an inside temperature between 15 and 18 °C (59 and 65 °F) and a relative humidity of about 95 per cent (Figure 31.10).

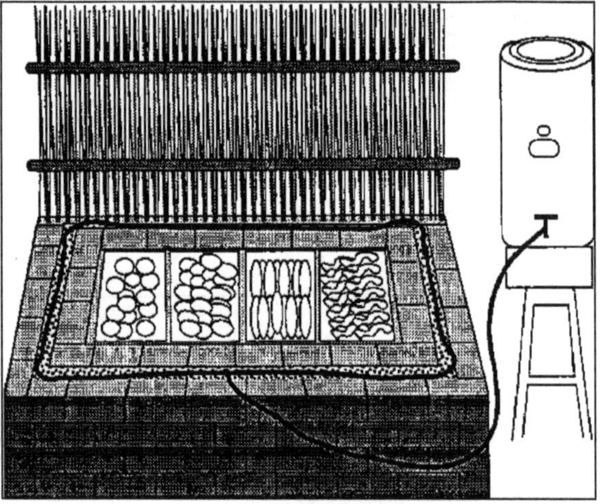

Figure 31.10

Packaging Machine/Sealing Machine

Pouch or laminated pouch packing machine was used for the purpose of sealing and packing of the fruits in food grade polyethylene pouches.

Blast Freezer

The blast freezer are used for frozen the fruits or fruits pieces in the blast freezer.

Deep Freezer

Deep freezer was used for storage of frozen products at minus 18°C or below temperature.

Hot Air Oven

A hot air oven (Range: 5°C – 250°C) was used for moisture content determination of different fruit crops.

NIR Spectrometer

A spectralyzer is a dual beam near infrared spectrometer suitable for compositional analysis of a wide variety of solids, semi solids, viscous, liquids *etc.* products.

Working

Both quantitative and qualitative near infra-red analysis is based on the interaction of near infra-red light with the molecules of a sample. The sample was irradiated with near infra-red light of specific wavelengths, selected from up to nineteen high precision interference filters. The light directed onto the sample was partly absorbed and partly reflected by the sample. A part of the wall of the gold coated interacting sphere contains a lead sulphide detector which measured the reflected portion of the incident light. The resulting spectral information was characteristic for the products and allowed a determination of product relevant properties (Figure 31.11).

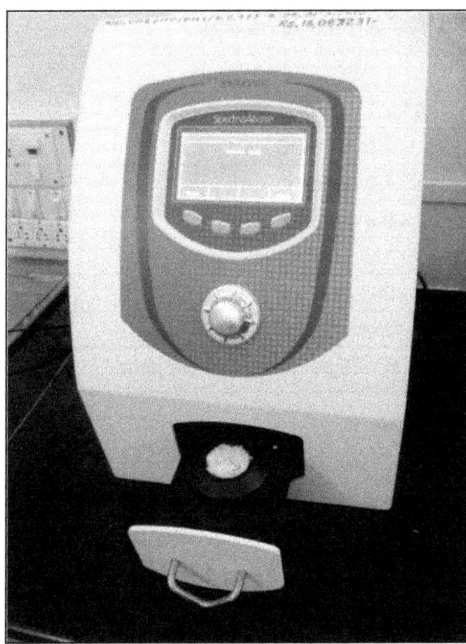

Figure 31.11: NIR Spectrometer Used for Measuring Compositional (Chemical) Analysis.

Figure 31.12: Brookfield Texture Analyzer (CT3) Used for Measuring Textural Properties.

Texture Analyzer

The texture analyzer was used for measurement of various textural properties of food masteials. Hardness is the force required to compress a substance between the molar teeth or between tongue and plate to a given deformation or penetration and designate as soft, firm or hard. Texture analyzer was equipped with a 50 kg load cell and TA3/100 probe operated at a test speed of 1 mm/s transversally. Hardness is the maximum force required to break the flake in Newton (N) as shown in Figure 31.12.

Viscometer

Brookfield viscometer was used for measuring viscosity of different food materials.

Brookfield Viscometer DV-II+Pro

The Brookfield viscometer DV-II+Pro uses the principle of rotational viscometers as shown in Figure 31.13a. It measures the torque required to rotate an immersed element (the spindle) in a fluid. The spindle is driven by a motor through a calibrated spring. The viscous drag of the fluid against the spindle is measured by the spring deflection. Spring deflection is measured with a rotary transducer. The drive motor is located at the top of the instrument inside the housing. Below the main case pivot shaft is provided to which spindles are attached. The Brookfield viscometer DV-II+Pro is provided with a set of four spindles and programmed with 10 possible speeds. The specifications of this viscometer are given below.

Specifications of Brookfield Viscometer DV-II+Pro

Sl.No.	*Parameter*	*Value*
1.	Net Weight	9 kg
2.	Temperature Sensing Range	-100°C to 300°C
3.	Viscosity Accuracy	+/- 1.0 per cent
4.	Viscosity Repeatability	+/- 0.2 per cent
5.	Temperature Accuracy	+/- 1°C : -100°C to +149°C
		+/- 2°C : +150°C to +300°C
6.	Operating Environment	Temperature : 0°C to 40°C
		R. H. : 20 per cent to 80 per cent

Thermostatic Water Bath

Thermostatic water bath (Figure 31.13b) was used to achieve and maintain the desired temperature of the samples. The water bath consists of an electrical heating oil element, which converts electrical energy into heat energy. This is used to heat the water inside the container of the water bath. The container is provided with a cover with circular holes which is used to support the beakers placed inside them. The holes are provided with covers of three different sizes to hold different size of utensils. The water bath has twelve such holes which can be covered when not

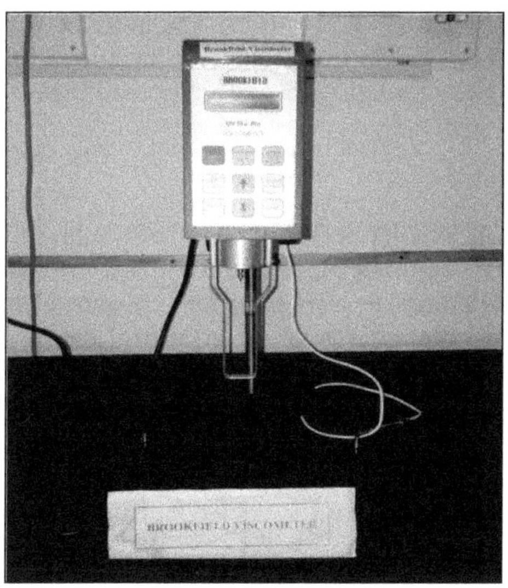

Figure 31.13a: Brookfield Viscometer DV-II+Pro Used for the Measurement of Viscosity.

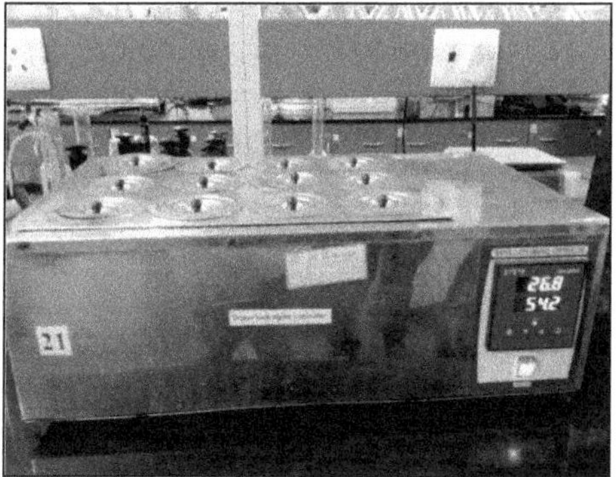

Figure 31.13b: Thermostatic Water Bath Used for Maintaining Temperature of Samples.

in use. The display gives the temperature which has been set. Three buttons are provided which are used to set the temperature; apart from these a main ON-OFF button is also provided.

Sample Holding Beaker

For measurements with standard viscometer a container with an inside diameter of 3*1/4 inches (83mm) was used. Use of smaller container will result in an increase in viscosity readings (DV-II+ Pro manual). The sample container used for this purpose was a 500 ml low form griffin beaker throughout the all viscosity measurements carried out.

Thermometer

The thermometer having least count of 1°C was used every time to check the temperature of oil sample kept in water bath.

Procedure for Viscosity Measurement

The sample taken into the sample holding beaker of 500 mL was placed in water bath. The temperature of water bath was set to the desired temperature and once the desired temperature was achieved the temperature of sample was checked (*i.e.* 23.4°C) manually by using thermometer and then the sample was taken for the viscosity measurement.

The sample obtained from the water bath was used for the measurement of viscosity. The auto zero adjustment was done in the equipment, and spindle (probe) no. 64 was used at 200 rpm taking into consideration the trial and error method followed for spindle and speed selection as mentioned earlier. The spindle was placed at the centre and inserted in the test material until the fluid level is at the immersion groove of the spindle's shaft. After stabilizing the viscosity reading was recorded from the display.

Boiling Pans/Pasteurizers

There are two types of boiling pans: at smaller scales of operation, a simple stainless steel (or less desirably aluminium) pan can be placed directly over the heat source. However, viscous products such as sauces, syrups, jams and chutneys, are likely to burn onto this type of pan unless great care is taken to control the heat and thoroughly stir the mixture as it is being heated. This reduces the quality of the product and also significantly slows down production because the pan must be cleaned between batches. At a larger scale, a 'double jacketed' stainless steel boiling pan is needed. Steam is produced in the space between the outer jacket and inner pan to give more uniform heating and avoid localised burning of the product (Figure 31.14).

Figure 31.14

Bottle Coolers

These are used to increase the rate at which glass containers cool, by allowing the containers to pass through an inclined water bath. There is a flow of cooling water in the opposite direction to the containers (Figure 31.15).

Figure 31.15

Bottle Washers

The time consuming part of bottle washing is rinsing out the detergent and bottle washers are used to reduce this time. They are made by soldering vertical pipes onto a larger base pipe and connecting the base pipe to a water supply. In use the bottles or jars are inverted over the vertical pipes and rinsed until free of detergent (Figure 31.16).

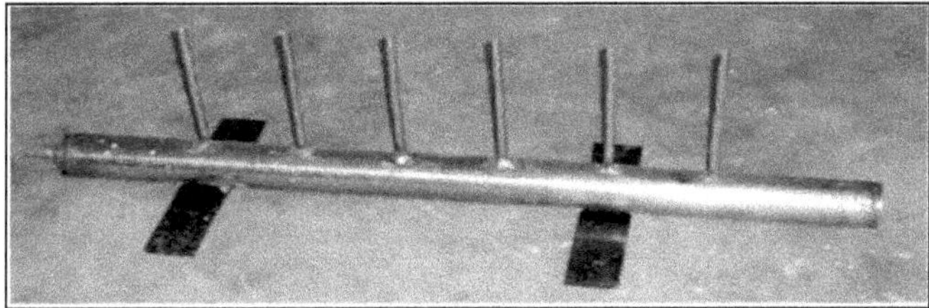

Figure 31.16

Crown Corking Machine

A corking machine compresses the cork and inserts it into the bottle. In operation a cork is inserted into hole in the top of the machine and a bottle is held in place by the sprung stand. The lever is then lowered to force the cork into the bottle (Figure 31.17).

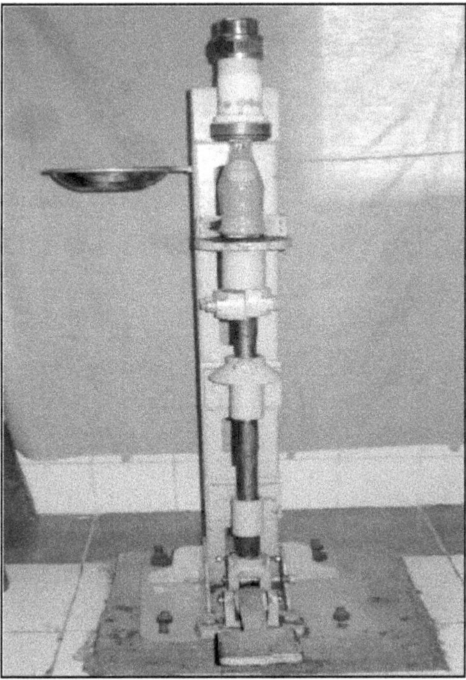

Figure 31.17

Crown Cappers

A hand operated crown capper is a simple die that is placed over the cap on a bottle and is then struck with a hammer to seal it. Another design has two handles that are lowered to force the cap onto the bottle (Figure 31.18).

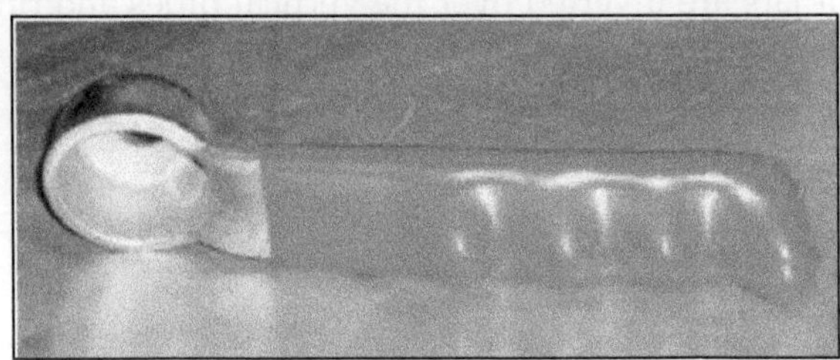

Figure 31.18

Deep Fat Fryers

The simplest fryer is a pan of oil over a fire, but there is little temperature control and a risk of burning both the oil and the product (Figure 31.19).

Figure 31.19

Dicers

Manual dicing machines first cut the material into strips and these are then cut into cubes. The machines produce uniform sized cubes, which are difficult to achieve using a knife. The blade should be sharpened as needed.

Fruit Crushers

Powered machines that are used to crush hard fruits consist of a rotating screw inside a casing. The casing has a screen in the base. Juice/pulp drains from an outlet in the base of the casing and skin is ejected from the end of the casing down a chute. The casing should be fitted with wing nuts to remove it easily (Figure 31.20).

Figure 31.20

Basket Press

Manual fruit presses have a stainless steel cage in which a press plate is raised and lowered by a screw. The screw should be made from a harder grade of steel than the bearing to prevent the more expensive screw from wearing down (the bearing can be replaced more cheaply than the screw). Layer plates made from stainless steel, nylon or painted mild steel may be needed in larger presses to promote efficient pressing. In operation a muslin or cotton bag is placed in the cage and pulp is poured in. The bag is closed and the press plate is lowered. After a few minutes, additional turns of the screw increase the pressure further. Juice is collected via a pipe. The screw is then raised and the bag is removed (Figure 31.21).

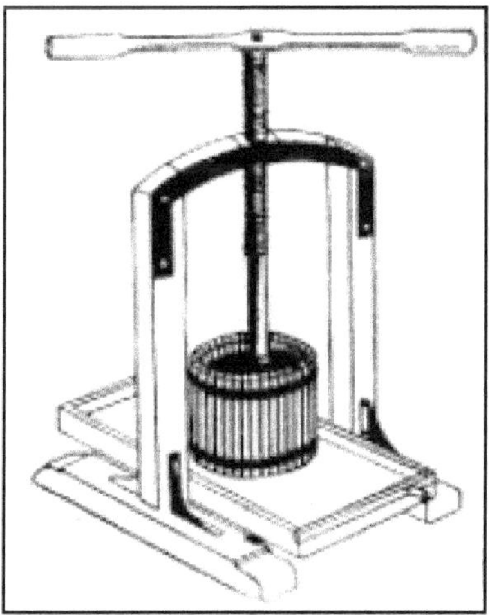

Figure 31.21

Pulper

The pulper is more suited to pulping different fruits and other softer fruits. It is supplied with a coarse and fine mesh screen (Figure 31.22).

Figure 31.22

Reamers

There are two types of reamer that are used in fruit processing: one is designed to extract coconut meat from the shell and the other is used to extract juice from halved citrus fruits. A citrus juice reamer attachment can also be obtained to fit an electric mixer (Figure 31.23).

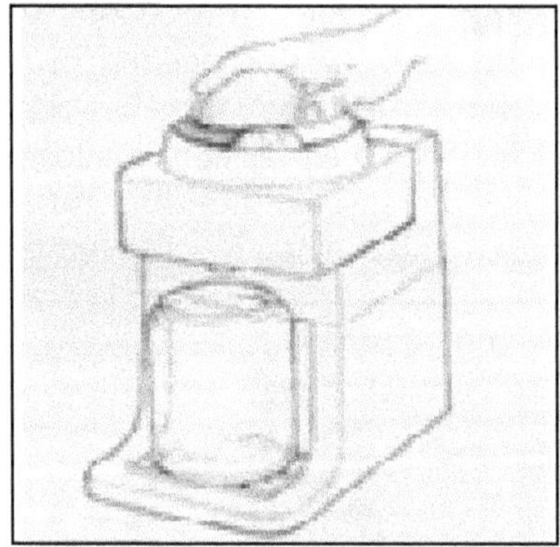

Figure 31.23

Sulfuring Box

A low cost sulfuring box can be constructed from a large cardboard box that is slashed in several places to allow adequate ventilation. Trays for drying be stacked using bricks and wooden spools as spacers. The trays must be made completely of wood, since sulfur fumes will corrode metal. The entire assembly must be located out of doors, preferably on bare soil. Use 2g sulfur per kg of fruit. Place the sulfur in a container well away from the side of the box since it will become quite hot. Seal the bottom edges of the box with soil (Figure 31.24).

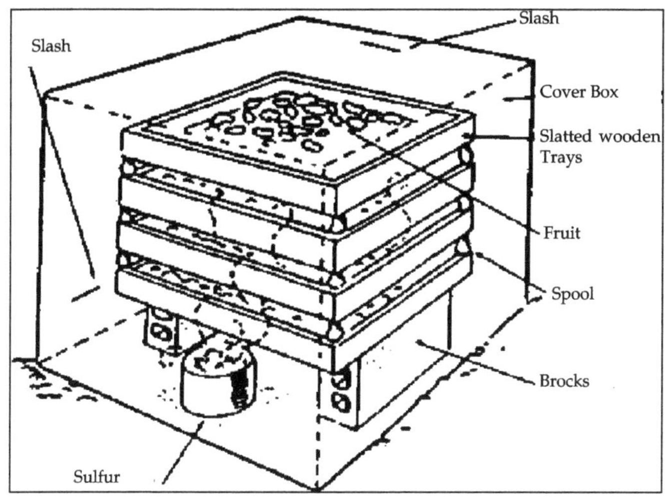

Figure 31.24

Direct Solar Drier

More complex models of solar driers have glass or clear plastic windows that cover the produce, providing some protection from insects while capturing more of the heat of the sun (Figure 31.25).

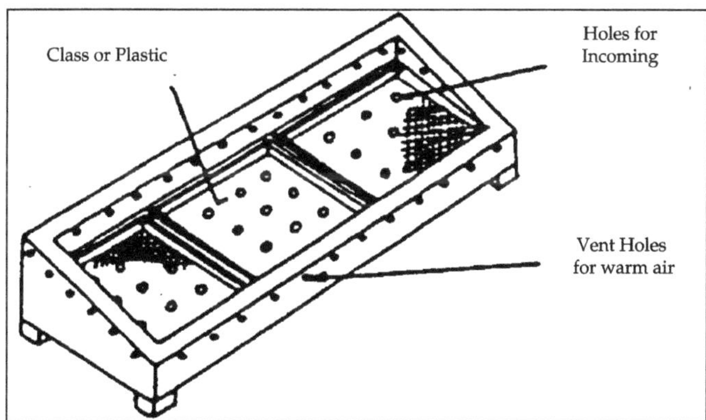

Figure 31.25

Indirect Solar Drier

Indirect driers are constructed so the sun shines upon a solar collector (a shallow box, the insides painted black, topped with a pane of glass) heating air which then moves upward through a stack of four to six trays loaded with produce (Figure 31.26).

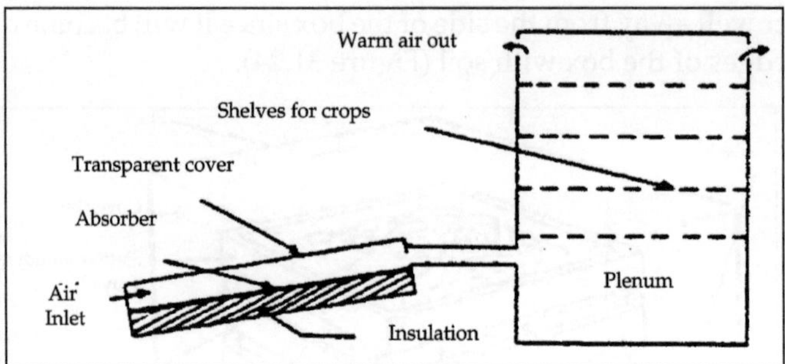

Figure 31.26

Forced-air Dehydrators

This dehydrator can be dried in bulk using a dehydrator that combines a steady stream of air with an external source of heat. The plenum chamber below the produce is covered with a floor of perforated sheet metal or wooden slats. A fan located between the furnace and the plenum chamber moves the hot air through the drying produce (Figure 31.27).

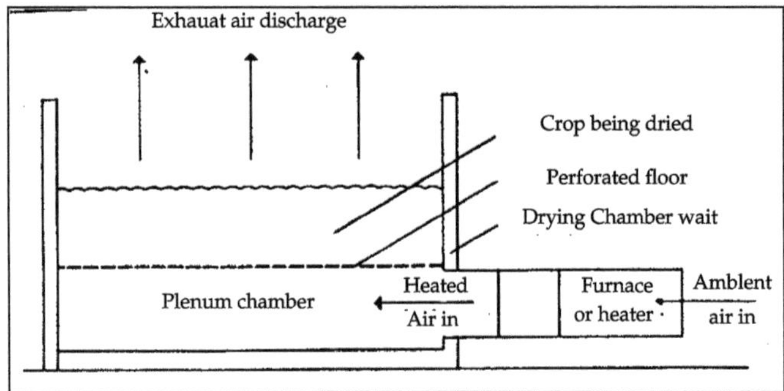

Figure 31.27

Tray or Cabinet Dryers

These types of dryers use trays or similar product holders to expose the product to heated air in an enclosed space. The trays holding the product inside a cabinet or similar enclosure are exposed to heated air so that dehydration will proceed. Air movement over the product surface is at relatively high velocities to ensure that heat and mass transfer will proceed in an efficient manner (Figure 31.28).

Fluidized-Bed Drying

In this system, the product pieces are suspended in the heated air throughout the time required for drying. As illustrated in Figure given below, the movement of product through the system is enhanced by the change in mass of individual particles as moisture is evaporated. The movement of the product created by fluidized particles results in equal drying from all product surfaces. The primary limitation to the fluidized-bed process is the size of particles that will allow efficient

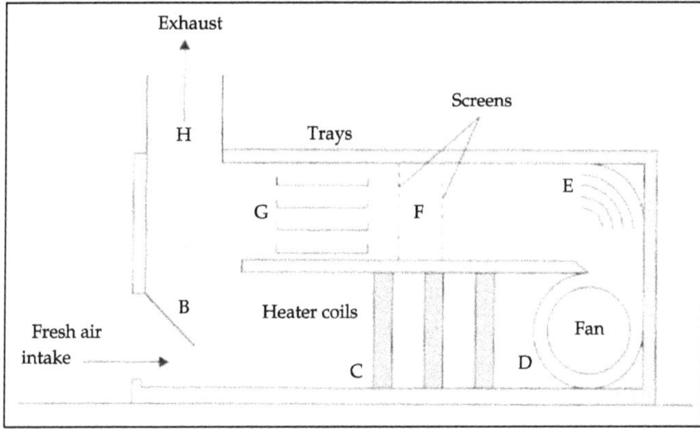

Figure 31.28

drying. As would be expected, smaller particles can be maintained in suspension with lower air velocities and will dry more rapidly (Figure 31.29).

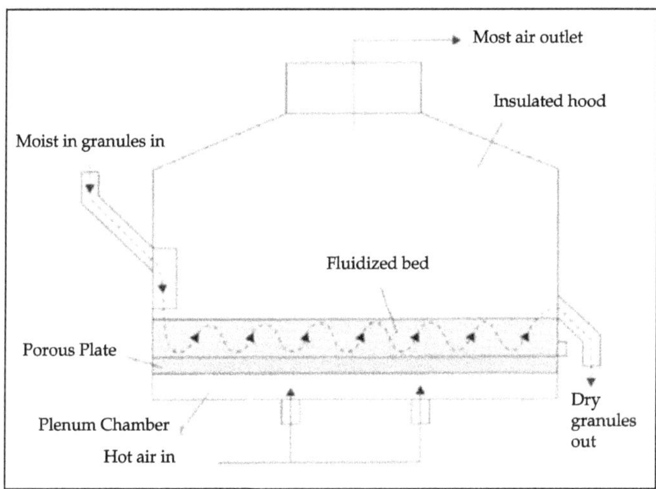

Figure 31.29

Spray Drying

The drying of liquid food products is often accomplished in a spray dryer (Figure 31.30). Moisture removal from a liquid food occurs after the liquid is atomized or sprayed into heated air within a drying chamber. While liquid food droplets are moving with the heated air, the water evaporates and is carried away by the air. Much of the drying occurs during a constant-rate period and is limited by mass transfer at the droplet surface. After reaching the critical moisture content, the dry food particle structure influences the falling-rate drying period. During this portion of the process, moisture diffusion within the particle becomes the rate-limiting parameter.

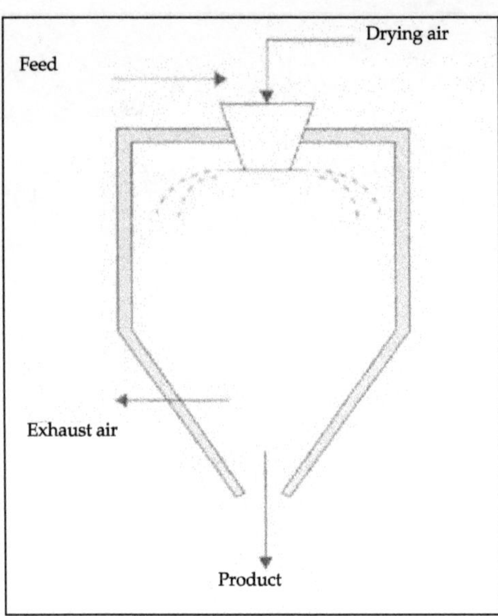

Figure 31.30

After the dry food particles leave the drying chamber, the product is separated from air in a cyclone separator. The dried product is then placed in a sealed container at moisture contents that are usually below 5 per cent. Product quality is considered excellent due to the protection of product solids by evaporative cooling in the spray dryer. The small particle size of dried solids promotes easy reconstitution when mixed with water.

Freeze-Drying

Freeze-drying is accomplished by reducing the product temperature so that most of the product moisture is in a solid state, and by decreasing the pressure

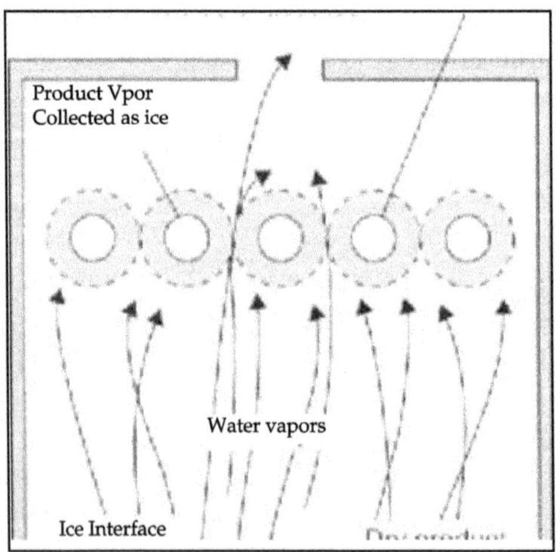

Figure 31.31

around the product, sublimation of ice can be achieved. When product quality is an important factor for consumer acceptance, freeze-drying provides an alternative approach for moisture removal (Figure 31.31).

Vacuum Chamber

The heat- and mass-transfer processes during freeze-drying are unique. Depending on the configuration of the drying system, heat transfer can occur through a frozen product layer or through a dry product layer. Obviously, heat transfer through the frozen layer will be rapid and not rate- limiting. Heat transfer through the dry product layer will be at a slow rate due to the low thermal conductivity of the highly porous structure in a vacuum. In both situations, the mass transfer will occur in the dry product layer. The diffusion of water vapor would be expected to be the rate limiting process because of the low rates of molecular diffusion in a vacuum.

Hot-Water Blanchers

There are a number of different designs of blancher, each of which holds the food in hot water at 70-100°C for a specified time and then removes it to a dewatering-cooling section. Developments in hot-water blanchers, based on the Individual Quick Blanching (IQB) principle, reduce energy consumption and minimize the production of effluent. For example, **the *blancher-cooler*** has three sections: a pre-heating stage, a blanching stage and a cooling stage. The food remains on a single conveyor belt throughout each stage and therefore does not suffer the physical damage associated with the turbulence of conventional hot water blanchers. The food is pre-heated with water that is re-circulated through a heat exchanger. After blanching, a second recirculation system cools the food. The two systems pass water through the same heat exchanger, and this heats the pre-heat water and simultaneously cools the cooling water. Up to 70 per cent of the heat is recovered. A recirculated water-steam mixture is used to blanch the food, and final cooling is by cold air (Figure 31.32).

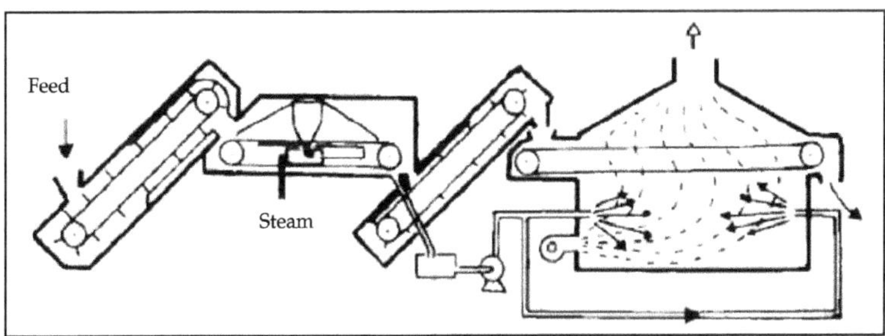

Figure 31.32

Belt Conveyors

A belt conveyor (Figure 31.33) is an endless belt operating between two pulleys with its load supported on idlers. The belt may be flat for transporting bagged material or V-shaped. The belt conveyor consists of a belt, drive mechanism and end pulleys, idlers and loading and discharge devices. On the belt conveyor baggage/product lie still on the surface of belt and there is no relative motion between the product and belt. This results in generally no damage to material. Belt can be run at higher speeds, so, large carrying capacities are possible. Horizontally the material can be transported to longer distance.

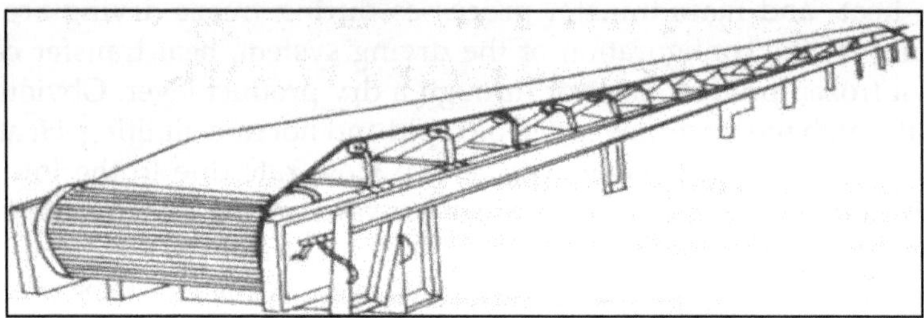

Figure 31.33

Bucket Elevator

A bucket elevator (Figure 31.34) consists of buckets attached to a chain or belt that revolves around two pulleys one at top and the other at bottom. The vertical lift of the elevator may range between few metres to more than 50 m.

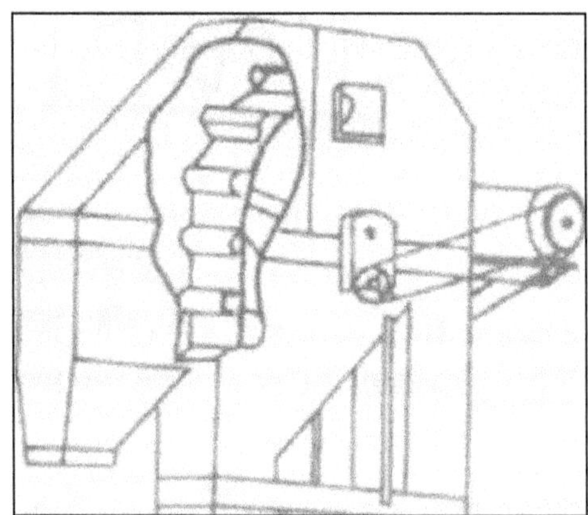

Figure 31.34

Screw Conveyor

The screw conveyor consists of a tubular or U-shaped trough in which a shaft with spiral screw revolves. The screw shaft is supported hanger bearings at ends. The rotation of screw pushes the grain along the trough. A typical screw conveyor is shown in the Figure 31.35. The screw conveyor is used in viscous or powder product handling facilities, animal feed industries and other installations for conveying of products generally for short distances. Screw conveyor requires relatively high power and is more susceptible to wear than other types of conveyors. The pitch of a standard screw which is the distance from the centre of one thread to the centre of the next thread is equal to its diameter. For example a 10 cm diameter screw has a pitch of 10 cm.

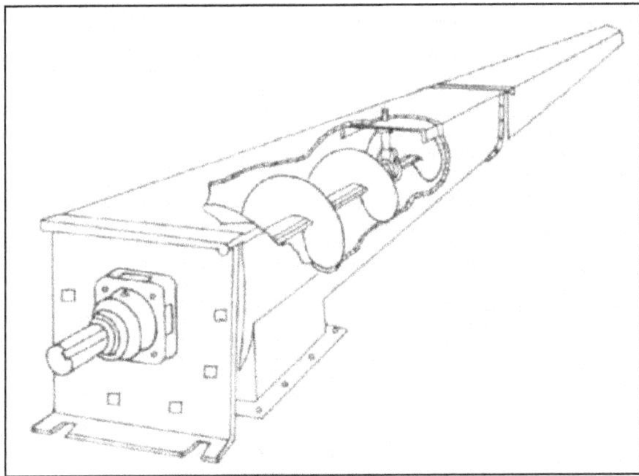

Figure 31.35

Belt-and-Roller Sorter

The particle size distribution of a material is expressed as either the mass fraction of material that is retained on each sieve or the cumulative percentage of material retained. The shape of some foods is important in determining their suitability for processing or their retail value. The separation of solids into two or more fractions on the basis of differences in size. This difference in size is adjusted accordingly by roller and belt (Figure 31.36).

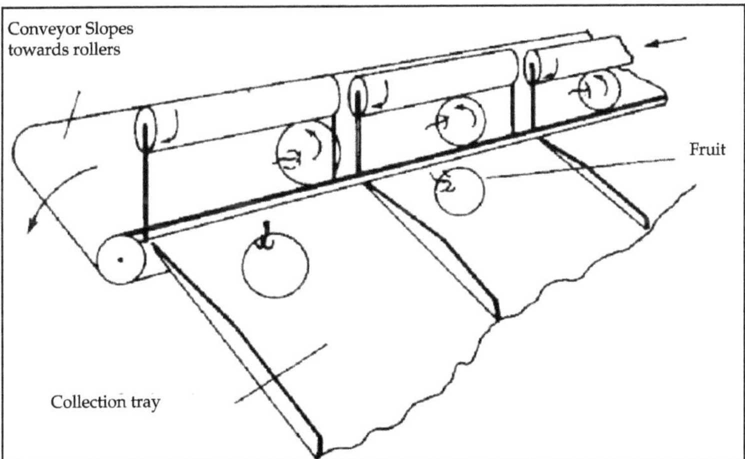

Figure 31.36

Image Processing

Image processing is used to sort foods on the basis of length, diameter, and number of surface defects and orientation of the food on a conveyor as well as colour. The images of the surface of the commodities are recorded and stored in the memory of a microprocessor. The information is then analyzed and compared with pre-programmed specifications for the product, and the commodities either rejected or moved into a group with similar characteristics.

Colour Sorting

Manual sorting by colour is still widely used but is increasingly expensive in both labour costs, operator training and the space required for sorting tables. There has therefore been considerable development of machine vision sorting systems which are said to have lower operating costs and greater accuracy than manual methods. These include monochrome (black and white), bichrome (4100 shades of red and green) and trichromatic or full colour (262 000 shades of red, green and blue, with optional infrared). Each is controlled by a programmable logic controller which has pre-set programs for different products that are easily changeable by operators using a video display (Figure 31.37). They are used for example, to sort potatoes for defects and blemishes by identifying dark areas on the potato surface. Light sensitive cells in the camera (termed 'pixels') produce a voltage that is proportional to the intensity of light received. An electronic circuit that receives a lower voltage than the pre-set value can thus detect darker objects or areas which reflect less light than normal. The voltage produced in the electronic circuit can be adjusted to alter the sensitivity of detection. Up to 10 tonnes of product per hour pass beneath the cameras on conveyors operating at 150-180 m per min. Defective items are removed

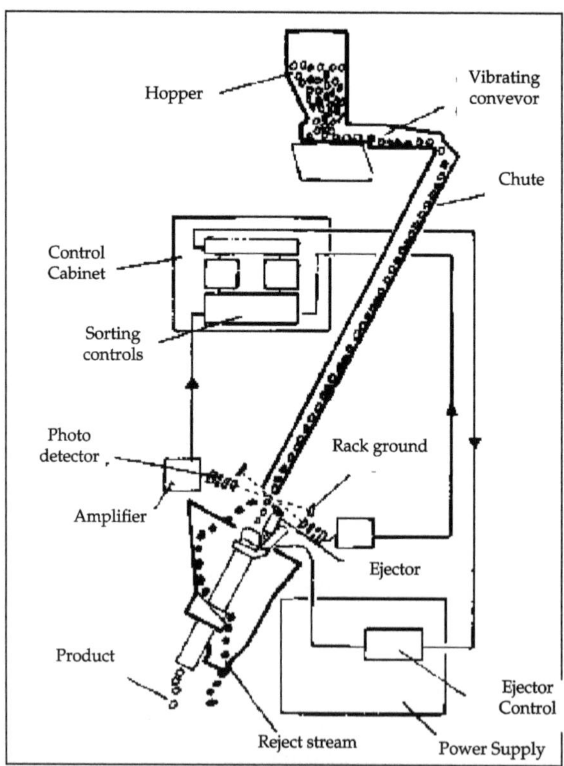

Figure 31.37

by electronically controlled air jets that can operate for 20 milliseconds, thus covering 50 mm of the belt length in a single blast. In another system, vegetables in free-fall are scanned 1000 times per second, as they leave a conveyor belt, using concentrated helium-neon or laser light beams and a high-speed rotating mirror. The machine detects differences in reflectivity between good product and unwanted material.

Small-particulate foods may be automatically sorted at high rates using microprocessor controlled colour sorting equipment. Particles are fed into the chute one at a time. The angle, shape and lining material of the chute are altered to control the velocity of the pieces as they pass a photodetector. The colour of the background and the type and intensity of the light used for illuminating the food (including infrared and ultraviolet options) are closely controlled for each product. Photo detectors measure the reflected colour of each piece and compare it with pre-set standards, and defective foods are separated by a short blast of compressed air. The computer can store 100 named product configurations to enable rapid changeover to different products using an operator touchpad.

Hammer Mills

These mills (Figure 31.38) all contain a high-speed rotor turning inside a cylindrical casing. Usually the shaft is horizontal. Feed dropped into the top of the casing is broken and falls out through a bottom opening. In a hammer mill, the particles are broken by sets of swing hammers pinned to a rotor disk. A particle of feed entering the grinding zone cannot escape being struck by the hammers. It shatters into pieces, which fly against a stationary anvil plate inside the casing and break into still smaller fragments. These in turn are rubbed into powder by the hammers and pushed through a grate or screen that covers the discharge opening.

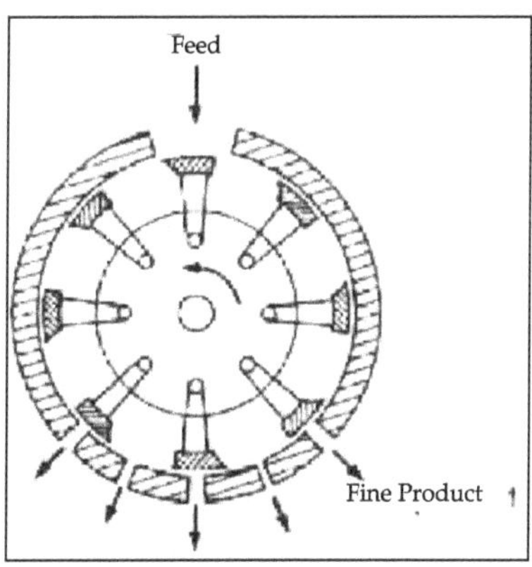

Figure 31.38

Transportation of Horticultural Crops

Temperature management is critical during long distance transport, so loads must be stacked to enable proper air circulation to carry away heat from the produce itself as well as incoming heat from the atmosphere and off the road. Transport vehicles should be well insulated to maintain cool environments for pre-cooled commodities and well ventilated to allow air movement through the produce. During transport, produce must be stacked in ways that minimize damage and then be braced and secured.

The small cooler as shown in Figure 31.39, uses a room window air conditioner to cool air inside the insulated box. The cool air inside the front of the box is forced through the produce by a pressure fan in a second wall inside. The return air passes under a false floor to the front of the box.

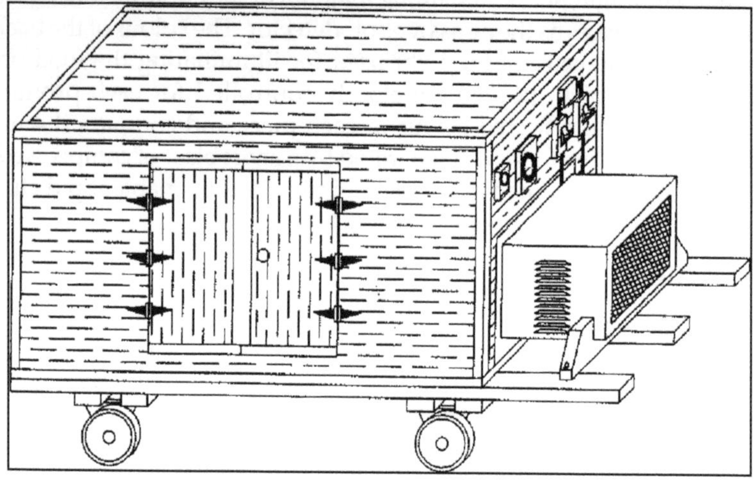

Figure 31.39

Refrigerated Trailers

For optimum transport temperature management, refrigerated trailers (Figure 31.40) need insulation, a high capacity refrigeration unit and fan, and an air delivery duct.

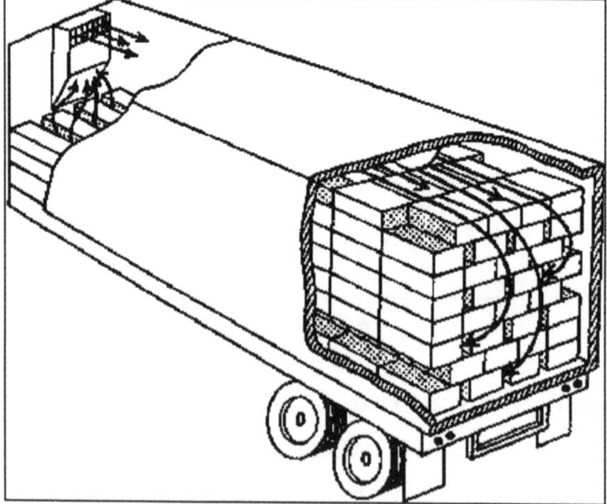

Figure 31.40

References

ASEAN-PHTRC. 1984. Village Level Handling of Fruits and Vegetables: Traditional Practices and Technological Innovations. Postharvest Horticulture and Training Center, College of Agriculture, University of the Philippines at Los Banos. Extension Bulletin No. 1

Fellows, P. and Hampton, A. 1992. Small-Scale Food Processing A Guide to Appropriate Equipment. London: Intermediate Technology Publications, pp. 158.

Kader, AA (ed). 2002. Postharvest Technology of Horticultural Crops (3rd Edition). UC Publication 3311. University of California, Division of Agriculture and Natural Resources, Oakland, California 94608, pp. 535.

Lisa K. and Kader A. A. 2003. Small-Scale Postharvest Handling Practices: A Manual for Horticultural Crops (4th Edition). University of California, Davis Postharvest Technology Research and Information Center.

Proctor, F.J. 1985. Postharvest handling of tropical fruit for export. The Courier 92: 83-86.

Reyes, M. U. 1988. Design Concept and Operation of ASEAN Packinghouse Equipment for Fruits and Vegetables. Postharvest Training and Research Center, University of the Philippines at Los Bahos, College of Agriculture, Laguna.

Shewfelt R.L. and Prussia, S.E. (eds). 1993. Postharvest Handling A Systems Approach. San Diego: Academic Press, pp. 358.

Talbot, M. T. and Fletcher, J.H 1993. Design and Development of a Portable Forced-Air Cooler. Proceedings of the Florida State Horticultural Society 106:249-255.

Thompson, J.F. *et al.,* 2001. Effect of cooling delays on fruit and vegetable quality. Perishables Handling Quarterly Issue No. 105: 2-5.

Chapter 32

Waste Management During Processing of Dryland Fruit Crops

P.M. Kotecha

Department of Food Science and Technology
Mahatma Phule Krishi Vidyapeeth, Rahuri – 413 722, Dist. Ahmednagar, M.S.

Fruits and vegetables are more prone to spoilage then cereals due to their nature and composition, and this spoilage occurs at the time of harvesting, handling transportation, storage, marketing and processing resulting in waste. Efficient management of these wastes can help in preserving vital nutrients of our foods and feeds, and bringing down the cost of production of processed foods, besides minimizing pollution hazards. The postharvest technologies for perishable horticultural produce serves as an effective tool for getting better return to the produce and also help in avoiding wastage both at production site and distribution centers.

The large scale facilities not only minimize the quantum of the fruit wastes for there profitable exploitation into value added byproducts, but also improves the overall economics of these units. There are about 4,000 processing units in India with a turn over of about 0.8 million tones of the fruit products. About 90 per cent of these units are in small-scale sector with an annual production limit between 50 to 250 tones. These units have no any organized method of waste disposal. Most of the solid wastes are dumped outside, which are taken care of by the stray cattle. An efficient disposal and recycling of these wastes can help to minimize pollution hazards, add vital nutritional components in our foods and feeds as well as bring down the cost of production of processed foods.

The more important wastes are the following:

1. Peel, rag and seeds in case of citrus fruits
2. Peel and stones in mangoes
3. Rind and seeds in jack fruit
4. Core and peel in guava
5. Cull fruits
6. Grape seeds, stone and skins
7. Pomace of fruits

Different Uses of Processing Waste

Source of Fat

The industrial processing of mangoes end up with the considerable proportion of the stone and peels, which are discarded and end up in wastes.

Source of Essential Oils

India is producing about 3 million tones of citrus fruits like mandarins, sweet oranges, limes, lemons *etc.* and generating large wastes around processing units. Citrus wastes are rich source of clouding agents, essential oils, pectin and a variety of the byproducts.

Source of Animal Feeds

Orange's wastes in the form of peels, rages and seeds are dried in California and Florida for the use of livestock. It contains lower amount of crude proteins, fiber and fat, but it is a rich source of the carbohydrates. Often, it is called as carbohydrate concentrate. The liquid concentrate made from the crude citrus waste is called molasses. Its use is similar to cane molasses and it is widely used for feeding dairy and beef cattle. Growing yeast on it and using it for animal feed is further helpful.

Ripe banana peels, which constitute abut 20 to 30 per cent, are found littered all around the streets causing pollution and skidding risks. In Vietnam, these peels are sun dried and used as poultry feed.

Source of Pectin

Majority of the pectin produce in the world is extracted from the citrus peels. Pectin is mainly used for jam making, pharmaceutical and several other industries. Mango peels, which are also available from processing factories have also been shown to be a good source of pectin.

Source of Starch

Banana pseudostem, having a yield of 20 to 25 tones from about 1,000 banana plants, when extracted gives abut 5 per cent edible starch.

Source of Natural Colours

Due to their carcinogenic effects, coaltar dyes are being fast replaced by natural colours or plant pigments for the use in food products. Natural colours of blue grapes skin, *kokum, phalsa, jamun* have thoroughly been investigated for there nature, concentration, extractability, stability and suitability as food colours. Methods for making colour concentrates and powders from them have been developed. A new fat soluble yellow pigment known as garcinol was isolated from the rind of *kokum*. Central Committee on Food Standards is actively considering permitting the use of *kokum* and safflower pigments in food industry.

Use of Waste in Biogas

Some of the fruit wastes (both solid and liquid) may usefully be diverted for biogas generation and making manures. Unlike vegetables, which have there pH's in natural region, fruits are not suited for active fermentation in generators due to there acidic nature. Neutralizing agents like lime slurry has to be added along with fruit waste for the continuous active fermentation and to maintain the efficiency of gas generators. The total biogas yield from the various fruit/vegetable wastes was 158 lit/kg dry matter which is three times more than using mixture of cowdung.

Other Uses of Processing Waste

Juice, wine, vinegar, sugar, alcohol, plastics, cosmetics, *etc.* can be made from fruit processing waste.

Future Thrust

The waste from fruit and vegetable processing factories are rich source of several useful materials. In the developed countries, many value added products are recovered from such wastes, but the scenario with developing countries is different where this waste is not utilized, resulting in loss of precious resources besides a great pollution problem. The future should see greater utilization of this waste for recovery of various by-products which could also stabilize the economy of the country.

Microbial Utilization of Food Processing Industry Waste

Products	*Waste*
Ethanol	Citrus industry waste, apple pomace, peach waste, cashew apple pomace, pineapple waste
Biogas	Waste from fruit and vegetable industry as a whole, fermentation (wine and beer) waste
SCP	Apple pomace, peach waste, cashew apple pomace, citrus waste extract, molasses, potato peels, cabbage waste
Cider, beer and vinegar	Apple pomace
Pectin	Citrus waste, apple pomace
Citric acid	Apple pomace
Baker's yeast/industrial yeast	Waste from wine, beer and distillery
Colour	Apple pomace
Flavours	Fruit and Vegetable waste, citrus waste
Animal feed	Apple pomace, peach waste, potato industry waste

Microbiac Fermentation Technology for Utilization of Processing Waste

- ☆ Fermented edible products
- ☆ Single cell proteins
- ☆ Animal feed
- ☆ Ethanol
- ☆ Enzyme
- ☆ Food additives
- ☆ Organic compounds
- ☆ Biogas

References

Anshu Singh, Arindam Kuila, Sunita Adak, Moumita Bishai, and Rintu Banerjee. 2012. Utilization of Vegetable Wastes for Bioenergy Generation. Agricultural Research. 1(3): 213–222.

Fellows P. 2007. Guidelines for Small-Scale Fruit and Vegetables Processors. FAO Agr. Ser. Bull., 127.

Lal G, Siddappa GS and Tandon GL. 1998. Preservation of Fruits and Vegetables. ICAR.

Nelson PE and Tressler DK.1980. Fruit and Vegetable Juice Processing Technology. Vol. III. AVI Publ.

Salunkhe, D.K. and Kadam, S.S. Handbook of Fruit Science and Technology: Production, Composition, Storage and Procesiing.1995. Marcel Dekker, Inc, New York. P 611.

Srivastava RP and Kumar S. 2003. Fruit and Vegetable Preservation -Principles and Practices. International Book Distributors.

Verma LR and Joshi VK. 2000. Postharvest Technology of Fruits andVegetables. Indus Publ.

Chapter 33

Role of Agriculture Education in Human Resource Development

B.R. Ulmek

Dean and Director of Instruction
Mahatma Phule Krishi Vidyapeeth, Rahuri – 413 722, Dist. Ahmednagar, M.S.

Agriculture is the main source of livelihood for over 80 per cent of the rural poor in India. Although, it employs about 52 per cent of the labor force, it contributes to only 14.4 per cent of GDP and 10.23 per cent of all exports. Global food demand is expected to be doubled by 2050 and at the same time natural resources are continuously reducing and deteriorating. Inadequate attention to agriculture has led to increase in prices of food products and thus making it inaccessible to poor people. Poverty is closely associated with malnutrition. India has large number of poor people (250 million out of approximately 1000 million worldwide). India's food security depends on producing cereal crops, as well as increasing its production of fruits, vegetables and milk to meet the demands of a growing population with rising incomes. To do so, a productive, competitive, diversified and sustainable agricultural sector will need to emerge at an accelerated pace.

The rate and level of development of any nation is dependent on her level of agricultural development. In other words, food security is important to any nation in dire need of development. This is because a nation that is dependent on other nations for food remains vulnerable to the manipulations of the supplier nations. A nation that lacks the capacity to feed her self is indeed underdeveloped. This can easily be overcome through the application of knowledge, skills and attitudes acquired from science and technology into agricultural production. Also the need for improvement in the state and condition of rural areas and its inhabitants is great. This is where agricultural education should come in.

India achieved spectacular agricultural growth since 1966. The increase in food grain production from a meager 51 million tonnes in 1950 to about 251 million tonnes in 2014-15 is a remarkable achievement unparallel in the history of world agriculture. Similar enhancement in production of milk, fish, oilseeds and fruit and vegetables has also been observed. Green, blue, yellow and white revolutions have been responsible for bringing in prosperity to the farming community. The cradle of the success, besides government policies and high receptivity of the fanning community, has been the establishment of institutions of higher agricultural education. These institutions developed new breed of skilled human resource that were instrumental in not only generating new technologies but also in their assessment, refinement and dissemination to the farming community. Because of challenges such as need to be globally competitive in markets, conserving natural resources in agriculture produce safe foods *etc.*, agriculture is increasingly becoming more knowledge intensive.

Human capital is the key source of comparative advantage in today's globalized and integrated world economy. Education is an important policy priority in creating a knowledgeable, skilled and adaptable workforce able to contribute to the country's competitiveness and social cohesion. Education has been accepted as one of the most impounded methods of developing human resource and increasing the rate of human capital formation. The more efficient is educational system higher will be the rate of capital formation and higher is the rate of capital formation faster will be the rate of economic growth.

HRD is development processes were it focuses on improving the existing capabilities of employees and helping them to acquire new capabilities required for the achievement of organizational and individual goals. Human Resource Development is the integrated use of training, organization, and career development efforts to improve individual, group and organizational effectiveness. HRD develops the key competencies that enable individual's in organizations to perform current and future jobs through planned learning.

Human Resource development is critical for sustaining, diversifying and realizing the potentials of agriculture. Agricultural human resource development is a continuous process being undertaken through partnership and efforts of the components of the Indian Council of Agricultural Research (ICAR) - Agricultural Universities (AUs) System comprising 53 State Agricultural Universities (SAUs), five Deemed to be universities (DUs), one Central Agricultural University (CAU) and four Central Universities (CUs) with Agriculture Faculty. The Agricultural Universities in India set up on the 'land grant' pattern of USA have contributed immensely to human resource development as well as enhancement of agricultural productivity and production in the country over the years. The system of education in Agricultural Universities was basically taken from USA pattern that greatly enabled incorporation of a number of diverse subjects in the courses as also provision of hands-on practical experience to the students. Agricultural universities impart education in the various disciplines of agriculture *viz.* Agriculture, Agricultural Engineering, Forestry, Horticulture, Veterinary and Animal Husbandry, Dairy Science, Food Technology, Fisheries Science, Agriculture Information Technology,

Agri. Business Management *etc.* It imparts education at the level of diploma, degree, masters and doctoral level. The principles of Agriculture education are providing up to date technical skills and knowledge in agriculture; conducting experiential learning activities in real agricultural careers and improving students in leadership and personal development activities at local, state and national level.

Agricultural education has now to evolve in tune with fast changing national and international scenario. Future agriculture is dominated by looming dangers of food insecurity originating from an unholy alliance of existing and emerging issues such as stagnating/declining productivity and profitability; degradation and depletion of natural resources; increased risks in the face of changing climate; unsafe livelihoods for millions of small and marginal farmers; regional imbalances in agricultural productivity; rising input costs, unsound profits and vulnerable markets; changing food habits and quality concerns; high postharvest losses and fragmented processing industry; globalization of trade and commerce; weakened technology transfer system; fossil fuel crisis and growing emphasis on bio-fuels encroaching upon good agricultural lands; poorly coordinated natural disaster management system, and the looming prospects of bioterrorism *etc.*

Of the critical issues facing the nation, few are more compelling than improving the academic performance of public schools and ensuring a stable, safe and affordable food supply. Today agricultural education is positioned to contribute substantially in these arenas through a major national initiative. The focus of the unprecedented effort is twofold: create new programs in communities not yet served by agricultural education and ensure the quality and high performance of current programs providing personal, academic and career education in agriculture. First, the public's expectations for higher student achievement are leading to dramatic increases in accountability, standards, rigor and relevance throughout education. Especially critical is the need to raise math and science proficiency. Second, the industry of agriculture, already concerned about meeting growing domestic and global demands for food and fiber, is eager to identify the future managers, leaders and workers who will ensure the future security and productivity of agriculture. A forecasted shortage of well-educated workers is adding urgency to the issue. Also, concerns about food safety, security and independence are registering at the highest levels of agribusiness and government. Lastly, local communities are intent on cultivating leadership and securing effective participation from their citizens. Through the intra-curricular programs of agricultural education students are developing skills in leadership, communication, team building and civic engagement. They will be prepared to provide for the social, economic and cultural well-being of small communities and large urban centers alike.

The quality and relevance of higher agricultural education enhanced to facilitate and undertake human capacity building for developing self-motivated professionals and entrepreneurs in view the changing scenario of globalization of education, emergence of new areas of specialization such as IPRs, other WTO-related areas, techno legal specialties *etc.*, and the cutting edge technologies such as biosensors, genomics and biotechnology, alternative sources of energy, nanotechnology, *etc.* The graduates are able to possess professional capabilities to deal with the concerns of sustainable development (productive, profitable and stable) of agriculture in all its aspects. Higher agricultural education comprising use of IT, Experiential Learning Module, use of Technology Mediated Learning Approaches, Participatory Approaches, Learning from farmers created potential agricultural human capital. Integrated Education system *viz.* Learning, Training, Workshop, Seminar resulted in innovative outcome in terms of Sustainable development of Agriculture.

The curriculum in undergraduate and post undergraduate agriculture education is being reformed to make it more compatible with contemporary developments in teaching and learning. 28 'Niche Areas of Excellence1 sanctioned to augment strategic strength of AUs in specific areas including those in new and emerging cutting-edge technologies. At the Mahatma Phule Krishi Vidyapeeth (MPKV), Rahuri, the units on biotechnology, biocontrol, nursery management, microirrigation system, biofertilizer production, mushroom production, protected cultivation of gerbera, rose and capsicum, dairy byproduct technology and postharvest technology centre were developed and utilized for skill development of UG students. About 200 units for Experiential Learning established in all the universities for providing skill oriented hands-on training to the students at undergraduate level. Revised Course Curricula and Syllabi approved for implementation by the AUs for under graduate education with emphasis on hands-on training, experiential learning for skill development, as recommended by the IV Deans Committee constituted by the Council. About 1000 National Talent Scholarships for undergraduate studies, 475 Junior Research Fellowships for post-graduate studies and 200 Senior Research Fellowships for Ph. provided every year towards the Human Resource Development, national integration and reducing inbreeding. Promoted Indian higher agricultural education among developing countries by providing opportunities to overseas students.

Further, Facilitating agricultural diversification to higher-value commodities: Encouraging farmers to diversify to higher value commodities will be a significant factor for higher agricultural growth, particularly in rain-fed areas where poverty is high. Moreover, considerable potential exists for expanding agro-processing and building competitive value chains from producers to urban centers and export markets. Agricultural graduates possesses knowledge, skills, ability and also entreprcneurship to provide a class of village-based services such as diagnostic laboratories, advisories on new innovations, markets and avenues of development assistance for corporate and contract farming. Thus, agriculture education plays vital role in human resource development creating the efficient human capital contributing to economy.

Index

A

B

T

U

V

W

Z

www.ingramcontent.com/pod-product-compliance
Ingram Content Group UK Ltd.
Pitfield, Milton Keynes, MK11 3LW, UK
UKHW021530300726
14060UKWH00011B/193